CARBOCYCLIC NON-BENZENOID
AROMATIC COMPOUNDS

Carbocyclic Non-Benzenoid Aromatic Compounds

by

DOUGLAS LLOYD

*Senior Lecturer in Chemistry, United College of St. Salvator and St. Leonard,
University of St. Andrews (Scotland)*

ELSEVIER PUBLISHING COMPANY

AMSTERDAM - LONDON - NEW YORK

1966

ELSEVIER PUBLISHING COMPANY
335 JAN VAN GALENSTRAAT, P.O.BOX 211, AMSTERDAM

AMERICAN ELSEVIER PUBLISHING COMPANY, INC.
52 VANDERBILT AVENUE, NEW YORK, N.Y. 10017

ELSEVIER PUBLISHING COMPANY LIMITED
RIPPLESIDE COMMERCIAL ESTATE, BARKING, ESSEX

LIBRARY OF CONGRESS CATALOG CARD NUMBER 66-11404

ALL RIGHTS RESERVED
THIS BOOK OR ANY PART THEREOF MAY NOT BE REPRODUCED IN ANY FORM,
INCLUDING PHOTOSTATIC OR MICROFILM FORM,
WITHOUT WRITTEN PERMISSION FROM THE PUBLISHERS

PRINTED IN THE NETHERLANDS

Preface

In 1865
Friedrich August Kekulé
first published his views on the structure of aromatic compounds

>(*Bull. Soc. chim. France*, [2] 3 (1865) 98)

In 1925
James Wilson Armit and Robert Robinson
published their suggestion of the concept of the aromatic sextet of electrons

>(*J. Chem. Soc.*, 127 (1925) 1604)

In 1965 it seemed appropriate to mark the centenary of Kekulé's hypothesis, and with it the fortieth anniversary of Armit and Robinson's concept by writing a text dealing with some aspects of the chemistry of aromatic compounds. It seemed particularly appropriate that such a text in the English language should come from the Department of Chemistry of the United College of St. Salvator and St. Leonard in the University of St. Andrews since this is one of the addresses from which the second of the two papers cited above was published.

Acknowledgements

To Dr. J. F. W. McOmie (University of Bristol) who most kindly read the complete manuscript and offered very valuable comments and criticisms.

To Dr. I. J. Worrall (University of Lancaster) and Dr. F. I. Wasson (University of St. Andrews) who also critically read sections of the manuscript.

To Professor K. Hafner (University of Munich) and Professor F. Sondheimer (University of Cambridge) who greatly assisted me by making their work readily available to me.

To Miss M. Smith who typed the manuscript.

To my wife, to Mrs. P. A. Sugden and to Dr. F. I. Wasson who assisted me in the chore of proof reading.

To Professor W. Baker (University of Bristol) to whom I owe my interest in this subject.

To the owners, the Bristol City Line, and the Master of M. V. Halifax City for hospitality on this ship on which the first portion of the book was written.

Apologies

To all persons whose work in this field I have overlooked, neglected, inadequately dealt with, or misinterpreted.

For errors which the diligent will inevitably find lurking in the text.

December 1965   D. Lloyd

Contents

Preface	v
Chapter I. *Aromaticity and aromatic character*	1
Introduction	1
The development of the concept of aromaticity	1
Molecular orbital theory and aromaticity	6
A note on the representation of non-benzenoid aromatic compounds	10
Coda – aromaticity, aromatic compounds and aromatic character.	11
References	14
Chapter II. *Derivatives of cyclopropene*	16
Introduction	16
Preparation of cyclopropenium salts	18
Properties of cyclopropenium salts	20
(a) Solubility	20
(b) Spectra	20
(c) pK Values and stability	21
(d) Action of nucleophiles	22
(e) Reduction	23
(f) Reaction with azides and diazo-compounds	23
(g) Formation of metal complexes	24
Preparation of cyclopropenone derivatives.	24
Properties of cyclopropenones	26
(a) Dipole moments	26
(b) Spectra	26
(c) Action of acid	26
(d) Action of alkali	27
(e) Thermal decomposition	27
(f) Reduction of cyclopropenones	28
(g) Other reactions of cyclopropenones	28
Quinocyclopropenes	29
Methylene cyclopropenes	31
Calicenes	33
References	33
Chapter III. *Derivatives of cyclobutadiene*	35
Introduction	35
Metal complexes of cyclobutadiene	38
Cyclobutenium salts	40
Cyclobutene-3,4-diones	41
Benzocyclobutadiene	42
Biphenylene.	45
Preparation of biphenylene	45
Structure of biphenylene	46
Ultra-violet spectra of biphenylenes	47
Properties of biphenylene	47

Benzobiphenylenes 50
Comparison of benzocyclobutadienes and biphenylenes 51
References 51

Chapter IV. *Derivatives of cyclopentadiene* 55
Introduction 55
Cyclopentadienide salts 57
　Preparation 57
　Properties of cyclopentadienide salts 57
　Chemical reactions of cyclopentadienide salts 58
　Substituted cyclopentadienide salts 59
　Salts derived from indene, fluorene, etc. 60
Diazocyclopentadienes 61
　Preparation 61
　Spectra of diazocyclopentadienes 62
　Chemical reactions of diazocyclopentadiene 63
Other cyclopentadienylides 66
　Preparation 66
　Physical properties 66
　Spectra 67
　Stability 68
　Chemical reactions 68
Indenylides and fluorenylides 69
　Indenylides 69
　9-Diazofluorene 69
　Other fluorenylides 72
Cyclopentadienylidene derivatives 75
　Preparation 75
　Properties of cyclopentadienylidenedihydropyridines 78
　Properties of cyclopentadienylidenepyrans 78
　Other cyclopentadienylidene derivatives 79
　Indenylidene and fluorenylidene derivatives 79
Sesquifulvalenes 80
Fulvenes 81
　Preparation 81
　Structure of fulvenes 81
　Spectra of fulvenes 82
　Chemical reactions 83
　Fulvenes with nitrogen or oxygen atoms attached to the 6-carbon atom . . 84
Ferrocene 86
　Preparation 87
　Structure 88
　Stereochemistry of ferrocene derivatives 88
　Chemical properties of ferrocene 89
　Electrophilic substitution reactions 90
　Analogues of ferrocene 91
References 92

Chapter V. *Tropylium salts* 98
Introduction 98
Preparation of tropylium salts 98
　(a) From cycloheptatriene 98

  (b) From cycloheptatrienecarboxylic acid 99
  (c) From benzene and halocarbenes 100
  (d) From cyclooctatetraene 101
  (e) From tropyl (cycloheptatrienyl) compounds 101
  (f) Other methods of formation 102
 Structure of the tropylium ion 103
 Spectra 104
 Properties of tropylium salts. 104
 Chemical reactions of tropylium salts 105
  (a) Reduction 105
  (b) Oxidation 106
  (c) Reaction with nucleophilic reagents 107
  (d) Attempted electrophilic substitution of the tropylium ion. 112
  (e) Reactions of the alkyl groups in alkyltropylium salts 112
 Metal complexes. 113
 Benzotropylium salts 113
 References 114

Chapter VI. *Tropones, tropolones and related compounds* 117
 Introduction 117
 Preparation of tropones 122
  (a) From benzene derivatives 122
  (b) From cycloheptanone and cycloheptenone 124
  (c) From tropinone derivatives 125
  (d) From cycloheptatriene 126
  (e) From tropylium salts 126
 Preparation of tropolones 126
  (a) From cycloheptanone and its derivatives 127
  (b) From benzene and other benzenoid compounds. 127
  (c) From tropone 128
  (d) From cyclopentadiene 129
 Structure of tropone and tropolone 129
 Spectra 131
  (a) Tropones 131
  (b) Tropolones 132
 Properties of tropones 133
  (a) Reactions of the carbonyl group 133
  (b) Addition reactions 134
  (c) Substitution reactions 135
  (d) Replacement reactions of substituent groups 136
  (e) Rearrangement reactions 138
 Properties of tropolones 139
  (a) Oxidation 139
  (b) Reduction 140
  (c) Reactions of the carbonyl group 141
  (d) Reactions of the hydroxyl group 141
  (e) Addition reactions 142
  (f) Electrophilic substitution reactions 143
   (i) Halogenation 143
   (ii) Nitration 143
   (iii) Sulphonation 144

CONTENTS

   (iv) Azo-coupling 144
   (v) Alkylation and acylation 144
   (vi) Nitrosation 144
   (vii) Reimer–Tiemann reaction 144
   (viii) Hydroxymethylation and aminomethylation. 144
  (g) Replacement reactions of substituent groups 145
  (h) Rearrangement reactions 146
 3- and 4-Hydroxytropones 147
 Thiotropolones, aminothiotropones and aminoiminocycloheptatrienes 150
 Benzotropones and benzotropolones 151
  Preparation 151
  Properties of benzotropones 151
  Properties of benzotropolones 152
 Heptafulvenes 153
 References 155

Chapter VII. *Medium and large ring compounds* 162
 Cyclooctatetraenides 162
 Cyclononatetraenides 163
 Annulenes 165
  Preparation of annulenes 167
  N.M.R. spectra and aromaticity of annulenes 170
  U.V. spectra 172
  X-ray crystallographic analysis 172
  Chemical properties 173
  Dehydroannulenes 174
  Bridged ring annulenes 176
 References 179

Chapter VIII. *Polycyclic compounds* 181
 Introduction 181
 Azulenes 182
  Syntheses of azulenes 183
  Syntheses not involving dehydrogenation 187
  Structure of azulene 190
  Reactions of azulenes 194
   (a) Addition reactions; reduction and oxidation 194
   (b) Substitution reactions 195
    (i) Electrophilic substitution 195
    (ii) Nucleophilic substitution 198
    (iii) Radical substitution 200
   (c) Rearrangement reactions 201
  Spectra of azulenes. 201
 Sesquifulvalenes 203
 Calicenes 204
 Pentalenes 205
 Heptalenes 208
 General comments on polycyclic non-benzenoid aromatic hydrocarbons 209
 References 212

Subject Index 217

CHAPTER I

Aromaticity and aromatic character

INTRODUCTION

During the nineteenth and twentieth centuries the meaning of the term "aromatic", as applied to organic compounds, has varied as the understanding of the chemistry of these compounds has extended and developed. From just a hundred years ago until comparatively recently "aromatic" was to all intents and purposes synonymous with "benzenoid" (meaning a derivative of benzene). But since it was never explicit whether "aromatic" referred to structural features (*i.e.* the presence of a benzene ring) or to chemical properties (*i.e.* properties resembling those of benzene), the recognition that cyclic compounds other than benzene derivatives might none the less resemble benzene in some respects, most notably in a tendency to undergo substitution rather than addition reactions with electrophilic reagents, led inevitably to such compounds also being described as "aromatic". In order to distinguish them from benzenoid compounds* they were known as *non-benzenoid aromatic compounds*. It is with these "aromatic" compounds, not derived structurally from benzene, that this book is concerned.

It seems worthwhile to consider first of all the development of the concept of aromaticity, as it applies to both benzenoid and non-benzenoid compounds

THE DEVELOPMENT OF THE CONCEPT OF AROMATICITY

The term "aromatic" was first applied in the early part of the nineteenth century to compounds of aromatic odour which had been found in many volatile vegetable oils. Examples of these compounds were toluene (from "balsam of Tolu"), benzaldehyde (from oil of bitter almonds), cymene (from oil of caraway), methyl salicylate (from oil of wintergreen), anethole (from aniseed oil) and vanillin (from vanilla). It was recognised that many (al-

---

* The term benzenoid compounds, which is unequivocal, is used throughout this book to describe derivatives of benzene.

References p. 14

though not all) of these odoriferous compounds contained relatively more carbon than the *aliphatic* organic compounds derived from fats and they were described by Kekulé[1] as "kohlenstoffreichere Verbindungen". When it was recognised that these carbon-rich compounds were all derivatives of benzene, the term "aromatic" came to have a structural meaning instead of its hitherto purely descriptive meaning and signified a benzene derivative. This is a meaning which is still frequently given to it in modern text-books of organic chemistry but, as suggested above, it would seem preferable to use the term *benzenoid* for this purpose.

In 1865 Kekulé[1] suggested the cyclohexatriene structure for benzene*. He commented on the particular stability of the benzene ring as follows[2]:

"If we wish to give an account of the atomic constitution of the aromatic compounds, we are bound to explain the following facts:

(1) All aromatic compounds, even the most simple, are relatively richer in carbon than the corresponding compounds in the class of fatty bodies.

(2) Among the aromatic compounds, as well as among the fatty bodies, a large number of homologous substances exist.

(3) The most simple aromatic compounds contain at least six atoms of carbon.

(4) All the derivatives of aromatic substances exhibit a certain family likeness; they all belong to the group of "aromatic compounds". In cases where more vigorous action takes place, a portion of the carbon is often eliminated, but the chief product contains at least six atoms of carbon. The decomposition stops with the formation of these products, unless the organic groups be completely destroyed.

These facts justify the supposition that all aromatic compounds contain a common group, or, we may say, a common *nucleus*, consisting of six atoms of carbon. Within this *nucleus* a more intimate combination of the carbon atoms takes place; they are more compactly placed together, and this is the cause of the aromatic bodies being relatively rich in carbon. Other carbon atoms can be joined to this *nucleus* in the same way, and according to the same law, as in the case of the group of fatty bodies, and in this way the existence of homologous compounds is explained."

This initial formulation of the structure of benzene by Kekulé assumed fixation of the single and double bonds. Shortly afterwards[3] he modified this

---

* The priority of Kekulé as author of the cyclohexatriene formula for benzene has been questioned; see ref. 33.

picture and assigned a "dynamic structure" to benzene, involving a continuous interchange between the two "Kekulé forms":

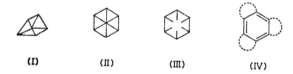

The only serious challenge to Kekulé's cyclohexatriene formula came from the prism formula (I)[4] but the latter was soon shown to be incompatible with the known facts concerning the isomerism of substituted benzene derivatives.

(I)   (II)   (III)   (IV)

Although the concept of a ring of six equivalent CH groups was generally accepted, alternative ways of expressing the "extra" valencies, instead of depicting them as alternating double and single bonds, were suggested by various authors. Many of these suggestions involved concepts having no parallel in the aliphatic series, for instance the "diagonal" formula (II)[5] which involved three para bonds linking carbon atoms on opposite sides of the ring and the "centric" formula (III)[6,7] in which "the remaining six (valencies) react upon each other — acting towards a centre as it were, so that the "affinity" may be said to be uniformly and symmetrically distributed".[6]

In 1899 the theory of "partial valencies" was extended to benzene, as a closed-chain triene[8]. The partial valencies were expressed as in (IV) and it was suggested that the neutralisation of the partial valencies around the ring led to the equalisation of all the C-C links. From the latter idea developed the concept of a fully symmetrical structure involving fractions of double-bonds[9].

Kekulé's use of the word "aromatic" was based on structural features, namely the presence of a benzene ring in the molecule, but very shortly afterwards the suggestion was made[10] that the concept of aromaticity should encompass substances of similar behaviour rather than those with a common structural feature; and the possibility of these alternative interpretations has subsequently bedevilled organic chemistry.

References p. 14

The special properties of aromatic compounds — regarded as compounds having properties similar to those of benzene — was associated with their having six "residual affinities" in the early 1890s[11]. It was stated that "a centric system can only exist as a hexacentric system" and the argument was developed that "hexacentric systems" could exist in ring-systems other than six-membered. Thus attention was drawn to the fact that pyrroles are exceptionally weak bases and it was suggested that this was because the salt-forming valencies of the nitrogen atom (in modern terminology, its lone pair of electrons) are required to form the six residual affinities. Pyridines, on the other hand, were relatively strong bases since the six residual affinities could be formed without calling on the latent valencies of the nitrogen atom, which therefore remained available for salt formation.

This idea very clearly foreshadows in pre-electron terms the concept of the "aromatic sextet" of electrons. The latter concept was specifically formulated by Armit and Robinson in 1925[12]. These authors were commenting on the fact that aromatic compounds were characterised by "reduced unsaturation and the tendency to retain the type", as illustrated, for example, by their tendency to undergo substitution rather than addition reactions. They observed: "The explanation is obviously that six electrons are able to form a group which resists disruption and may be called the aromatic sextet. It is not supposed that the existence of the sextet involves a change in the total covalency exerted by the carbon atoms of the ring nor does the theory as employed... require any particular assumption in regard to the position of the electrons or their orbits in space." They suggested the inscribed circle formula (V) for benzene, stating "The circle in the ring symbolises the view that 6 electrons in the benzene molecule produce a stable association which is reponsible for the *aromatic* character of the substances."

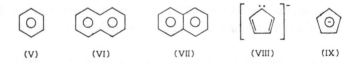

(V)　　(VI)　　(VII)　　(VIII)　　(IX)

Here again aromatic is clearly being thought of in terms of the properties of compounds rather than in structural terms. In order to extend this type of formulation to include naphthalene and other polycyclic compounds the central bridge bond had to be eliminated, as in (VI) in order to produce two complete sextets of electrons[12]. This is obviously not satisfactory and the problem of representing polycyclic compounds by means of inscribed circle formulae (*e.g.* VII) has now been solved by saying that the symbol represents a sextet of π-electrons which may in part be shared with an adjacent sextet or sextets. The use of inscribed circle formulae has some drawbacks; thus, for example, it may give the false impression that the rings in naphthalene are totally symmetrical, whereas in fact the 1,2-bond has much more double-bond character and is appreciably shorter than the 2,3-bond. Nowadays it is general practice to use Kekulé type formulae for benzenoid compounds, remembering, of course, that these are symbols and not precise representations of the structures of the molecules. (For a discussion and recommendations on the use of inscribed circle formulae see ref. 13. Dotted circles are used for delocalised systems of other than six π-electrons; see also p. 10).

The concept of the aromatic sextet at once explained the aromatic character of heterocycles such as thiophen and pyrrole; it also gave a rational explanation of the stability of the cyclopentadienide anion $[(C_5H_5)^-]$, (VIII)[12, 14,15]. This ion has six electrons which can be shared among five equivalent CH groups in the ring. In the inscribed circle notation it is depicted as in (IX) wherein the circle symbolises the sextet of π-electrons and the negative sign within the circle indicates that the negative charge, like the π-electrons, is shared over the whole ring.

Armit and Robinson[12] not only recognised that six was a specially favourable group of electrons but also that other groups made up from other numbers of electrons might not have this same stability: "The unsaturated nature of cyclooctatetraene suggests that a stable group of eight electrons analogous to the aromatic sextet is not formed."

It was also realised that aromatic character was not a uniform property, and that considerable variations might be evident between different aromatic compounds. Thus pyrrole and furan undergo electrophilic substitution more readily than benzene but on the other hand are more easily decomposed by oxidising agents and undergo addition reactions more readily as well. Even among benzene derivatives aromatic character is not constant; one reaction often used to demonstrate the aromaticity of benzene is its relative inertness towards oxidising agents such as potassium permanganate, yet resorcinol, an indisputably aromatic compound which undergoes electrophilic substitution more readily than benzene itself, is so easily oxidised that it very easily reduces Fehling's solution or ammoniacal silver nitrate. Similar comments apply to physical properties such as bond lengths. The difference between the lengths of the 1,2 and 2,3 bonds in naphthalene has already been mentioned above; irregular bond length can also be found in simple monocyclic benzene derivatives, *e.g.* salicylic acid.

MOLECULAR ORBITAL THEORY AND AROMATICITY

The basic rule underlying all present day ideas on aromaticity is known as *Hückel's rule*[16]. It states that "amongst fully conjugated, planar, monocyclic polyolefins only those possessing $(4n + 2)$ π-electrons, where n is an integer, will have special aromatic stability." Thus benzene has a resonance energy of 36 kcal/mole, *i.e.* it is more stable by this amount than it would be if it were made up from three isolated double-bonds.

In Hückel's treatment, the π-electrons of the conjugated system are regarded as common to all the carbon atoms of the system and are considered to occur in common molecular orbitals. Three types of molecular orbital are possible, namely (1) *bonding orbitals*, in which the energy of the electrons is less than their energy in the atomic orbitals; consequently the occurrence of the electrons in these orbitals makes the system more stable, (2) *non-bonding orbitals*, in which the energy of the electrons is equal to their energy in the atomic orbitals, and (3) *anti-bonding orbitals* in which the energy of the electrons is greater than their energy in the atomic orbitals; consequently the occurrence of electrons in anti-bonding orbitals is energetically disadvantageous. The numbers of each of these kinds of orbitals and their energies for any particular conjugated system depend on the number of atoms which make up the system and also on the symmetry of the system. They may be determined by solving the wave equation of the system. The Pauli principle requires that not more than two electrons can be allocated to any one molecular orbital.

Application of such molecular orbital calculations to benzene shows that it has three bonding and three anti-bonding orbitals. In the ground state, the six π-electrons occupy the lowest energy levels, *i.e.* the three bonding orbitals. This can be illustrated by the diagram at the top of page 7 in which the vertical scale represents increasing energy, the horizontal lines the energy levels of the different molecular orbitals and the arrows electrons occupying particular orbitals; two arrows ↑↓ in an orbital indicate electrons of opposing spin.

This elementary treatment shows that benzene, in the ground state, contains a stable "closed" system of six π-electrons, a "closed system" of electrons being one in which either removal of electrons or addition of further electrons increase the total energy of the system and consequently diminish its stability. Thus in the case of benzene there are no unoccupied positions in the bonding orbitals, hence entry of a further electron requires expenditure

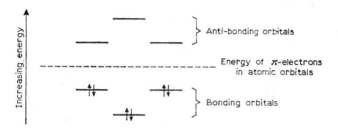

of energy, and furthermore all the π-electrons are in bonding orbitals and hence removal of any of them also requires expenditure of energy. This closed system of six π-electrons in benzene may be likened to the especially stable electron shells of the inert gases.

Hückel also successfully applied this concept of closed shells of π-electrons to five- and seven-membered rings. For example, it is found that a C_5H_5 ring made up from five sp^2 hybridised carbon atoms has three bonding and two anti-bonding orbitals. If, therefore, a hydrogen atom is removed from the methylene group of cyclopentadiene together with either two, one, or none of its bonding electrons, leaving respectively $C_5H_5^+$, $C_5H_5^{\cdot}$, or $C_5H_5^-$, the occupancy of the molecular orbitals in these three species can be represented by the diagrams:

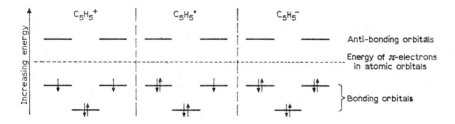

Thus of the three the cyclopentadienide anion alone has a closed shell of π-electrons and in consequence is the most stable of the three. This is in complete accord with experimental findings, for whereas salts of the cyclopenta-

References p. 14

dienide anion may be isolated, this is not the case with either the $C_5H_5^+$ ion or the $C_5H_5\cdot$ radical.

Similarly the seven-membered C_7H_7 ring is found to have three bonding and four anti-bonding orbitals. Thus of the species $C_7H_7^+$, $C_7H_7\cdot$ and $C_7H_7^-$ only the cation has all its electrons in bonding orbitals and a closed electron shell, and thus should be the stable species of the three.

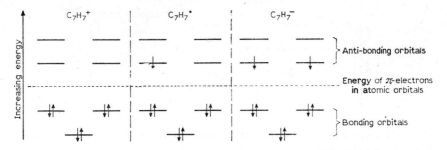

Hückel[16] wrote in 1931: "It may be expected that a seven-membered ring will tend to lose its unpaired electrons with the formation of a positively charged ion, so that a seven-membered ring will behave in relation to benzene in the same way as an atom of an alkali metal in relation to an atom of the neighbouring inert gas." This prediction was completely fulfilled by the preparation and characterisation, in 1954, of such a salt, known as a tropylium salt[17]. (See also Chapter V.) Both the cyclopentadienide and tropylium salts are examples of non-benzenoid aromatic compounds.

Molecular orbital theory thus gave a mathematical expression to Armit and Robinson's concept of the aromatic sextet, and, for that matter, the even earlier ideas of the special importance of *six* residual valencies, both of which had been conceived solely on the basis of organic chemical intuition.

The new theory took matters considerably further however by its postulate that a shell not only of six π-electrons but of any number of π-electrons which fitted the expression $(4n + 2)$ where $n =$ an integer was a closed shell, whereas cyclic polyolefins which did not fit this formula would not have the same aromatic stabilisation.

Thus Hückel's rule predicts that cyclobutadiene and cyclooctatetraene should not be aromatic. This is not evident from any of the earlier theories although it was foreshadowed by Armit and Robinson[12] (see above, p. 5).

Cyclobutadiene has one bonding, two non-bonding and one anti-bonding orbitals and its four π-electrons would be expected to be distributed as shown in the following diagram:

MOLECULAR ORBITAL THEORY

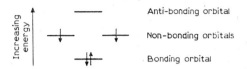

There is no closed shell of electrons in the sense we have been discussing. The total energy of the π-electrons in cyclobutadiene proves to be the same as that for two unconjugated double-bonds, *i.e.* its conjugation energy is zero[18,19]. It is likely in fact that cyclobutadiene would be a diradical; calculations show that the molecule should be rectangular rather than square[19,20] (see also Chapter III).

Similarly, cyclooctatetraene does not have a closed shell of π-electrons:

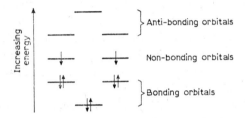

In consequence there is no special stability associated with this octet of electrons. If there were to be electronic interaction between the double-bonds the molecule would have to be planar but this would involve considerable angle-strain. This strain would not be compensated for by any appreciable gain in energy due to delocalisation and the molecule therefore takes up a non-planar form in which angle strain is relieved. In this form there is no overlap between the orbitals of the separate double bonds and the molecule therefore has a polyolefinic character with alternating single and double bonds of lengths 1.466 Å and 1.334 Å respectively.

Hückel's rule has also been used to explain the aromaticity of heterocyclic compounds. In this case there are also symmetry requirements to be fulfilled — for instance pyridine does not in fact fulfil these requirements on account of the different electronegativity of the nitrogen atom compared with the other ring atoms, but the deviation is not great and the "closed shell" theory in fact retains some force. The rule has also been applied to inorganic aromatic compounds such as borazole (X)[21].

Perhaps even more spectacularly it forecast new non-benzenoid aromatic systems wherein the closed shell of electrons was made of 2 or 10 or 14, etc.

References p. 14

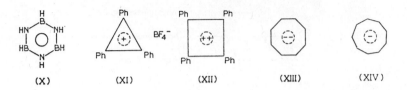

π-electrons. Until 1957 no such monocyclic systems were known but in that year a stable cyclopropenium salt (XI) (see Chapter II) was prepared having two π-electrons delocalised over three ring atoms[22]. Since then the cyclobutadiene dication (XII)[23] with two delocalised π-electrons has probably been obtained in solution (see Chapter III), and the cyclooctatetraenide dianion (XIII)[24] and cyclononatetraenide anion (XIV)[25], each having closed shells of ten π-electrons, have been prepared (see Chapter VII), and shown to have properties in accord with the presence of such delocalised systems of π-electrons. More strikingly still since 1959 a series of completely conjugated cyclic polyolefins or annulenes has been prepared[26] and studies, particularly of their nuclear magnetic resonance spectra, again testify to the validity of Hückel's rule (see Chapter VII).

A recent succinct restatement[27] of Hückel's rule which accommodates both uncharged cyclic polyolefins and the aromatic ions such as the tropylium, cyclopentadienide or cyclopropenium ions runs as follows: "The π-electron molecular orbitals of monocyclic conjugated molecules vary with the number of carbon atoms such that $(4n + 2)$ numbers of electrons completely fill the bonding molecular orbitals and produce the largest electron delocalisation energy. $(4n + 1)$ and $(4n + 3)$ electron molecules (radicals) tend to form a "more stable" system through the gain or loss of an electron to form the $(4n + 2)$ configuration."

A NOTE ON THE REPRESENTATION OF NON-BENZENOID AROMATIC COMPOUNDS

In the formulae of the cyclopentadienide anion given above (IX) and in the representation of the tropylium ion (see Chapter V) the inscribed circle in the formulae is an unbroken one, whereas in formulae (XI)–(XIV) above a dotted circle is used. This conforms with a suggestion that the unbroken circle be used only to represent an aromatic sextet of π-electrons and that for delocalised electron systems of other than six π-electrons the dotted circle should be used. (For a more full discussion see ref. 13.) It should be noted that although either the unbroken or dotted inscribed circles infer a generalised delocalisation of electrons, and, where appropriate, of charge, over the ring they do not necessarily imply that this delocalisation is equal over all the ring. For instance the presence of substituent groups will inevitably distort the delocalisation.

CODA – AROMATICITY, AROMATIC COMPOUNDS AND AROMATIC CHARACTER

Earlier in this discussion attention was drawn to the fact that the word "aromatic" has had two distinct interpretations, one based on structural features and the other on the properties, chemical and physical, of the compound under consideration. Thus Kekulé's original use of the word involved a structural definition — Armit and Robinson on the other hand were thinking in terms of "reduced unsaturation and the tendency to retain the type", in other words, chemical reactivity. Hückel's rule again concerns structural features, although in this case of electrons as well as of atoms.

Unfortunately the two interpretations are not entirely mutually interchangeable. Thus the delocalisation energy of larger aromatic molecules increases roughly in proportion to the number of π-electrons, yet these molecules are often more reactive in addition type reactions than their smaller homologues. For instance naphthalene has a larger delocalisation or resonance energy than benzene yet it also participates in addition reactions more readily. It follows therefore that energetic stabilisation of a molecule and low reactivity do not necessarily go hand in hand.

Nor is there any real reason why they should do so[28]. The reactivity of a molecule in any particular reaction depends on the difference in free energy between that of the initial molecule and that of the transition state of the reaction which ensues. Thus a knowledge solely of the energy of the starting material cannot define its reactivity. (There is the accidental connexion between the energy of the ground state and the reactivity in some particular reaction, namely that the lower the energy of the ground state, the larger the activation energy *if* the energy of the transition state is constant. There seems no reason, however, why, in the general case, the energy of the transition state should be even roughly constant. When the ground-state resonance energy can be successfully used in predicting aromaticity such a relation must occur, but it has no theoretical foundation and may break down without warning)[28a].

The situation is further complicated in the case of the aromatic ions, for these ions owe their stability to a favourable free-energy difference between the neutral form and the ion rather than to any low chemical reactivity. Furthermore the neutral form may itself decompose very readily and for such an ion to have any prolonged existence, conversion into the neutral form must be prohibitively slow.

All these considerations lead to the conclusion that chemical reactivity, or lack of it, is not a very sound guide to the inherent stability of a molecular or ionic species. A molecule might be both stable, and yet too reactive to be easily isolable. Even were it a qualitative guide a classification based on chemical properties would have the disadvantage that it is based on a number of properties (*e.g.* ease or difficulty of undergoing addition or substitution reactions, oxidations, etc.) which might be mutually conflicting. (Note for example the cases of benzene and pyrrole mentioned above, p. 5.) Even when such evidence was mutually consistent some artifical limit to aromaticity would have to be drawn, just where normally depending on the predisposition of the person drawing the line.

The aromaticity predicted by Hückel's rule concerns only the ground state of the molecule and refers to a particularly effective contribution by π-electron delocalisation to the total molecular stability. It would seem wiser therefore to use the term *aromaticity* to refer only to the ground state properties of the molecule and to deem a molecule or ion *aromatic* if it is stabilised in this particular way[29,30]. Although π-electron delocalisation represents a particularly effective factor which contributes to the overall stability of a molecule, it should be borne in mind that other contributing factors may in some circumstances nullify or override its effect. Thus it seems likely that steric factors would prevent the monocyclic polyolefin cyclodecapentaene, $C_{10}H_{10}$, from taking up the planar form necessary for complete cyclic delocalisation of its π-electrons, although it "fits" Hückel's rule.

The simplest way of detecting delocalisation of the π-electrons in an aromatic compound is by determining its nuclear magnetic resonance (n.m.r.) spectrum. The magnetic field which is applied in determining such spectra causes induced circulation of the delocalised π-electrons and in consequence, ring currents are set up in the molecule. These ring currents are in turn responsible for deshielding protons attached to the outside of the aromatic rings and this is shown up experimentally by the magnitude of the chemical shift in the n.m.r. spectra of these protons[29]. Based on this mode of experimental investigation the following definition of an aromatic compound has been suggested: "The essential feature is a ring of atoms so linked that π-electrons are delocalised right round the ring. We can define an aromatic compound, therefore, as a compound which will sustain an induced ring current. The magnitude of the ring current will be a function of the delocalisation of π-electrons around the ring and therefore a measure of aromaticity".[29] Because of the stability conferred on an aromatic molecule by the delocalised system of π-electrons there is a tendency for this system to be retained, or to quote Armit and Robinson[12] yet again "a tendency to retain the type". This is exemplified by the reaction of aromatic compounds with electrophilic reagents to proceed by substitution rather than addition. This type of behaviour, *i.e.* tendency to undergo substitution rather than addition reactions, resistance to oxidation, etc. can be described by the term *aromatic character*.

Aromatic character varies considerably however from one indisputably aromatic compound to another. Thus whereas benzene is stable to air and moisture the cyclopentadienide salts inflame on exposure to these conditions. Even within one series of aromatic compound there is considerable gradation of chemical properties — for example substituted cyclopentadienide salts wherein the substituent groups are electron withdrawing are stable in moist air. Even the classical aromatic compounds, the benzenoid ones, undergo reactions generally considered more typical of normal unsaturated compounds. Thus for example resorcinol decolorises potassium permanganate solution immediately, toluic acid may be reduced by sodium and ethanol, and even benzene itself reacts readily with bromine by addition under the appropriate conditions. One may say indeed that the chemical differences from simple olefins which are frequently regarded as evidence of aromatic character are in reality differences of degree rather than of kind. A roughly quantitative criterion of the aromatic character of a molecule may be obtained from a consideration of its localisation energies[31] for electrophilic, nucleophilic and free radical attack at the most reactive positions in the molecule[28a].

Nor is the "tendency to retain the type" restricted to aromatic compounds only. For example the carbonyl group frequently reacts by substitution, as in the case of carboxylic acid derivatives, although it also reacts by addition in aldehydes or ketones. A recent classification of compounds having multiple bonds[32] suggests that such compounds may be succinctly classified as (1) *aromatic* (as defined above), (2) *pseudoaromatic*[34] (including benzo derivatives), describing molecules which have cyclic formally conjugated

References p. 14

systems of double bonds but which are not aromatic as defined above (*e.g.* cyclic polyolefins with $4n$ π-electrons), (3) *quasi-aromatic* compounds, being compounds which contain acyclic or incomplete cyclic conjugated π-electron systems which show chemical reactions akin to those of aromatic compounds, especially in a tendency to react by substitution with retention of type, and (4) *olefinic* compounds. The first three of these classes are mutually exclusive, and the fourth contains compounds not in the first three. As with many classifications, a compound may fall into one or another class depending on the conditions to which it is subjected. Thus all molecules containing double bonds may react by addition under appropriate conditions making the distinction between (3) and (4) imprecise. None the less this classification does appear to provide a general basis for the classification of compounds having multiple bonds.

REFERENCES

1 F. A. KEKULÉ, *Bull. soc. chim. France*, [2]3 (1865) 98; *Ann.*, 137 (1866) 129; *Ber.*, 2 (1869) 362.
2 Quoted in English by W. BAKER in A. R. TODD (Editor), *Perspectives in Organic Chemistry*, Interscience, New York, 1956, p. 29.
3 F. A. KEKULÉ, *Ann.*, 162 (1872) 77.
4 A. LADENBURG, *Ber.*, 2 (1869) 140, 272.
5 A. CLAUS, *Theoretische Betrachtungen und deren Anwendungen zur Systematik der organischen Chemie*, Freiburg, 1867, p. 207.
6 H. E. ARMSTRONG, *J. Chem. Soc.*, 51 (1887), 258.
7 L. MEYER, *Die moderne Theorie der Chemie*, Second edition, 1872, p. 183; A. BAEYER, *Ann.*, 269 (1892) 145.
8 J. THIELE, *Ann.*, 306 (1899) 87.
9 J. J. THOMSON, *Phil. Mag.*, 27 (1914) 784.
10 E. ERLENMEYER, *Ann.*, 137 (1866) 327.
11 E. BAMBERGER, *Ber.*, 24 (1891) 1758; 26 (1893) 1946; *Ann.*, 273 (1893) 373.
12 J. W. ARMIT AND R. ROBINSON, *J. Chem. Soc.*, 127 (1925) 1604.
13 W. BAKER, *Proc. Chem. Soc.*, (1959) 75.
14 See also W. O. KERMACK AND R. ROBINSON, *J. Chem. Soc.*, 121 (1922) 437; J. W. ARMIT AND R. ROBINSON, *J. Chem. Soc.*, 121 (1922) 827.
15 F. R. GOSS AND C. K. INGOLD, *J. Chem. Soc.*, (1928) 1268.
16 E. HÜCKEL, *Z. Physik*, 70 (1931) 204; 72 (1931) 310; *Grundzüge der Theorie ungesättigter und aromatischer Verbindungen*, Verlag Chemie, Berlin, 1938.
17 W. VON E. DOERING AND L. H. KNOX, *J. Am. Chem. Soc.*, 76 (1954) 3203.
18 J. D. ROBERTS, A. STREITWIESER AND C. M. REGAN, *J. Am. Chem. Soc.*, 74 (1952) 4759.
19 J. E. LENNARD-JONES AND J. TURKEVITCH, *Proc. Roy. Soc.*, A 158 (1937) 297.
20 S. SHIDA AND Z. KURI, *J. Chem. Soc. Japan, Pure Chem. Sect.*, 76 (1950) 322; Y. OSHIKA, *Busseiron Kenkyu*, No. 33 (1955) 95.
21 A. STOCK AND E. POHLAND, *Ber.*, 59 (1926) 2255.
22 R. BRESLOW, *J. Am. Chem. Soc.*, 79 (1957) 5318.

23 D. G. FARNUM AND B. WEBSTER, *J. Am. Chem. Soc.*, 85 (1963) 3502; H. H. FREEDMAN AND A. M. FRANTZ, *J. Am. Chem. Soc.*, 86 (1964) 734.
24 T. J. KATZ, *J. Am. Chem. Soc.*, 82 (1960) 3784.
25 T. J. KATZ AND P. J. GARRATT, *J. Am. Chem. Soc.*, 85 (1963) 2852; E. A. LaLANCETTE AND R. E. BENSON, *J. Am. Chem. Soc.*, 85 (1963) 2853.
26 For a summary see F. SONDHEIMER, *Pure Appl. Chem.*, 7 (1963) 363.
27 R. BRESLOW AND C. YUAN, *J. Am. Chem. Soc.*, 80 (1958) 5991.
28 For further discussion, see *inter alia* (*a*) D. PETERS, *J. Chem. Soc.*, (1960) 1274; (*b*) W. VON E. DOERING, *Tetrahedron*, 11 (1960) 183.
29 J. A. ELVIDGE AND L. M. JACKMAN, *J. Chem. Soc.*, (1961) 859; J. A. ELVIDGE, *Chem. Comm.*, (1965) 160.
30 F. SONDHEIMER, R. WOLOVSKY AND Y. AMIEL, *J. Am. Chem. Soc.*, 84 (1962) 274; L. M. JACKMAN, F. SONDHEIMER, Y. AMIEL, D. A. BEN-EFRAIM, Y. GAONI, R. WOLOVSKY AND A. A. BOTHNER-BY, *J. Am. Chem. Soc.*, 84 (1962) 4307.
31 G. W. WHELAND, *J. Am. Chem. Soc.*, 64 (1942) 900.
32 D. LLOYD AND D. R. MARSHALL, *Chem. and Ind.*, (1964) 1760.
33 C. MENTZER, *Bull. Soc. chim. France*, (1964) 2671.
34 D. P. CRAIG, *J. Chem. Soc.*, (1951) 3175.

CHAPTER II

Derivatives of cyclopropene

INTRODUCTION

According to Hückel's rule the cyclopropenium cation, derived from cyclopropene by loss of a hydride ion, should have aromatic character and stabilisation; the isolation of a stable cyclopropenium salt in 1957[1] provided excellent evidence for the validity of this rule.

This first example of a cyclopropenium salt was prepared by carbene addition to an acetylenic bond to give a cyclopropene derivative, which was treated with boron trifluoride etherate in the presence of a trace of water:

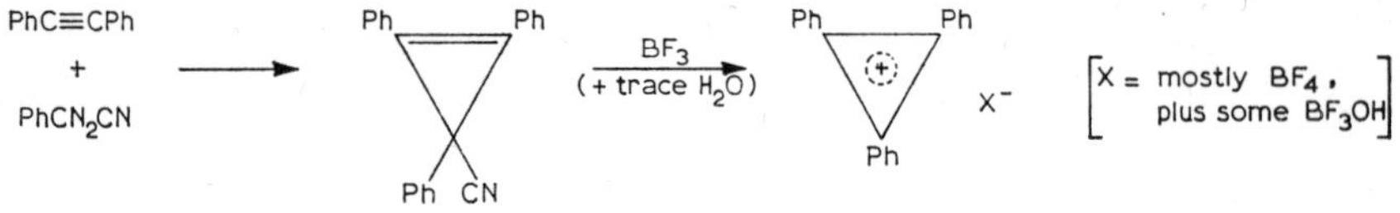

PhC≡CPh + PhCN$_2$CN → [Ph, Ph, Ph, CN triangle] $\xrightarrow{BF_3 \text{ (+ trace } H_2O)}$ [Ph, Ph, Ph triangle with (+)] X$^-$ [X = mostly BF$_4$, plus some BF$_3$OH]

The product had salt-like properties and its n.m.r. spectrum[2] showed the equivalence of the three phenyl groups.

Attention was drawn[2] to the contrast between triphenylcyclopropenium bromide which exists as an ionic salt and triphenylmethylbromide in which the bromine atom is covalently bound. Yet "B-strain"[3] should favour the dissociation of the triphenylmethyl compound more than the triphenylcyclopropenium compound; furthermore the increased electronegativity of the ring carbon atoms in the cyclopropene derivative, due to the altered hybridisation in the strained three-membered ring, should make cation formation less easy than in the case of triphenylmethylbromide. The stable existence of the cyclopropenium cation is thus evidence of the extra stability conferred on a molecule by the aromatic delocalisation of the π-electrons.

In order to demonstrate that the presence of phenyl groups was not essential to the stability of the cyclopropenium ion, dipropyl-[4] and tripropylcyclopropenium[5] salts were prepared. These were again shown to be ionic by their solubility properties and infra-red spectra. The n.m.r. spectrum of the dipropyl compound[5] is consistent with the presence of two equivalent propyl groups attached to a strongly electronegative group. The ring hydrogen shows up as a singlet at very low field.

Attempts to prepare the unsubstituted cyclopropenium cation by hydride extraction from cyclopropene have not as yet been successful, however[6]. It has been calculated[6] that the difference between the heat of formation of the allyl and cyclopropenium cations is only *ca.* 10 kcal/mole, which is considerably less than might be expected from simple molecular orbital calculations. This value may however be reasonable, if the probable increase in strain in going from cyclopropene to a cyclopropenium cation is considered, and if account is taken of the fact that inclusion of electron repulsion integrals will reduce the calculated delocalisation energy of the cyclopropenium ion more than that of the allyl cation.

A crystallographic examination of triphenylcyclopropenium perchlorate confirms that the compound is ionic[23]. It shows that the phenyl groups are not coplanar with the three-membered ring being, on average, at angles of 21° to it. This lack of planarity is probably to accommodate the *o*-hydrogen atoms of adjacent phenyl groups which would otherwise be too close to one another. The bond lengths in the cyclopropenium ring are identical with those in the benzene ring, *i.e.* 1.40 Å; the bonds linking the phenyl groups to the cyclopropenium ring are each 1.45 Å in length.

Owing to the conjugative effect of the carbonyl group, it is possible to have a canonical form of a cyclopropenone molecule with two π-electrons delocalised over three trigonal carbon atoms:

In 1959 diphenylcyclopropenone was isolated as a stable compound[7,8]. This is all the more remarkable when one considers that it has not proved possible to isolate any cyclopropanone derivative with a free carbonyl group present, very possibly owing to the strain that would be present in such a molecule. Once again it seemed possible that stabilisation of this cyclopropene derivative might be due in large measure to the presence of the phenyl groups. That this is not in fact the case is shown by the subsequent isolation of alkyl substituted cyclopropenones[9]. The latter are moderately sensitive to oxygen but in other respects are remarkably stable. Thus whereas diphenylcyclopropenone is completely destroyed by heating at 190° under nitrogen, the dipropyl analogue is only 75% destroyed after 15 min. under the same conditions. Similarly on treatment with alkali dipropylcyclopropenone is recovered unchanged after one hour under conditions in which the cyclopropene ring of diphenylcyclopropenone is 90% cleaved after only three minutes. These facts do not however necessarily signify that the propyl groups confer greater stabilisation on such a molecule than do the phenyl groups. On the contrary it is likely that these results reflect a greater stabilisation by the phenyl groups of the transition states in the decompositions[9].

In accord with the predictions of Hückels rule, no stable cyclopropenide anion or cyclopropyl radical has been isolated, although both may exist as transitory intermediates[10,11,38].

PREPARATION OF CYCLOPROPENIUM SALTS

The majority of preparative methods which have been used for obtaining cyclopropenium salts involve carbene addition to an acetylenic bond.

An example has already been given (p. 16) of the use of a carbene obtained by decomposition of a diazo compound in the preparation of the triphenylcyclopropenium cation[1,2].

Chlorocarbenes, got by the action of potassium tertiary butoxide on benzal chloride have also been used, for example in an alternative preparation of the triphenylcyclopropenium cation[12]:

$$PhCHCl_2 + KOBu^t \longrightarrow Ph\ddot{C}Cl \xrightarrow{PhC\equiv CPh} \underset{Ph\ \ Cl}{\overset{Ph\ \ \ \ \ Ph}{\triangle}} \xrightarrow{OBu^{t-}} \underset{Ph\ \ OBu^t}{\overset{Ph\ \ \ \ \ Ph}{\triangle}} \xrightarrow{HBr} \underset{Ph}{\overset{Ph\ \ \ \ \ Ph}{\triangle^{(+)}}} Br^-$$

The same method has also been used to prepare the diphenylcyclopropenium cation but in this case it proved desirable to isolate a dicyclopropenyl ether as an intermediate; direct treatment of the carbene adduct with hydrobromic acid did not give a satisfactorily pure cation[13].

$$PhC\equiv CH + PhCHCl_2 + KOBu^t \longrightarrow \underset{Ph\ \ OBu^t}{\overset{Ph}{\triangle}} \xrightarrow{H_2O} \underset{Ph}{\overset{Ph}{\triangleright}}-O-\underset{Ph}{\overset{Ph}{\triangleleft}} \xrightarrow{HBr} \underset{Ph}{\overset{Ph}{\triangleright^{(+)}}} Br^-$$

Oxidation of 1,2-dipropyl- and 1,2-diphenyl-cyclopropene carboxylic acids (themselves obtained by carbene addition to the appropriate alkyne) by means of acetyl perchlorate or fluoroborate has afforded a method for obtaining dipropyl-[4,5] and diphenyl-cyclopropenium[14] salts:

This method of preparation is comparable to a general method of obtaining tropylium salts from cycloheptatriene carboxylic acid and its derivatives. (See Chapter V.)

Tripropylcyclopropenium perchlorate has been obtained by reaction of a dipropylcyclopropenium salt with propyl lithium; or by means of the reaction of propyl magnesium bromide with 1-methoxy-2,3-dipropylcyclopropene (obtained by reaction of dipropylcyclopropenium salts with methanol, see below)[5]:

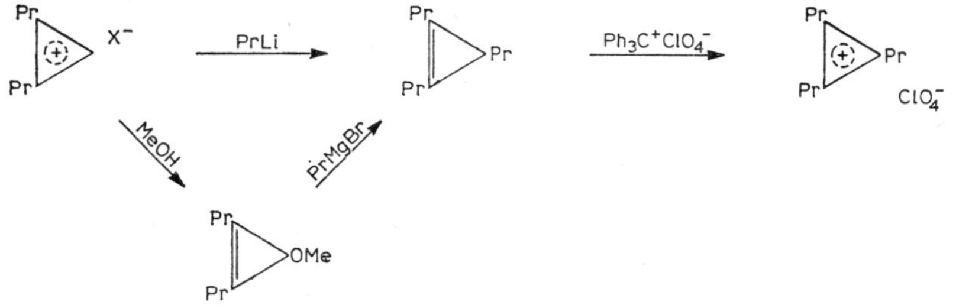

The last step involves hydride extraction from a cyclopropene, a method which has not proved successful when applied to cyclopropene itself[6]. Hydride ion abstraction from 1,2,3-triphenylcyclopropene has also been achieved using 2,3-dichloro-5,6-dicyano-1,4-benzoquinone in the presence of perchloric acid[26].

The reaction of propylmagnesium bromide with a cyclopropenylether, followed by hydride extraction with triphenylmethyl perchlorate has also been used to obtain diphenylpropylcyclopropenium salts:

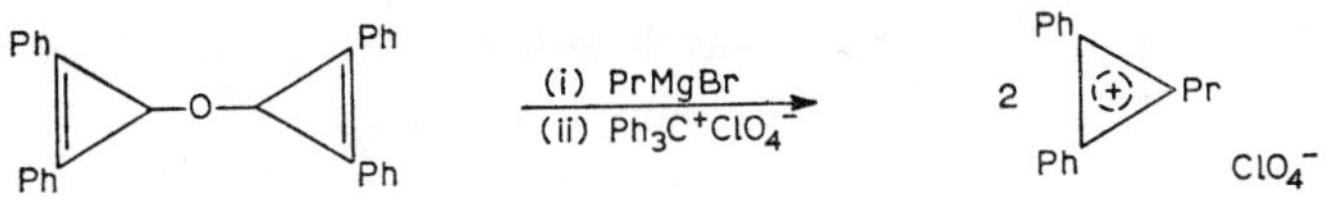

References p. 33

Trichloropropenium salts have been prepared by the action of Lewis acids on tetrachlorocyclopropene[24]:

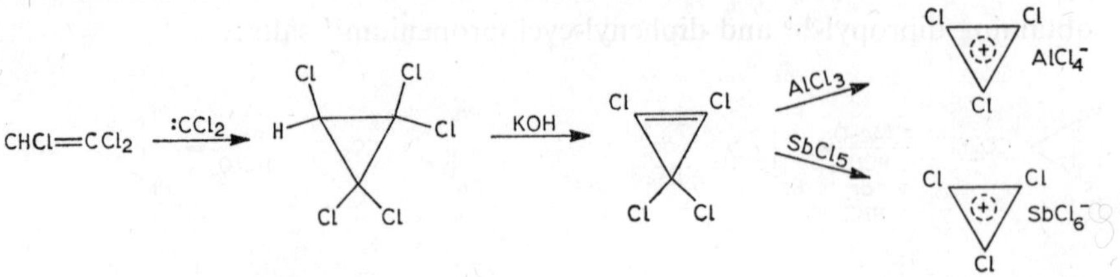

$$CHCl=CCl_2 \xrightarrow{:CCl_2} \text{(pentachlorocyclopropane)} \xrightarrow{KOH} \text{(tetrachlorocyclopropene)} \xrightarrow[SbCl_5]{AlCl_3} \text{trichlorocyclopropenium salts (AlCl}_4^-\text{ or SbCl}_6^-\text{)}$$

The infra-red spectra of the two salts are essentially identical with the exception of bands which may be assigned to the anions, and indicate the presence of a highly symmetrical system. On treatment of either salt with water, tetrachlorocyclopropene is regenerated in fairly good yield (60+%).

Ethoxy-, anilino- and dimethylamino-diphenylcyclopropenium salts have been prepared[36,45] (see below, p. 28).

PROPERTIES OF CYCLOPROPENIUM SALTS

(a) Solubility

As might be expected of salts, cyclopropenium compounds are insoluble in non-polar solvents such as ether, chloroform or benzene. They are soluble in polar solvents such as acetone, acetonitrile and dimethylformamide; they dissolve in alcohols but react with them to form covalent ethers. (see below).

(b) Spectra

An infra-red absorption peak at *ca.* 7.0μ seems to be characteristic of the cyclopropenium ion[5]. Alkyl substituted cyclopropenium ions show only end absorption in the ultra violet but aryl substituted ones show intense absorption in the ultra-violet (*e.g.* diphenyl[14], λ_{max} 305,292,246 mμ; log ε_{max} 4.52, 4.30, 4.03; triphenyl[2], λ_{max} 320, 307, 255 mμ; log ε_{max} 4.62, 4.67, 4.20). The presence of *p*-methoxy groups in the phenyl rings shifts the position of the long wavelength peak to longer wavelength[12], *e.g.* triphenyl λ_{max} 307 mμ; *p*-anisyldiphenyl, λ_{max} 341 mμ; di-*p*-anisylphenyl, λ_{max} 352 mμ; tri-*p*-anisyl, λ_{max} 359 mμ). The n.m.r. spectra are mentioned above (p. 16).

(c) pK Values and stability

Cyclopropenium salts are soluble and stable in acid solution but neutral aqueous solutions become turbid owing to equilibration with a covalent carbinol:

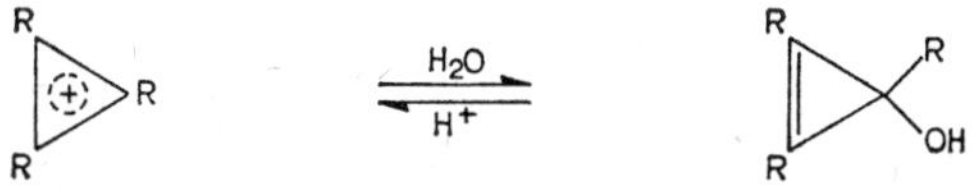

It has been suggested[5] that the pH value at which there is 50% ionisation of the carbinol to cyclopropenium ion, described as pK_R^+, can be taken as a criterion of the stability of the cation.

Some values for pK_R^+ which have been determined experimentally, some by potentiometric titration and others spectrophotometrically are as follows[5]:

Substituent groups	pK_R^+
Dipropyl	2.7
Tripropyl	7.2
Diphenyl	0.32–0.67
Triphenyl	3.1, 2.8
p-Anisyldiphenyl	4.0
Di-p-anisylphenyl	5.2, 5.2
Tri-p-anisyl	6.5, 6.4
Diphenylpropyl	3.8

A number of factors other than π-electron stabilisation will affect the position of equilibrium so that the actual values of pK_R^+ are much less easy to interpret than their relative values.

The above figures make it appear that propyl groups stabilise the cation form more effectively than do phenyl groups. This could be due to two factors, viz. (1) that the propyl groups do, in fact, stabilise the cation more than phenyl groups, and (2) that the covalent carbinol is more effectively stabilised by phenyl than by propyl groups. If one compares the diphenyl and diphenylpropyl compounds the second of these possible effects should be about the same in each compound, yet the pK_R^+ value is greater in the case of the latter. This suggests that the propyl group does indeed have a stabilising effect on the cyclopropenium cation; it seems likely that this is due to an inductive rather than a hyperconjugative effect[5].

(d) Action of nucleophiles

Not surprisingly the cyclopropenium ring is attacked by nucleophilic reagents. An example has just been given in the reaction with water to form a covalent carbinol. Alcohols and cyanide ions react similarly to give, respectively, covalent ethers[2,5] and nitriles[1] e.g.

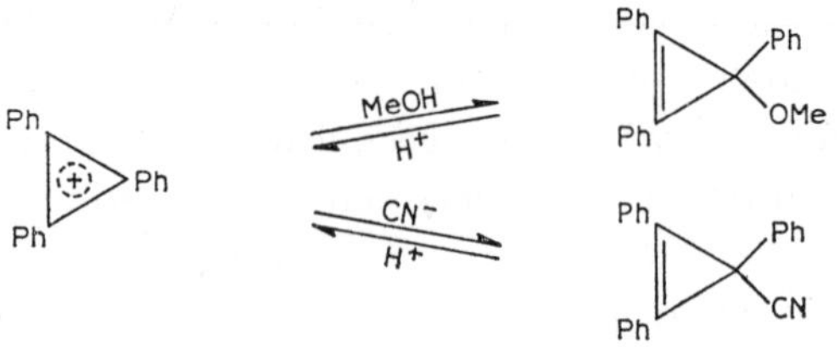

Aqueous sodium bicarbonate reacts with diphenylcyclopropenium perchlorate to form 3,3-bis(1,2-diphenylcyclopropenyl)ether[14].

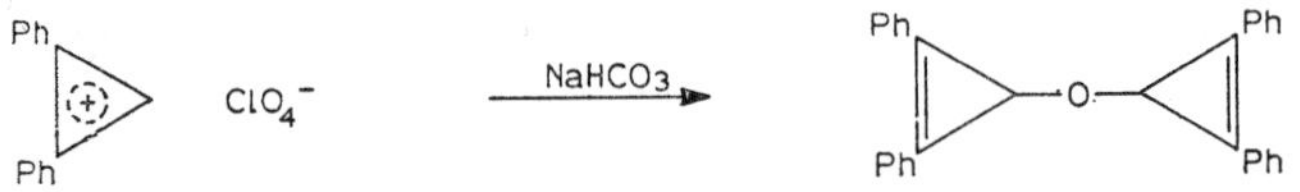

Presumably the carbinol is formed first and then makes a nucleophilic attack on another diphenylcyclopropenium cation. Aqueous hydroxide causes ring-opening and formation of an unsaturated carbonyl compound, presumably by nucleophilic attack followed by arrangement to a keto form:

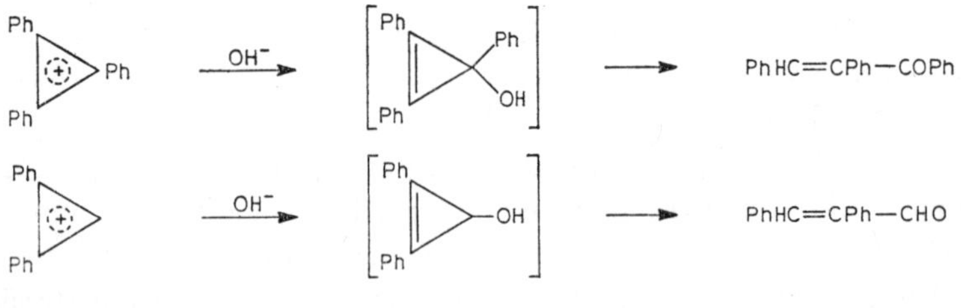

$$PhHC=CPh-COPh$$

$$PhHC=CPh-CHO$$

The cyclopropenium ring can be alkylated be means of lithium alkyls[5] or Grignard reagents[38].

Activated aromatic compounds may be alkylated in the benzene ring by reaction with cyclopropenium salts[27]:

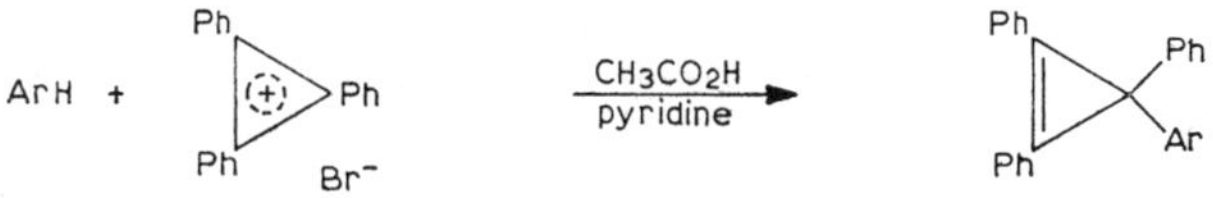

Ar = p-dimethylaminophenyl; p-anisyl; 2,4-dimethoxyphenyl; 2-methoxynaphthyl; anthron-10-yl

The trichlorocyclopropenium cation has also been shown to effect electrophilic substitution in a benzene ring[28].

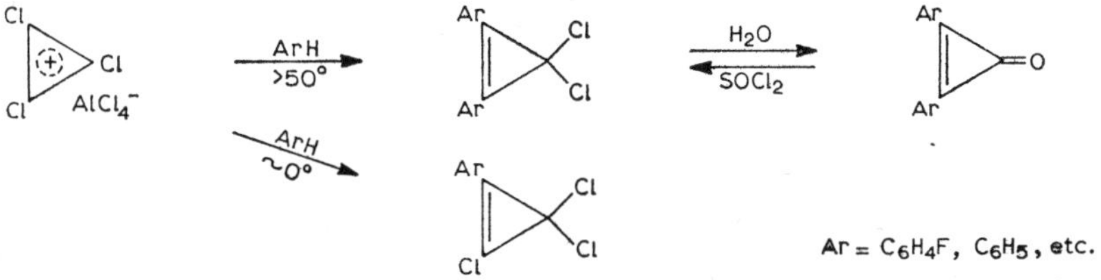

Ar = C_6H_4F, C_6H_5, etc.

The 1,2-diaryl-3,3-dichlorocyclopropene is hydrolysed to a diarylcyclopropenone during aqueous work up of the reaction[28].

(e) Reduction

The triphenylcyclopropenium cation is reduced to triphenylcyclopropene in good yield by means of lithium aluminium hydride[38]. Zinc brings about a bimolecular reduction[11]:

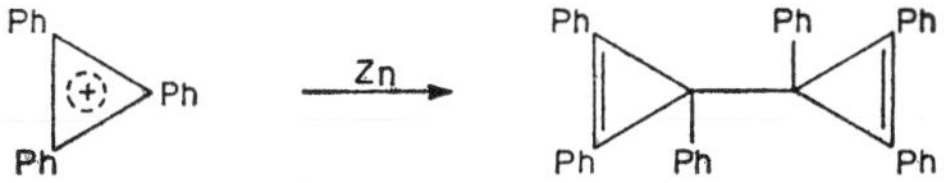

The product is converted into hexamethylbenzene on heating. The diphenylcyclopropenium cation is also reduced to a bimolecular product by this means[11].

(f) Reaction with azides and diazo-compounds

Sodium azide reacts with the triphenylcyclopropenium cation to give an azidotriphenylcyclopropene, which slowly rearranges to 4,5,6-triphenyl-1,2,3-triazine[39].

References p. 33

The product of the reaction between triphenylcyclopropenium bromide and phenyldiazomethane is 1,2,3-triphenylazulene[40]. It is suggested that the first product is again a cyclopropene derivative which loses nitrogen and undergoes a series of intramolecular rearrangements to provide the final product.

(g) Formation of metal complexes

Triphenylcyclopropenium bromide reacts with the cobalt tetracarbonylate anion to form a complex[15] which probably has the structure:

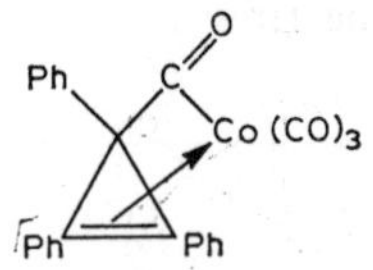

The $[Fe(CO)_3NO]^{3-}$ ion reacts similarly.

Nickel tetracarbonyl also reacts with triphenylcyclopropenium bromide to form a complex; carbon monoxide is evolved in the reaction[41].

PREPARATION OF CYCLOPROPENONE DERIVATIVES

The preparation of cyclopropenone derivatives has also usually involved carbene addition to multiple bonds. The first recorded examples involved halocarbene addition to diphenylacetylene in one case[8] and to phenylketene dimethylacetal in the other[7] (see also ref. 36):

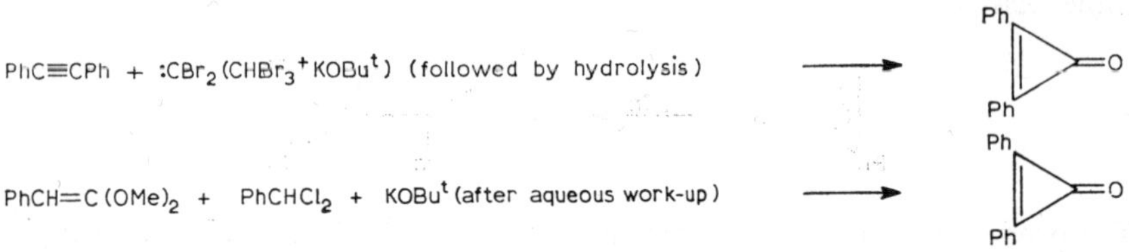

$PhC\equiv CPh$ + $:CBr_2$ $(CHBr_3 + KOBu^t)$ (followed by hydrolysis) ⟶

$PhCH=C(OMe)_2$ + $PhCHCl_2$ + $KOBu^t$ (after aqueous work-up) ⟶

The second example involves a dehydrohalogenation step as well as carbene addition. Diphenylcyclopropenone has also been obtained by reaction of 1-methoxy-2-phenylacetylene with phenylchlorocarbene (from benzal chloride and potassium tertiary butoxide)[36] and by a Favorskii reaction on di-(bromobenzyl)ketone (see below)[36]. The latter is probably the best method of preparation of diphenylcyclopropenone[36].

Dipropylcyclopropenone has been prepared similarly by addition of dichlorocarbene to oct-4-yne, the dichlorocarbene being generated either from the decarboxylation of sodium trichloroacetate or from the reaction between sodium methoxide and ethyl trichloroacetate[9]:

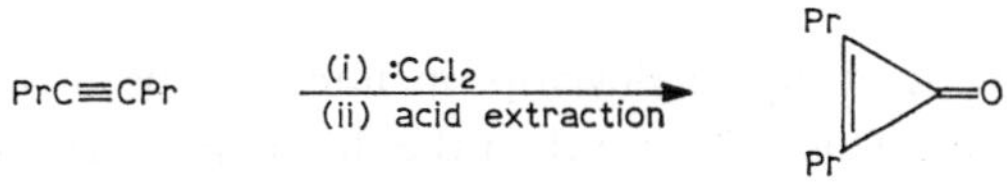

$$PrC{\equiv}CPr \xrightarrow[\text{(ii) acid extraction}]{\text{(i) :CCl}_2} \text{[2,3-dipropylcyclopropenone]}$$

More recently the Favorskii reaction[16] has been used to prepare cyclopropenone derivatives from certain $\alpha\alpha'$-dibromoketones[17,36,37] *e.g.*

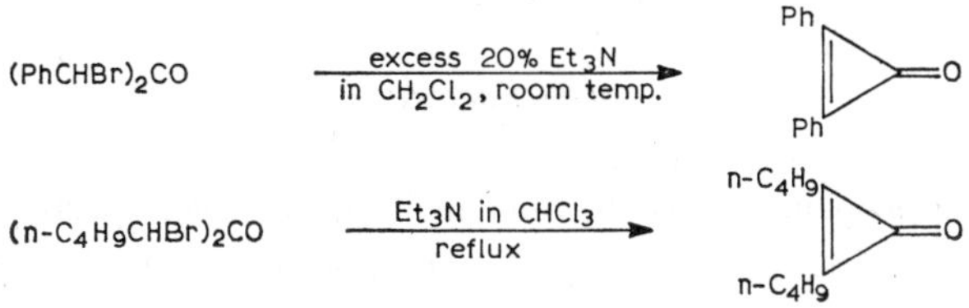

$$(PhCHBr)_2CO \xrightarrow[\text{in CH}_2\text{Cl}_2, \text{room temp.}]{\text{excess 20\% Et}_3\text{N}} \text{[2,3-diphenylcyclopropenone]}$$

$$(n\text{-}C_4H_9CHBr)_2CO \xrightarrow[\text{reflux}]{\text{Et}_3\text{N in CHCl}_3} \text{[2,3-di-}n\text{-butylcyclopropenone]}$$

By the same method a bicyclic cyclopropenone derivative has been obtained from 2,8-dibromocyclooctanone:

[Structural scheme: 2,8-dibromocyclooctanone $\xrightarrow{\text{Et}_3\text{N}}$ bicyclic cyclopropenone $\xrightarrow[\text{reflux}]{\text{aq. KOH}}$ cycloheptene carboxylic acid]

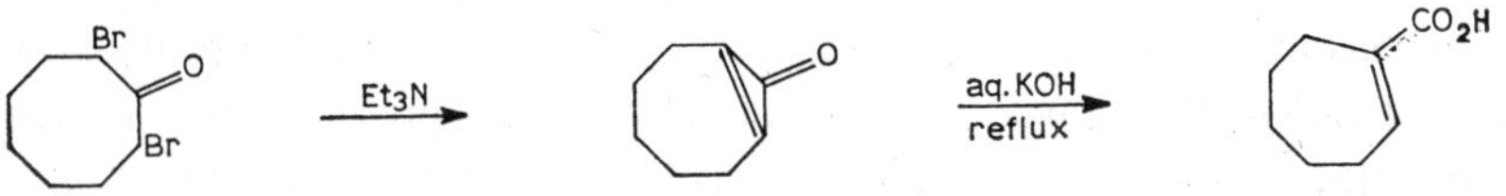

On being refluxed with aqueous alkali this product is converted into the normal product expected from a Favorskii ring-contraction reaction. Other bicyclic cyclopropenones have been prepared in the same way[37].

So far the method has failed to yield either unsubstituted cyclopropenone or a cyclohexenocyclopropenone; this may be due to their instability and immediate decomposition on formation. It was also not possible to obtain an analytically pure sample of dimethylcyclopropenone, although the product had the expected n.m.r. and i.r. spectra[37]. Some 1-diethylamino-2-phenylcyclopropenone is obtained when 1,1,3-trichloro-3-phenylpropan-2-one is treated with a mixture of diethylamine and triethylamine[37].

2-Hydroxy-3-phenylcyclopropenone has been prepared by the reaction of 1,1,3,3-tetrachloro-2-phenylpropene with potassium tertiary butoxide in ether[29]; presumably an intramolecular carbene addition reaction is involved. The product is strongly acidic and gives colours with ferric chloride (*cf.* tropolone, p. 141); it decolorises bromine water and reduces both potassium permanganate and silver nitrate[29].

The preparation of diarylcyclopropenones from trichlorocyclopropenium salts[28] mentioned above (p. 23) may also be noted.

References p. 33

PROPERTIES OF CYCLOPROPENONES

(a) Dipole moments

The dipole moment of diphenylcyclopropenone has been recorded as 5.08 D[8,30], and 5.14 D[36]. This is higher than the dipole moment of tropone. The dipole moments of dialkylcyclopropenones are slightly lower (*e.g.* dipropylcyclopropenone, 4.78 D) presumably since interaction by the phenyl groups increases the length of the dipole[37].

(b) Spectra

Diphenylcyclopropenone has maxima in its ultra-violet spectrum at λ 310 (shoulder), 297, 282, 226, 220 mμ (log ε 4.04, 4.3, 4.25, 4.13, 4.16)[7] but dipropylcyclopropenone shows only end absorption in the ultra-violet region[9,37] (*cf.* cyclopropenium salts, above).

The infra-red spectrum (in carbon disulphide) of the diphenyl compound[7,30,31,36] has in addition to bands due to the phenyl groups other peaks at 5.4, 6.1 and 7.45 μ; strong peaks at 5.4 and 6.1 μ are also present in the spectrum of the dipropyl compound[9,37]. There has been some disagreement about the assignment of the peaks at 5.4 and 6.1 μ, but it is suggested that according to the solvent dependence of the positions of these peaks the one at 5.4 μ is associated with the C=C bond in the three-membered ring and the one at 6.1 μ with C=O stretching[42]. This value for the C=O stretching band indicates a considerable contribution from the dipolar form. The peak at 7.45 μ has been attributed to C=O bending. In hydrobromic acid the carbonyl frequencies disappear and a broad band at 3.36 μ due to a hydroxyl group appears instead[31,32].

The n.m.r. spectrum of dipropylcyclopropenone has the form expected for propyl groups attached to electron withdrawing substituents[9,37]. It does not strongly resemble the spectrum of a cyclopropenium compound but resembles rather more nearly that of a cyclopropene derivative[37].

(c) Action of acid

Cyclopropenones are reversibly converted into hydroxycyclopropenium salts by the action of strong acids[8,9,18]:

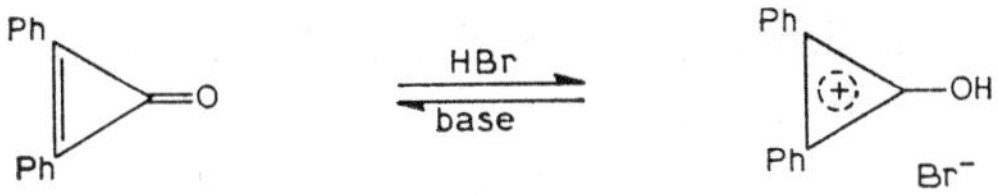

Diphenylcyclopropenone is less basic than tropone[18]. The dipropyl compound is more basic than the diphenyl one; for example, unlike the latter, the former is completely extracted from carbon tetrachloride by 12N hydrochloric acid[9]. Addition of sodium bicarbonate regenerates the ketone.

(d) Action of alkali

The three-membered ring in diphenylcyclopropenone is readily cleaved by alkali to give α-phenylcinnamic acid, but dipropylcyclopropenone is not affected under identical conditions[9,37]. For a further discussion see above, p. 18. In the presence of alkoxide ions the methylene groups adjacent to the ring in dipropylcyclopropenone readily undergo deuterium exchange[19].

(e) Thermal decomposition

Cyclopropenones melt reversibly but decompose on stronger heating with evolution of carbon monoxide[7–9,17,36,37]. Thus diphenyl cyclopropenone breaks down to carbon monoxide and diphenylacetylene on heating to 160° or above[36]. At 145–150° an alternative product is also formed, which appears to be a dimer of diphenylcyclopropenone; this product does not give diphenylacetylene on pyrolysis showing that the formation of the acetylene and of the dimer are competing reactions and that formation of the dimer is not an intermediate step in acetylene formation[36]. Dipropylcyclopropenone is similarly broken down at 190° to give oct-4-yne and carbon monoxide[37]. Cycloundecenocyclopropenone is pyrolysed at 210° to carbon monoxide and cycloundecyne but cycloheptenocyclopropenone gives carbon monoxide and tricycloheptenobenzene[37]:

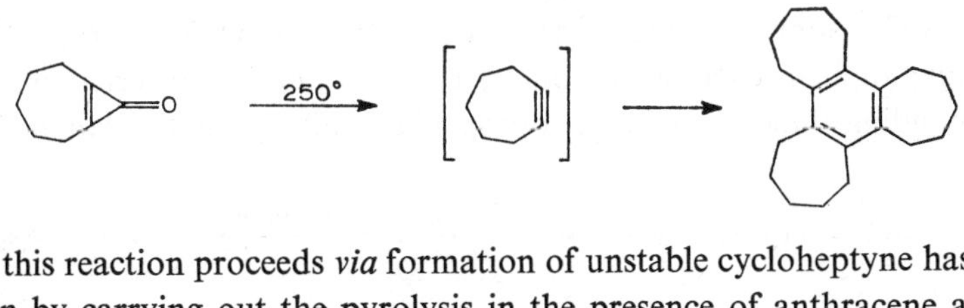

That this reaction proceeds *via* formation of unstable cycloheptyne has been shown by carrying out the pyrolysis in the presence of anthracene and of tetraphenylcyclopentadienone, when adducts of these compounds and cycloheptyne have been obtained[37].

(f) Reduction of cyclopropenones

Diphenylcyclopropenone absorbs two moles of hydrogen on reduction[8,36]. According to one group of workers the product obtained is 2,3-diphenylcyclopropanol[8]; other workers have obtained dibenzyl ketone as the reduction product[36]. Dipropylcyclopropenone is reduced to di-n-butyl ketone in the presence of platinum, and to α-phenylcinnamaldehyde in the presence of palladium on carbon[37]. Both the carbonyl group and carbon–carbon double bond of diphenylcyclopropenone are reduced by lithium aluminium hydride[8].

(g) Other reactions of cyclopropenones

The carbonyl group in diarylcyclopropenones has been converted into a *gem*-dichloro group by means of thionyl chloride[28], by phosgene[45] and by phosphorus pentachloride in chloroform[44]. The ketone is regenerated from the dichlorocompound by reaction with water[44]. Phosphorus pentasulphide reacts with diphenylcyclopropenone to give the corresponding thioketone[44].

It has been claimed that diphenylcyclopropenone forms a 2,4-dinitrophenylhydrazone[8] but the derivative has also been described as ill-defined[36]. The formation of an oxime has also been described[44] but attempts by other workers to obtain the oxime have produced instead a mixture of deoxybenzoin oxime and 3,4-diphenylisooxazolone, reaction presumably proceeding in the first step by conjugate addition of the hydroxylamine to the unsaturated ketone[36].

The reaction of phenyl magnesium bromide with diphenylcyclopropenone gives a bis (triphenylcyclopropenyl)ether which on treatment with perchloric acid is cleaved to give triphenylcyclopropenium perchlorate[36]. Similar reactions with some aliphatic Grignard reagents do not work as satisfactorily[36].

Diphenylcyclopropenone reacts with triethyloxonium fluoroborate to form an ethoxy-diphenylcyclopropenium salt; the latter is converted into the corresponding dimethylaminocyclopropenium salt by dimethylamine[36]:

The ethoxy derivative is hydrolysed very readily but the dimethylamino derivative is very stable and can be recrystallised from water without decomposition[36]. The stability of the latter compound has been ascribed to the delocalisation of the positive charge over the exocyclic nitrogen atom as well as over the three-membered ring. The corresponding anilino and N-methylanilino derivatives have also been made, by the action of the appropriate amine on 3,3-dichloro-1,2-diphenylcyclopropene[45], but the N,N-dimethylamino derivative could not be obtained in this way. When ethoxy-diphenylcyclopropenium fluoroborate is treated with N,N-dimethylaniline, a p-dimethylaminophenyl-diphenylcyclopropenium salt results[45].

An interesting reaction of dipropylcyclopropenone which has been investigated is that with diazomethane[19]. Unlike keten or cycloalkanones, which react by ring expansion to give a cycloalkanone of larger ring size, dipropylcyclopropenone gives a pyridazone. Diphenylcyclopropenone reacts similarly[36]. Such a product could be formed by attack at either the carbonyl group or the olefinic bond but the structure of the product indicates the latter:

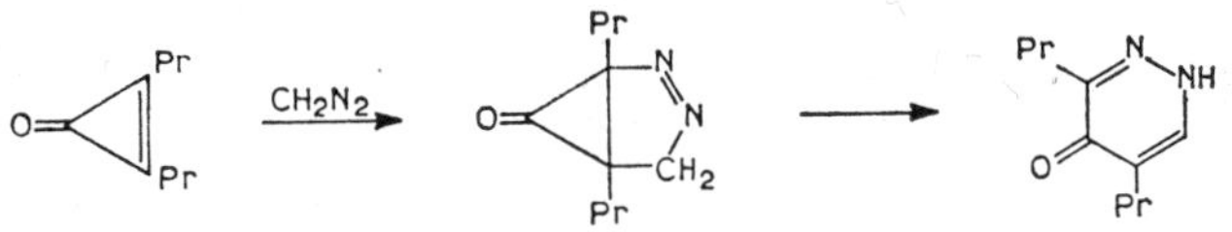

The reaction is thus identical to that which takes place with a simple $\alpha\beta$-unsaturated ketone. A further example of behaviour analogous to that of an $\alpha\beta$-unsaturated ketone is provided by the fact that the α-methylene groups in dipropylcyclopropenone undergo rapid alkoxide deuterium exchange in perdeuteromethanol as solvent[19]. Diphenylcyclopropenone is reported to form a nickel carbonyl complex[25].

Diphenylcyclopropenone reacts with 1-diethylaminobuta-1,3-diene to give 2,7-diphenyltropone in good yield. Reaction probably proceeds by a Diels–Alder reaction followed by elimination of the diethylamino group[46].

QUINOCYCLOPROPENES

Quinocyclopropenes have the structural formula:

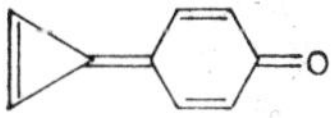

They can therefore be regarded as "extended" cyclopropenones and will more properly be represented as resonance hybrids of this covalent form and a dipolar one:

Examples of this type of compound have been synthesised as shown in the following chart[20]:

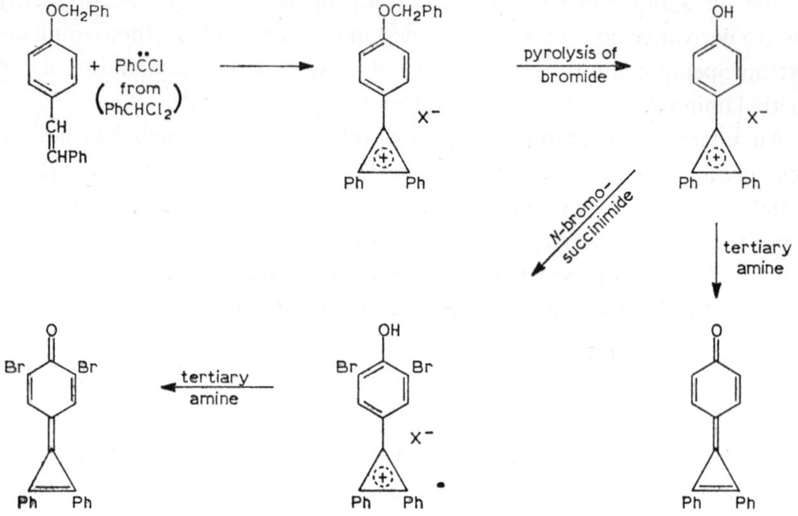

The quinocyclopropenes were reconverted into cyclopropenium salts by acid. The unbrominated compound could not be isolated as a solid but the dibromocompound was obtained as orange needles which were stable to light and to prolonged heating at 140°. (Decomposition set in above 230°). It was readily destroyed by hydroxylic solvents and, as suggested by molecular orbital calculations, was more basic than diphenylcyclopropenone.

An alternative method of preparing a quinocyclopropene utilised the reaction between a cyclopropenium salt and anthrone[27]:

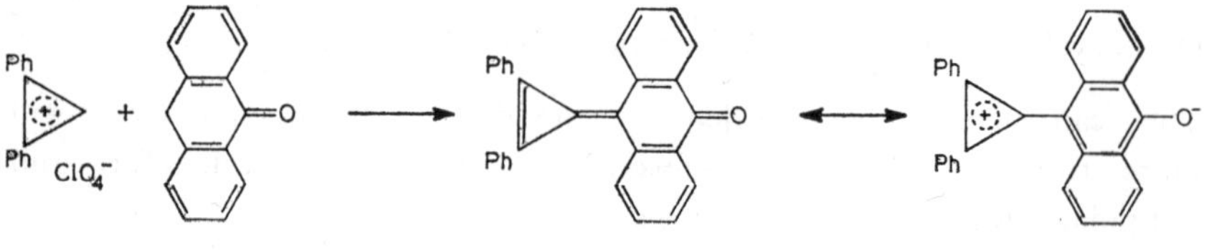

Reaction probably proceeds by alkylation of anthrone (*cf.* p. 23) to give a 9-cyclopropenylanthrone which is then dehydrogenated by reaction with further cyclopropenium cations. This quinocyclopropene is thermally very stable. The contribution of the dipolar form is shown by the infra-red spectrum, the absorption due to the carbonyl group being at much lower wavelength than in the case of 9-methyleneanthrone.

More recent work[45] has however cast doubts on the identity of this product and an alternative product, obtained by treating 1-ethoxy-2,3-diphenylcyclopropenium fluoroborate or 3,3-dichloro-1,2-diphenylcyclopropene with anthrone, followed by triethylamine, has been assigned this formula. The newer product has a dipole moment of 9.4 D; this and its i.r. spectrum indicate that there is a relatively large contribution from the dipolar form.

METHYLENE CYCLOPROPENES

An example of a methylene cyclopropene has been synthesised as follows[21]:

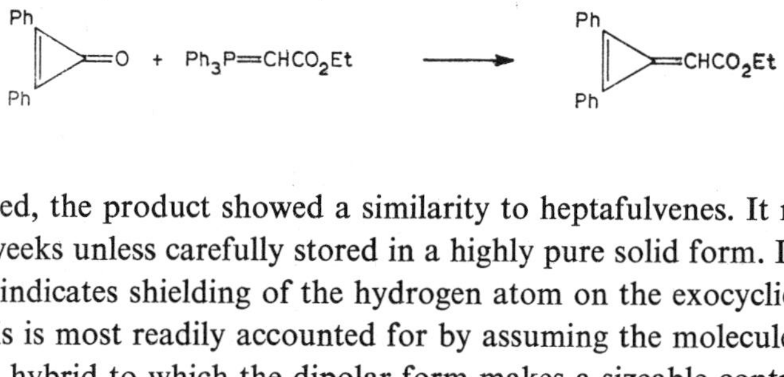

As expected, the product showed a similarity to heptafulvenes. It resinified in a few weeks unless carefully stored in a highly pure solid form. Its n.m.r. spectrum indicates shielding of the hydrogen atom on the exocyclic carbon atom. This is most readily accounted for by assuming the molecule to be a resonance hybrid to which the dipolar form makes a sizeable contribution:

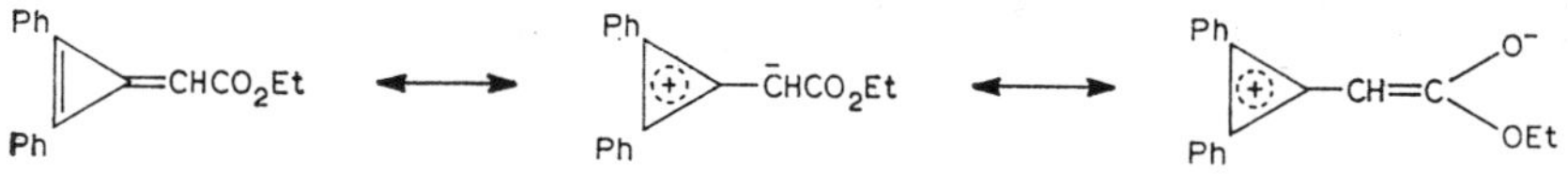

The effect of solvent on the long wavelength absorption band in the ultraviolet spectrum also supports the idea of a highly polar ground state. This compound has also since been prepared by the following methods[33]:

References p. 33

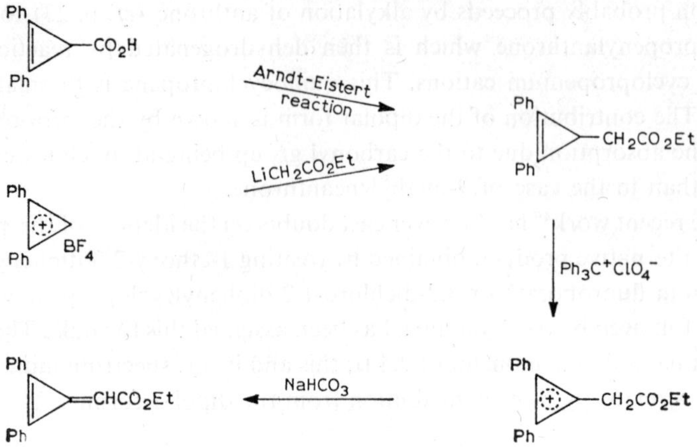

Another example of a methylenecyclopropene which has been synthesised is the following molecule[22]:

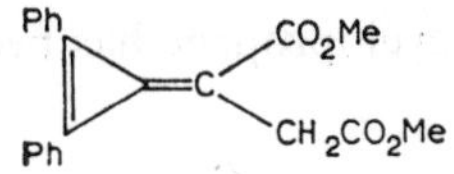

Both of these methylenecyclopropenes react with acids to form cyclopropenium salts, *e.g.*

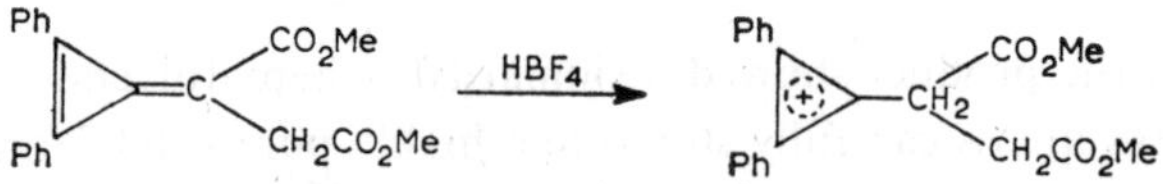

Dicyanomethylenecyclopropenes have been prepared by condensation of cyclopropenones with dicyanomethane[34,35,44,47]:

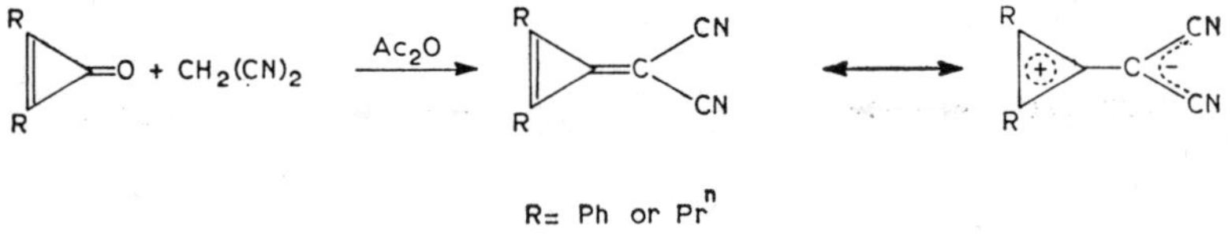

R = Ph or Pr^n

Both compounds are unprotonated in trifluoroacetic acid[34,35]. The contribution of the dipolar form in the case of the diphenyl compound is indicated by the dependence of the wavelength of u.v. light absorption on the nature of the solvent and by the dipole moment (= 7.9 D)[34]. The n.m.r. spectrum of the di-n-propyl compound indicates that the cyclopropene ring bears a positive charge[35]. The latter compound does not react with bromine in methylene chloride in the cold[35]. A similar product is obtained from diphenylcyclopropenone and cyanoacetic ester[47]. It also is not protonated in strong acid. It has a dipole moment of 5.9 D and this and its i.r., u.v. and n.m.r. spectra again clearly accord with considerable charge separation in the molecule.

The relative stability of all these methylenecyclopropenes can be associated with the presence of substituent groups at the exocyclic carbon atom which can delocalise the charge on this atom. Theoretical calculations suggest that methylenecyclopropene itself should polymerise readily and be highly reactive towards free radicals[43].

CALICENES

Calicene is a trivial name for cyclopropenylidenecyclopentadiene:

Calicene should exist as a hybrid of dipolar and covalent forms. The parent hydrocarbon has not yet been prepared but derivatives of it are known. They are discussed in Chapter VIII (pag. 204).

REFERENCES

1 R. Breslow, *J. Am. Chem. Soc.*, 79 (1957) 5318.
2 R. Breslow and C. Yuan, *J. Am. Chem. Soc.*, 80 (1958) 5991.
3 H. C. Brown and R. S. Fletcher, *J. Am. Chem. Soc.*, 71 (1949) 1845; H. C. Brown, H. Bartholomay and M. D. Taylor, *J. Am. Chem. Soc.*, 66 (1944) 435.
4 R. Breslow and H. Höver, *J. Am. Chem. Soc.*, 82 (1960) 2644.
5 R. Breslow, H. Höver and H. W. Chang, *J. Am. Chem. Soc.*, 84 (1962) 3168.
6 K. B. Wiberg, W. J. Bartley and F. P. Lossing, *J. Am. Chem. Soc.*, 84 (1962) 3980.
7 R. Breslow, R. Haynie and J. Mirra, *J. Am. Chem. Soc.*, 81 (1959) 247.
8 M. E. Vol'pin, Y. D. Koreshkov and D. N. Kursanov, *Izv. Akad. Nauk S.S.S.R., Otdel. khim. Nauk*, (1959) 560; *Zh. Obshch. Khim.*, 30 (1960) 2877.
9 R. Breslow and R. Peterson, *J. Am. Chem. Soc.*, 82 (1960) 4426.
10 R. Breslow and M. Battiste, *Chem. and Ind.*, (1958) 1143; R. Breslow, W. Bahary and W. Reinmuth, *J. Am. Chem. Soc.*, 83 (1961) 1763.
11 R. Breslow and P. Gal, *J. Am. Chem. Soc.*, 81 (1959) 4747.
12 R. Breslow and H. W. Chang, *J. Am. Chem. Soc.*, 83 (1961) 2367.
13 R. Breslow, J. Lockhart and H. W. Chang, *J. Am. Chem. Soc.*, 83 (1961) 2375.

14 D. G. FARNUM AND M. BURR, *J. Am. Chem. Soc.*, 82 (1960) 2651.
15 C. E. COFFEY, *J. Am. Chem. Soc.*, 84 (1962) 118.
16 A. S. KENDE, *Org. Reactions, XI*, Wiley, New York, 1960, p. 261.
17 R. BRESLOW, J. POSNER AND A. KREBS, *J. Am. Chem. Soc.*, 85 (1963) 234.
18 B. E. ZAITSEV, Y. D. KORESHKOV, M. E. VOL'PIN AND Y. N. SHEINKER, *Dokl. Akad. Nauk S.S.S.R.*, 139 (1961) 1107.
19 P. T. IZZO AND A. S. KENDE, *Chem. and Ind.*, (1964) 839.
20 A. S. KENDE, *J. Am. Chem. Soc.*, 85 (1963) 1882.
21 M. A. BATTISTE, *J. Am. Chem. Soc.*, 86 (1964) 942.
22 W. M. JONES AND J. M. DENHAM, *J. Am. Chem. Soc.*, 86 (1964) 944.
23 M. SUNDARALINGAM AND L. H. JENSEN, *J. Am. Chem. Soc.*, 85 (1963) 3302.
24 S. W. TOBEY AND R. WEST, *J. Am. Chem. Soc.*, 86 (1964) 1459.
25 C. W. BIRD AND E. M. HOLLINS, *Chem. and Ind.*, (1964) 1362.
26 D. H. REID, M. FRASER, B. B. MOLLOY, H. A. S. PAYNE AND R. G. SUTHERLAND, *Tetrahedron Letters*, (1961) 530.
27 B. FÖHLISCH AND P. BÜRGLE, *Angew. Chem.*, 76 (1964) 784.
28 S. W. TOBEY AND R. WEST, *J. Am. Chem. Soc.*, 86 (1964) 4215.
29 D. G. FARNUM AND P. E. THURSTON, *J. Am. Chem. Soc.*, 86 (1964) 4206.
30 Y. G. BOROD'KO AND Y. R. SYRIKIN, *Dokl. Akad. Nauk S.S.S.R.*, 134 (1960) 127.
31 B. E. ZAITSEV, Y. N. SHEINKER AND Y. D. KORESHKOV, *Dokl. Akad. Nauk S.S.S.R.*, 136 (1961) 1090.
32 V. D. ZAITSEV, Y. N. SHEINKER, Y. D. KORESHKOV AND M. E. VOL'PIN, *Fiz. Probl. Spektroscopii. Akad. Nauk S.S.S.R. Materialy 13-go [Trinadtsatogo] Soveshch. Leningrad*, 1 (1960) 442; *Chem. Abs.*, 59 (1963) 12303.
33 W. M. JONES AND R. S. PYRON, *Tetrahedron Letters*, (1965) 479.
34 E. D. BERGMANN AND I. AGRANAT, *J. Am. Chem. Soc.*, 86 (1964) 3587.
35 A. S. KENDE AND P. T. IZZO, *J. Am. Chem. Soc.*, 86 (1964) 3587.
36 R. BRESLOW, T. EICHER, A. KREBS, R. A. PETERSON AND J. POSNER, *J. Am. Chem. Soc.*, 87 (1965) 1320.
37 R. BRESLOW, L. J. ALTMAN, A. KREBS, E. MOHACSI, I. MURATA, R. A. PETERSON AND J. POSNER, *J. Am. Chem. Soc.*, 87 (1965) 1326.
38 R. BRESLOW AND P. DOWD, *J. Am. Chem. Soc.*, 85 (1963) 2729.
39 E. CHANDROSS AND E. SMOLINSKY, *Tetrahedron Letters*, No. 13 (1960) 19; R. BRESLOW, R. BOIKESS AND M. BATTISTE, *Tetrahedron Letters*, No. 26 (1960) 42.
40 R. BRESLOW AND M. MITCHELL, unpublished work quoted by R. BRESLOW in P. DE MAYO (Editor), *Molecular Rearrangements*, Vol. I, Interscience, New York, 1963, p. 276.
41 E. W. GOWLING AND S. F. A. KETTLE, *Inorg. Chem.*, 3 (1964) 604.
42 A. W. KREBS, *Angew. Chem.*, 77 (1965) 10.
43 J. D. ROBERTS, A. STREITWIESER AND C. M. REGAN, *J. Am. Chem. Soc.*, 74 (1952) 4579.
44 Y. KITAHARA AND M. FUNAMIZU, *Bull. Chem. Soc. Japan*, 37 (1964) 1897.
45 B. FÖHLISCH AND P. BÜRGLE, *Tetrahedron Letters*, (1965) 2661.
46 J. CIABOTTONI AND G. A. BERCHTOLD, *J. Am. Chem. Soc.*, 87 (1965) 1404.
47 S. ANDREADES, *J. Am. Chem. Soc.*, 87 (1965) 3940.

CHAPTER III

Derivates of cyclobutadiene

INTRODUCTION

The cyclobutadiene system has four π-electrons and thus does not conform to the requirements of Hückel's rule for aromaticity. Two of these electrons are in a bonding orbital and two are in non-bonding orbitals and have unpaired spins. The molecule may thus be a diradical. The conjugation energy of cyclobutadiene is zero[1,2,86]. Molecular orbital calculations suggest that the molecule should not have four sides of equal length, but be rectangular with two single bonds and two double bonds[73,86].

In view of these considerations it is not altogether surprising that cyclobutadiene has never yet been isolated. It seems probable that this molecule is unlikely to be stable enough to be isolable, although there is strong evidence that it has been generated as a transient intermediate.

As far back as 1894 an attempt was made to prepare cyclobutadiene-1,2-dicarboxylic acid by the action of alkali on 1,2-dibromocyclobutane-1,2-dicarboxylic acid[3]; only cyclobutene derivatives were obtained.

The first attempt to prepare cyclobutadiene itself involved the action of bases on 1,2-dibromocyclobutane[4]. When the latter compound was heated with potassium hydroxide at 100–105° 1-bromocyclobutene was formed. This compound is a vinyl bromide and proved very stable, only reacting with potassium hydroxide at 210° to give acetylene in 30% yield.

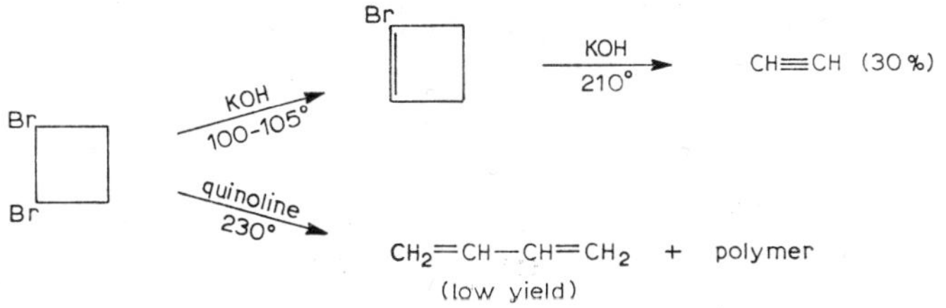

On heating the dibromocyclobutane with quinoline at 230° a small yield of 1,3-butadiene was obtained, together with coloured polymeric material[4].

It has frequently been assumed that cyclobutadiene is in fact formed in the alkaline decomposition of bromocyclobutene and that it at once splits into

References p. 51

two molecules of acetylene. Should this be the case the conversion cannot be a straightforward process but would have to involve a migration of a hydrogen atom. Such a rearrangement is not impossible at the elevated temperature involved but the formation of acetylene in this experiment certainly cannot be regarded as proof of the formation of cyclobutadiene as an unstable intermediate[5]. It has also been suggested[6] that the intermediate cyclobutene was not in fact the 1-bromo-compound but rather the 3-bromo compound, which would be dehydrobrominated to cyclobutadiene without any molecular rearrangement. This suggestion seems unlikely to be correct in view of the difficulty with which hydrogen bromide is eliminated from the molecule.

The experimental work has been confirmed by more recent workers[7]. They argued that if cyclobutadiene is indeed an intermediate in the formation of acetylene, then acetylene should also be produced in other reactions wherein cyclobutadiene might be formed as a transient intermediate. They therefore investigated the thermal decomposition of cyclobutane-1,3-bistrimethylammonium hydroxide and obtained 1,3-butadiene in about 20% yield rather than acetylene. In consequence they deduced that cyclobutadiene was not obtained from 1,2-dibromocyclobutane by the action of potassium hydroxide but that it was produced as a transient species by the action of quinoline, and that it was also formed as an intermediate in the decomposition of the quaternary ammonium hydroxide. If cyclobutadiene is indeed a highly reactive diradical it is likely that it would react at once with solvent or other neighbouring molecules by hydrogen abstraction thereby producing 1,3-butadiene. Recent semi-empirical calculations confirm that cyclobutadiene should not be unstable with respect to dissociation into acetylene[86].

(I)

Another attempted synthesis of cyclobutadiene involved the thermal decomposition of cyclobutane-1,2-bistrimethylammonium hydroxide[8]. No cyclobutadiene was obtained nor any product which might have derived from it.

INTRODUCTION

The dechlorination of 3,4-dichloro-1,2,3,4-tetramethylcyclobutene has also been investigated[9,10,33,75]. Lithium amalgam in ether gave a crystalline compound[9], $C_{16}H_{24}$. This molecular formula corresponds to a dimer of tetramethylcyclobutadiene, which is considered to have the tricyclic structure (I).

The action of zinc on the same cyclobutene derivative has also been studied[10]. When the reaction was carried out in the presence of either dimethylacetylene or dimethyl acetylenedicarboxylate it was possible to isolate o-xylene and dimethyl phthalate from the respective reaction mixtures. This suggests that a diradical has been formed which reacts with the alkyne which is present to give a benzene derivative, but it does not show whether this diradical has a cyclic or an open-chain structure viz.[10].

There is also evidence that tetramethylcyclobutadiene is generated by the gasphase reaction between 3,4-dichloro-1,2,3,4-tetramethylcyclobutene and sodium-potassium vapour in helium[33]. Among the products isolated were the dimer (I), octamethylcyclooctatetraene and 3-methylene-1,2,4-trimethylcyclobutene. All attempts to prepare diphenylcyclobutadiene by debromination of 3,4-dibromo-1,2-diphenylcyclobutene gave only intractable products[76].

Abstraction of HX (X = Cl, OH or NMe$_2$) from 3-chloro-, 3-hydroxy- and 3-dimethylamino-1,2,3,4-tetramethylcyclobutene gave 1,2,4-trimethyl-3-methylenecyclobutene; pyrolysis of the same compounds produced butadienes or their rearrangement products[77].

Strong evidence has been produced very recently (1965) that cyclobutadiene is formed by the action of ceric ion on cyclobutadieneiron tricarbonyl[87]. If the reaction is carried out in the presence of methyl propiolate a methoxycarbonylbicyclo-[2,2,0]-hexadiene is formed, which would result from a Diels–Alder reaction between the ester and cyclobutadiene. It was shown that this product is not formed directly from the original metal complex.

It was at one time suggested that the instability or reactivity of cyclobutadiene and its consequent non-isolation could be attributed to the angle-strain in the molecule, but this cannot be an over-riding factor since dimethylenecyclobutenes such as (II)[11] and tetramethylenecyclobutane (III)[12] have been prepared. Both of these compounds polymerise rapidly at room temperature. Compound (II) was allowed to react with tetracyanoethylene in the hope that normal diene synthesis would result in the formation of the cyclobutadiene derivative (IV), but instead the spiran (V) was obtained[16]. 1,2-Dimethylenecyclobutane (VI)[13], 1,3-dimethylenecyclobutane (VII)[17], and 1-methylenecyclobut-3-ene (VIII)[14,15] have also been prepared. It is notable that

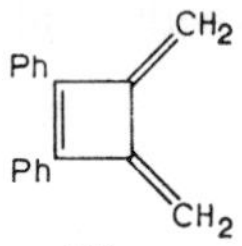

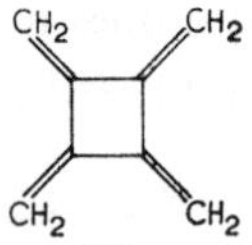

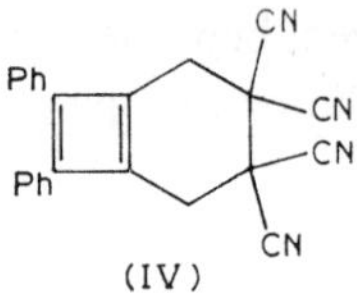

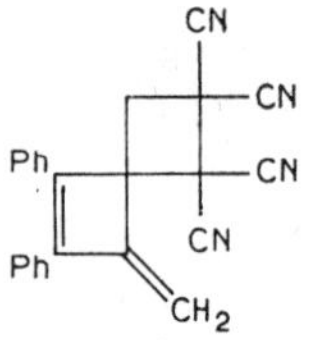

none of these compounds shows any tendency to undergo prototropic rearrangement to give cyclobutadiene derivatives. Similar angle-strain must be present in various cyclobutenediones which have been prepared (see below) yet these are stable compounds. It thus appears that angle-strain plays little part in the apparent instability of cyclobutadiene and its derivatives, and that the electronic factors discussed at the beginning of the chapter are of far greater significance.

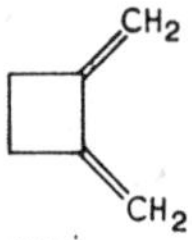

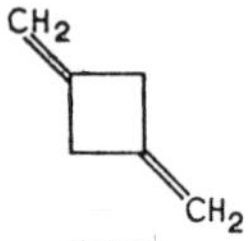

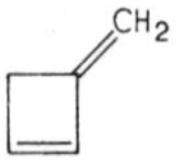

METAL COMPLEXES OF CYCLOBUTADIENE

Following the suggestion[18] made in 1956 that cyclobutadiene should form stable complexes with transition metals the preparation of a number of such complexes has been achieved.

The first such complex to be described[19] (in 1959) was a tetramethylcyclobutadiene nickel complex prepared as follows:

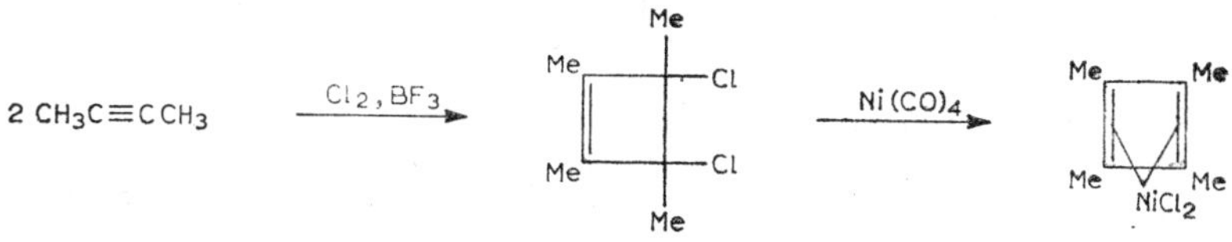

2 CH₃C≡CCH₃ →(Cl₂, BF₃) [tetramethyl-dichlorocyclobutene] →(Ni(CO)₄) [tetramethylcyclobutadiene·NiCl₂ complex]

This compound forms reddish-violet crystals soluble in water and in chloroform. Its n.m.r. spectrum shows the presence of twelve equivalent hydrogen atoms. It is proved to be monomeric in solution by its reaction with sodium nitrite to give the known 1,2-dihydroxy-1,2,3,4-tetramethylcyclobut-3-ene. The complex decomposes on heating under reduced pressure, but only at 250°, forming a dimer of tetramethylcyclobutadiene:

[Me₄-cyclobutadiene·NiCl₂] →(250°, reduced pressure) [octamethyl tricyclic dimer]

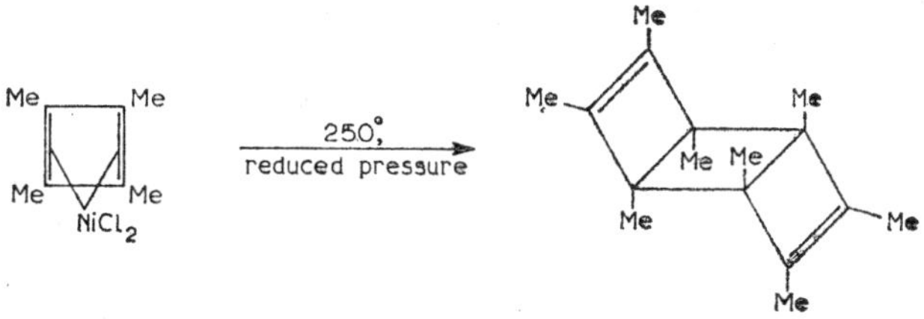

Oxidation of this dimer to a cyclobutanetetracarboxylic acid proves that the rings are *trans* or *anti* to one another[20].

In the same year the preparation of a cyclobutadiene silver complex was also claimed[21], as follows:

[tetrabromocyclobutane] →(i) Li/Hg (ii) AgNO₃ → [cyclobutadiene·AgNO₃] → (IX)

This has more recently been shown to be in fact a silver complex of the tricyclic dimer of cyclobutadiene (IX) in which the rings are *cis* or *syn* to one another[22]. Since then various other substituted cyclobutadiene metal complexes have also been prepared[23]. An unsubstituted cyclobutadiene iron carbonyl complex has been obtained; its n.m.r. spectrum shows a single sharp peak at 6.09 τ[78]. It undergoes electrophilic substitution, for example with acid chlorides in the presence of aluminium chloride[88].

References p. 51

CYCLOBUTENIUM SALTS

Removal of two electrons from a molecule of cyclobutadiene would leave it with two electrons in a bonding orbital and the resultant doubly charged cation meets the requirements of Hückel's rule for aromaticity, with two π-electrons delocalised over the ring carbon atoms.

The preparation of such a salt from a cyclobutadiene metal complex was claimed in 1962:

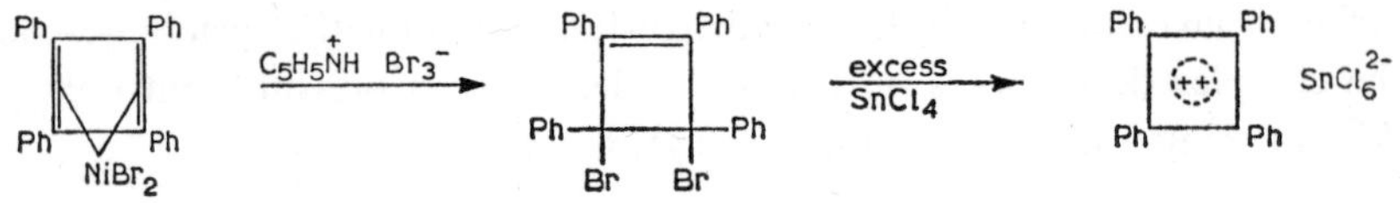

$$\text{Ph}_4\text{C}_4\text{NiBr}_2 \xrightarrow{C_5H_5\overset{+}{N}H\ Br_3^-} \text{1,2-dibromo-1,2,3,4-tetraphenylcyclobut-3-ene} \xrightarrow{\text{excess } SnCl_4} [\text{Ph}_4C_4]^{2+}\ SnCl_6^{2-}$$

It has recently been shown, however, by X-ray diffraction that this salt in fact has structure (X) in the solid state[25]. It is interesting to note that it had already been shown that the action of silver hexafluoroantimonate on 1,2-dichloro-1,2,3,4-tetramethylcyclobut-3-ene produced the monocation (XI) and not the doubly charged cation (XII)[26].

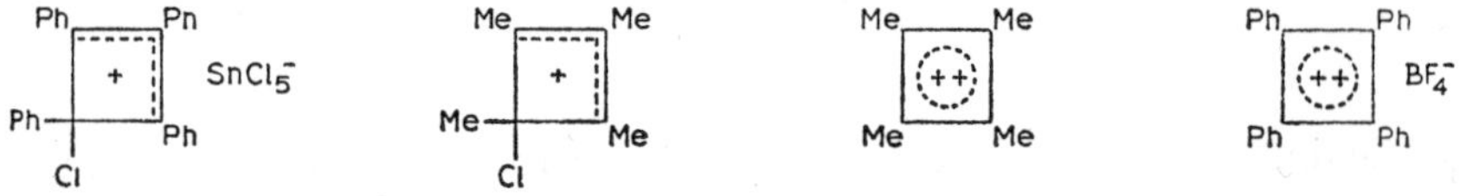

(X) Ph₄C₄Cl⁺ SnCl₅⁻   (XI) Me₄C₄Cl⁺   (XII) Me₄C₄²⁺   (XIII) Ph₄C₄²⁺ BF₄⁻

On the other hand the action of silver fluoroborate on 1,2-dibromo-1,2,3,4-tetraphenylcyclobut-3-ene produces a deep red solution which definitely appears to contain the doubly charged cyclobutenium cation (XIII)[27]. This is shown by the ^{19}F n.m.r. spectrum and by the fact that the solution reacts with cycloheptatriene giving tropylium fluoroborate and 1,2,3,4-tetraphenylcyclobutene.

A doubly charge cyclobutenium cation also appears to be formed when 2-bromo-2,4-diphenyl-3-hydroxy-cyclobut-3-enone is dissolved in 96% sulphuric acid[28]. Its structure follows from its n.m.r. spectrum. If the deep red solution in sulphuric acid is poured into water a dihydroxydiphenylcyclobutenone is obtained:

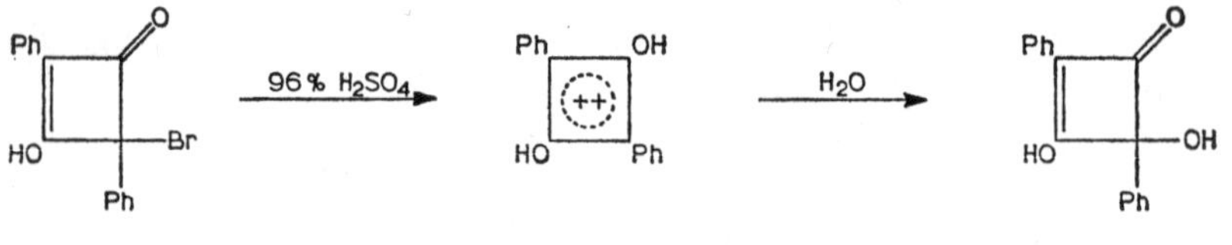

2-bromo-2,4-diphenyl-3-hydroxy-cyclobut-3-enone $\xrightarrow{96\%\ H_2SO_4}$ dication $\xrightarrow{H_2O}$ dihydroxydiphenylcyclobutenone

CYCLOBUTENE-3,4-DIONES

In the same way that one canonical form of cyclopropenone (XIV) involves a structure having two π-electrons delocalised over three trigonal carbon atoms so it is possible to write a canonical form of cyclobutene-3,4-diones (XV) in which two π-electrons are delocalised over the four ring carbon atoms, all of which are trigonally hybridised. Since such a form meets the requirements of Hückel's rule it will be energetically favoured and will undoubtedly contribute to the overall structure of the molecule.

An example of such a diketone is phenylcyclobutene-3,4-dione (XVI, R=H)[29,30]. This is a remarkably stable compound. It cannot be reduced to the corresponding diol, does not take part in diene syntheses and reacts with bromine by substitution to give the 2-bromo compound (XVI, R=Br). The latter is hydrolysed rapidly to give the 2-hydroxy compound (XVI, R=OH).

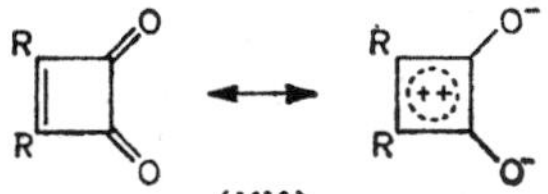

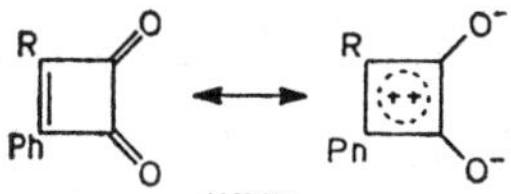

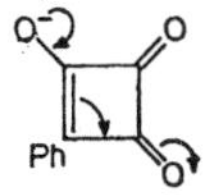

This hydroxy compound is a remarkably strong acid, with $pK_a \sim 1$. Two factors which contribute to its acidity are the contribution of the dipolar form in which the positive charge on the ring facilitates the separation of the hydrogen atom of the hydroxyl group as a proton, and the delocalisation of charge in the anion so produced (see XVI a).

The dihydroxycyclobutene (XVII)[31] is an even stronger acid, which has been appropriately named *squaric acid*. This gives rise to a completely symmetrical dianion with delocalisation over all the atoms in the ion.

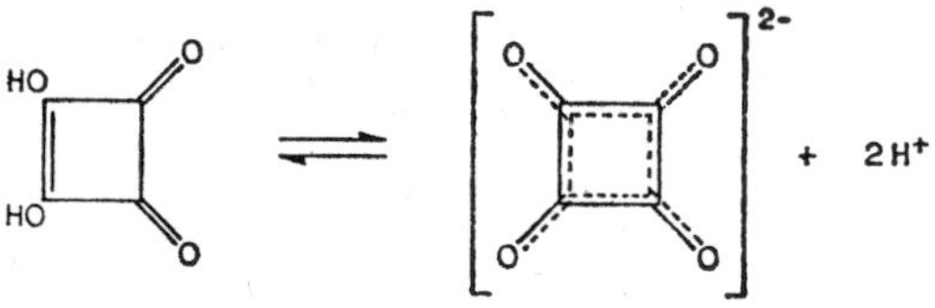

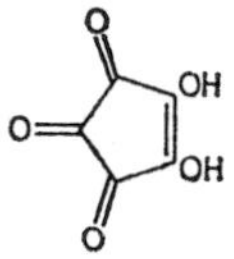

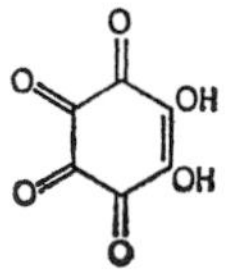

References p. 51

It has been suggested[32] that this anion and those derived from croconic acid (XVIII) and rhodizonic acid (XIX) might represent a series of aromatic anions with totally delocalised systems of π-electrons.

BENZOCYCLOBUTADIENE

The preparation of benzocyclobutadiene by the action of quinoline on *o*-xylylene dibromide was attempted as far back as 1907[34]. Shortly afterwards the cyclisation of tetrabromo-*o*-xylene by means of sodium iodide to give a dibromobenzocyclobutene was achieved[35].

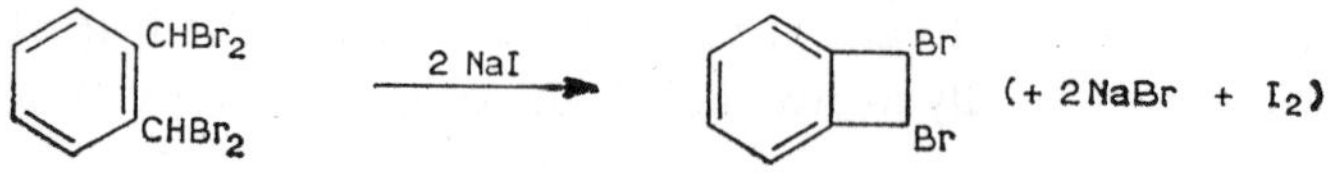

$$\text{o-C}_6\text{H}_4(\text{CHBr}_2)_2 \xrightarrow{2\text{ NaI}} \text{dibromobenzocyclobutene} \quad (+ 2\text{NaBr} + \text{I}_2)$$

This latter work has been confirmed and extended in more recent years[36,37]. Further treatment of the dibromobenzocyclobutene with sodium iodide failed to produce elimination of the two bromine atoms and concomitant double-bond formation; instead a substitution reaction took place to give the corresponding di-iodo compound[36].

Other debrominating agents gave rise to dimeric products. Thus the action of zinc and of sodium or lithium amalgam resulted in the formation of the hydrocarbon (XX) (plus polymeric material)[37,39], whereas sodium amalgam in the presence of tetracarbonyl nickel produced the isomeric compound (XXI)[38]. With potassium hydroxide the bromo-compound (XXII) was formed[35,37].

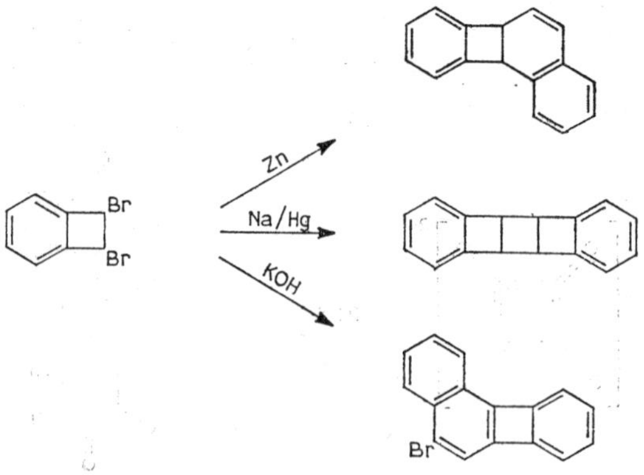

(XX)

(XXI)

(XXII)

Yet another dimer of benzocyclobutadiene has been prepared by oxidative cleavage of a benzocyclobutadiene iron carbonyl complex[78] (see p. 45). The hydrocarbons could have been formed by dimerisation of benzocyclobutadiene, and it has indeed been shown that the latter compound is probably formed as an intermediate by carrying out the debrominations in the presence of a reactive diene, whereby adducts of benzocyclobutadiene and the diene have been isolated[39,40,85]:

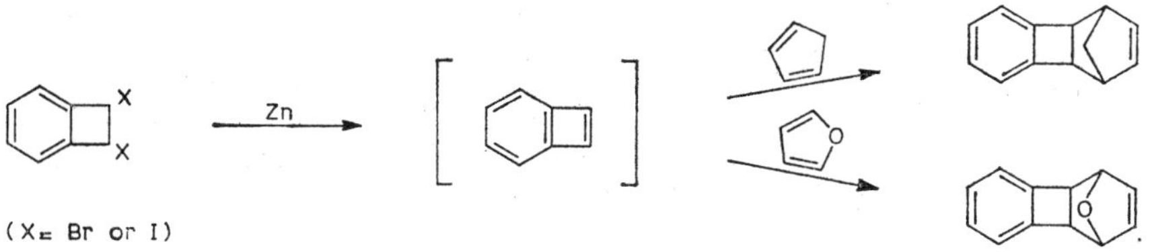

(X = Br or I)

Dehydrobromination of a monobromobenzocyclobutene by means of potassium tertiary butoxide also leads to the dimer (XX)[41].

It thus appears that, like cyclobutadiene, benzocyclobutadiene is far too reactive a molecule to have any prolonged existence. Once again this reactivity cannot be solely due to angle strain in the molecule since the diketobenzocyclobutene (XXIII), which has similar angle strain, has been prepared and is a stable compound[42].

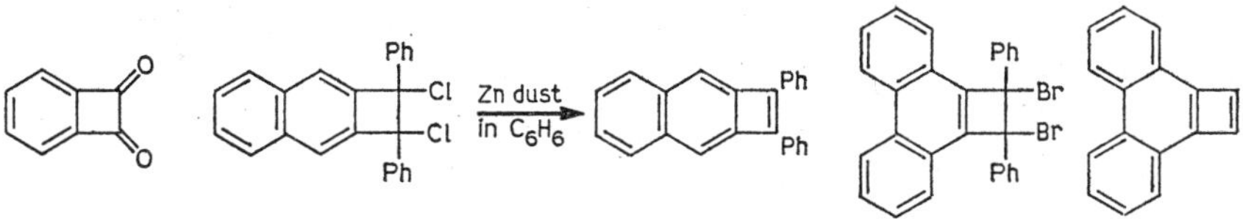

(XXIII)   (XXIV)   (XXV)   (XXVI)   (XXVII)

On the other hand the 2,3-naphthocyclobutadiene (XXV) has been prepared by dehalogenation of the dichloro-compound (XXIV)[43], and forms red crystals which remain unchanged when exposed to air and light for several weeks. This stability has been attributed to the fact that the $\beta\beta$ bonds in naphthalene have less double bond character than the $\alpha\beta$ bonds (or the ring bonds of benzene) and that in consequence the four-membered ring is fused to bonds which have a very low bond order. The four-membered ring thus has more cyclobutene than cyclobutadiene character. The n.m.r. spectrum[43] indicates a remarkably fixed double bond structure in the adjacent $\alpha\beta$ bonds, for in addition to showing 14 protons in the normal aromatic region there are two protons giving a sharp peak very near to the position of the peak given by the vinyl hydrogen atoms in *cis*-stilbene.

If the stability of this compound is connected with the low bond order of the bond common to the four- and six-membered rings then it follows that a molecule wherein the four-membered ring is fused to a position of high bond-order should not be stable. This has been confirmed by the attempted conversion of the dibromo-compound (XXVI) to 9,10-phenanthrocyclobutadiene (XXVII)[44]. It proved impossible to isolate either a dimer or a Diels–Alder adduct of (XXVII). When the debromination was carried out by means of zinc in the presence of oxygen, some 9,10-dibenzoylphenanthrene was isolated, however, which could have arisen by oxidation of (XXVII). If the latter is a diradical it would be expected to react at once with oxygen in this way.

The low bond order of the $\beta\beta$ bond is probably not the only contributing factor to the stability of (XXV) however, a further factor being the stabilisation afforded to the entire system by the phenyl substituents[89]. Significantly the isomer of (XXV) having phenyl groups substituted at the 1 and 4 positions of the naphthalene ring instead of in the four-membered ring is apparently unstable[89,90], but replacement of the hydrogen atoms attached to the four-membered ring in this latter compound by bromine atoms results in a stable compound[89].

Dibromobenzocyclobutene reacts with triphenylphosphine to form a bis-triphenylphosphonium salt, and this salt, on treatment with base (n-butyl lithium or lithium hydroxide), gives a dark red solution which is thought to contain a bis-triphenylphosphoniumylide[82]:

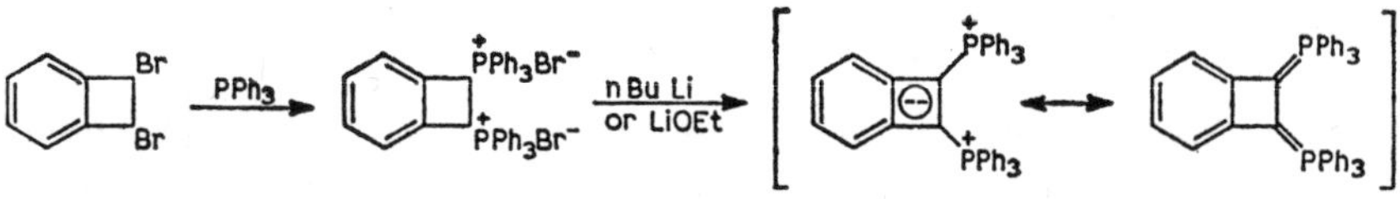

The product quickly decomposes to a brown tar at room temperature but appears to be stable at $-40°$ for at least several hours. In the dipolar canonical form six π-electrons are delocalised over the four-membered ring (or ten π-electrons over both rings) and this ring should therefore be aromatic in character.

A benzocyclobutadiene iron carbonyl complex has been prepared by the reaction of dibromobenzocyclobutene with $Fe_2(CO)_9$[78]. Oxidative cleavage of this complex by means of silver nitrate produces a dimer of benzocyclobutadiene:

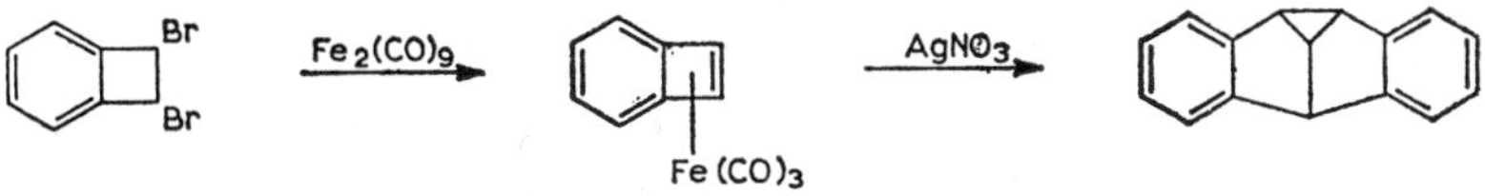

BIPHENYLENE

In apparent contrast to cyclobutadiene or benzocyclobutadiene, dibenzocyclobutadiene or biphenylene (XXVIII) is a very stable compound.

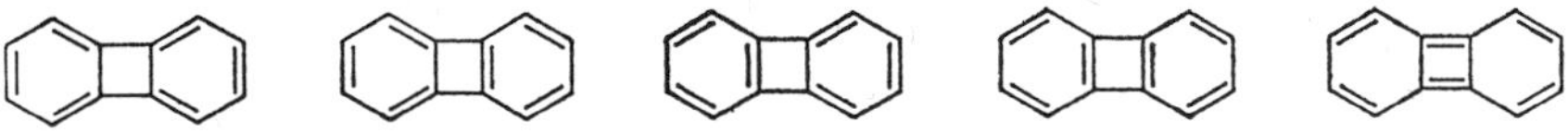

(XXVIII)   (XXIX)   (XXX)   (XXXI)   (XXXII)

Various canonical forms (*viz.* (XXVIII)–(XXXII)) can be written for biphenylene; chemical evidence[45,46] and X-ray diffraction studies[47] suggest that (XXVIII) is the predominant form. In this form the central ring has in fact no cyclobutadiene character.

Preparation of biphenylene

Biphenylene was first prepared in 1941, by distillation of 2,2'-dibromo- or 2,2'-diiodo-biphenyl with cuprous oxide[48].

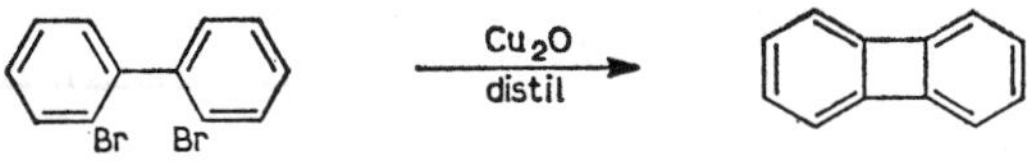

2,7-Dimethylbiphenylene was made by the same method. It was found that it could be made either from 2,2'-diiodo-4,4'-dimethylbiphenyl or from 2,2'-diiodo-5,5'-dimethylbiphenyl, thus demonstrating the symmetry of the biphenylene molecule.

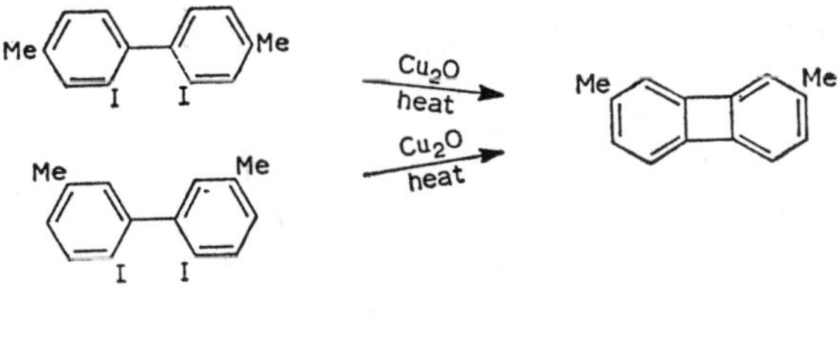

This method has been examined in detail and improved by later workers[49] and has been used to prepare a variety of substituted biphenylenes[49,50]. It is probably the most satisfactory route for the preparation of biphenylene derivatives.

Other preparative routes include the reaction between 2,2′-biphenylene-mercury and silver powder[52], and the dimerisation of benzyne[53]. It is also obtained in very poor yield by oxidation by means of cupric chloride of the Grignard reagent from 2,2′-dibromobiphenyl[51].

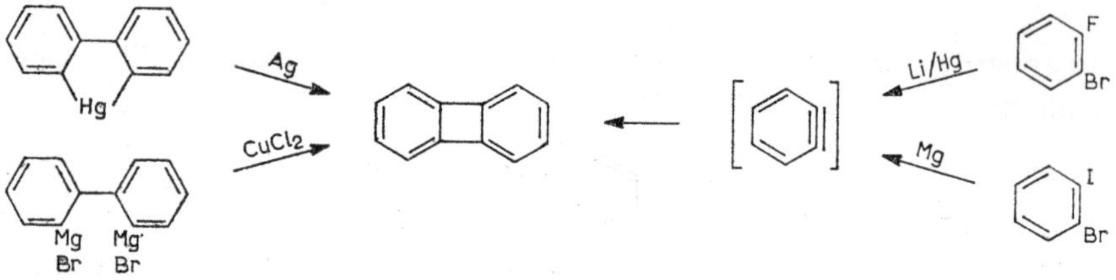

Oxidation of 1-aminobenzotriazole with lead tetraacetate produces biphenylene in high yield[91].

Structure of biphenylene

The structure of biphenylene has been confirmed by means of electron diffraction measurements[54] and X-ray crystallographic methods[47,55].

Simple resonance ideas would suggest that formula (XXXI) above, with a cyclobutadienoid central ring, was the best representation of the molecule, but molecular orbital calculations[56] suggest that formula (XXVIII), which consists of two benzene rings linked by two sp^2 hybridised σ-bonds, is a better representation.

An X-ray diffraction study of biphenylene[47] gives the following lengths (Å) for the various bonds in the molecule:

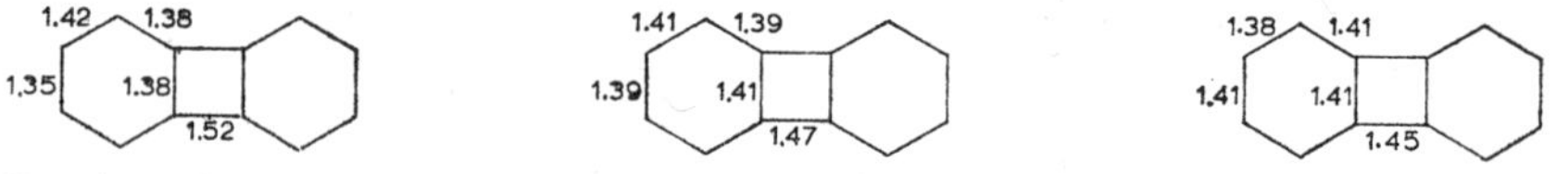

Experimental values

Calculated values (molecular orbital)

Calculated values (simplest resonance theory)

These values accord well with those obtained by molecular orbital calculations[72] but there are marked discrepancies between the measured values and those calculated by simplest resonance theory. The measured values indicate that the preferred structure is best represented by formula (XXVIII), but that this formula alone represents too severe a fixation of the double-bonds. The best agreement is obtained by considering the molecule as a hybrid of forms (XXVIII)–(XXXII), with a major contribution from (XXIII), lesser contributions from (XXIX) and (XXX) and a very small contribution from (XXXI). The values for the *ortho* proton–proton coupling constants in the n.m.r. spectrum accord with such an arrangement[79]. In all the theoretical models the bonds common to both the four- and six-membered rings are calculated to have a greater length than that observed by experiment. The divergence may well result from no allowance being made in these theoretical models for the strain involved in forming the four-membered ring.

The n.m.r. spectrum of biphenylene has been interpreted as showing ring currents only in the six-membered rings[57] and as indicating electronic interaction between these rings[79]. There is ample evidence, both chemical (described below) and physical to show that the two six-membered rings are not entirely independent. Thus there is ultra-violet light absorption at much longer wavelengths than in the case of biphenyl, suggesting quite extensive conjugation between the two benzenoid rings[49].

Ultra-violet spectra of biphenylenes

In contrast to biphenyl which has only one absorption maximum, at 250 mμ, the ultra-violet spectrum of biphenylene is considered to be made up of two overlapping band systems[58]. In ethanolic solution there are the following peaks[49]:

mμ	239	248	326	330	339	343	348	358
log ε_{max}	4.77	5.05	3.47	3.49	3.79	3.76	3.58	3.97

The split peak at 330–400 mμ appears to be characteristic of biphenylene derivatives. In general the ultra-violet spectra of biphenylene derivatives resemble that of the parent hydrocarbon except for loss of fine structure and bathochromic shift (for details see refs. 49,59).

Properties of biphenylene

Biphenylene forms pale yellow prisms, m.p. 112°. It is very stable, even at high temperatures, and may be kept for years without change. It is volatile

References p. 51

in steam and can be sublimed, and has a faint odour resembling that of the xylenes.

It forms molecular complexes, e.g. with picric acid and other nitro-compounds[48,49]. Its molecular complex with tetracyanoethylene has been shown to be more stable than the corresponding fluorene complex[60].

Since the four-membered ring in biphenylene must be strained, it might be expected that the reactions leading to the opening of this ring would be a feature of the chemistry of biphenylene. This is not in fact the case although catalytic reduction proceeds very readily to give biphenyl[49]. Reduction can also be brought about by sodium in liquid ammonia; in both cases the biphenyl formed undergoes further reduction[49]. Bromination of substituted biphenylenes has been shown to lead to opening of the four-membered ring[45,81].

Ring-enlargement of the four-membered ring is brought about by the action of hexacarbonylchromium on biphenylene, when fluorene is produced, presumably by carbonyl insertion, together with some bifluorenylidene, probably by reduction of fluorenone[65].

Biphenylene is oxidised to phthalic acid by chromic acid[48].

It was predicted[61] that electrophilic, nucleophilic and homolytic substitution in biphenylene should all take place at the 2-position.

The first such reaction to be investigated was the Friedel–Crafts acetylation with acetyl chloride/aluminium chloride[49]. This indeed gave 2-acetylbiphenylene; its orientation was established since on reduction in the presence of Raney nickel, the products obtained were 3- and 4-acetylbiphenyl. (The other possible acetylation product, 1-acetylbiphenylene, would give 2- and 3-acetylbiphenyl on reduction). This result does not, however, unequivocally confirm the theoretical predictions since Friedel–Crafts acylation is reversible and the nature of the product may therefore be equilibrium controlled.

Other substitution reactions to which this reservation does not apply have since been investigated and substitution invariably takes place at the 2-position[59]. Some of these reactions are illustrated in the chart on page 49. When biphenylene is treated with excess bromine without any catalyst present, two addition products, $C_{12}H_8Br_4$ and $C_{12}H_8Br_6$ are formed. The tetrabromo compound has been shown to be the tetrabromobenzocyclooctatriene (XXXIII)[62]. This may be converted into a benzocyclooctatetraene.

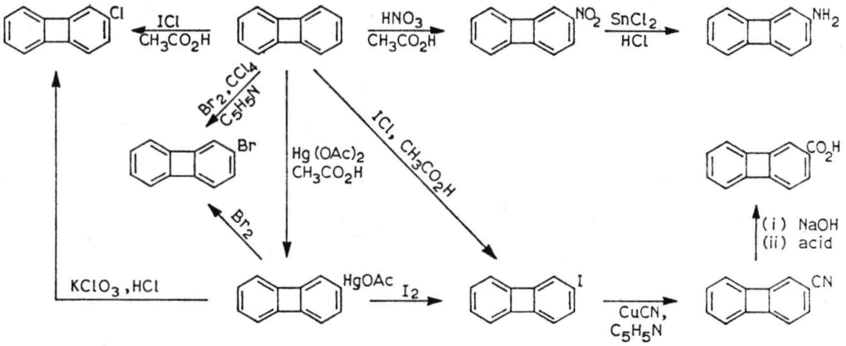

Biphenylene is unaffected by N-bromosuccinimide, even in the presence of benzoyl peroxide. This illustrates its low reactivity towards homolytic reagents, which is in agreement with theoretical predictions[61].

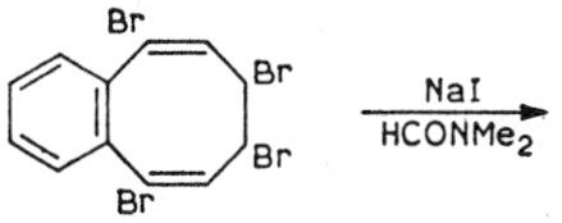

Diacetylation and dinitration of biphenylene give 2,6-disubstituted products[59]; in the case of the former a very small amount of 2,7-diacetylbiphenylene has also been identified[83]. This result indicates that there must be conjugation between the two benzene rings. With no such conjugation, substitution of the second group should take place at a β-position of the unsubstituted and hence not deactivated ring, and should occur equally readily at the 6- or 7-positions. In fact the 2,6-isomer is obtained predominantly. The acetyl group deactivates the 1,3,5- and 7-positions (see XXXIV, above). Of the remaining positions substitution is more likely to occur at the 6- than the 8-position since the β-positions are intrinsically more reactive than the α-positions. The small amount of 2,7-isomer is formed owing to this innate reactivity of both β-positions. The 2,6- and 2,7-diacetyl isomers are most probably formed directly by competing reactions since they cannot be interconverted under the conditions of the acetylation. It may be noted that by contrast 4-acetylbiphenyl gives 4,4'-diacetylbiphenyl and not the 4,3'-isomer.

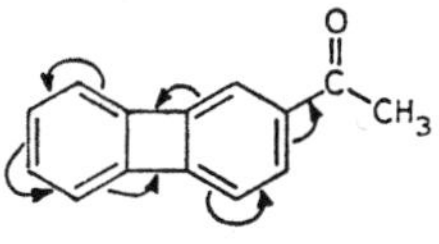

Molecular orbital calculations[56] suggested that an *o.p*-directing group at

References p. 51

the 2-position of biphenylene should direct a second substituent group to the 3-position, whereas simple resonance theory suggested that the 1-position should be attacked. Experimental evidence accords entirely with the molecular orbital calculations. Thus it has been shown that bromine reacts with 2-acetamidobiphenylene to give the 3-bromoderivative[45,63], and that 2-aminobiphenylene[46] and 2-hydroxybiphenylene[64] couple with benzene diazonium salts at the 3-position.

The 3-phenylazo-derivative (XXXV) has been converted, as shown below, to the o-quinone (XXXVI)[64]. This o-quinone is very stable and shows no tendency to polymerise. Its stability may be associated with the fact that another canonical form (XXXVII) can be drawn for this molecule, involving a delocalised decet of π-electrons.

There is evidence that biphenylene forms a dianion by the action of sodium or potassium in tetrahydrofuran[80]. This dianion will have a system of 14 π-electrons.

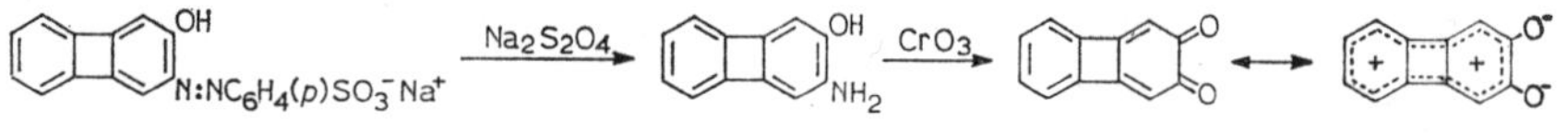

(XXXV)   (XXXVI)   (XXXVII)

BENZOBIPHENYLENES

As mentioned above (p. 42) the 3-bromo-1,2-benzobiphenylene (XXII) is formed by the action of potassium hydroxide on dibromobenzocyclobutene[35,37]. It was first prepared in 1910, although its structure was not recognised at the time. Other derivatives of this 1,2-benzobiphenylene (including the parent hydrocarbon) have been prepared, and also the linear 2,3-benzobiphenylene (XXXVIII)[67].

Three of the five possible dibenzobiphenylene isomers have been prepared, the linear isomer (XXXIX)[68,69] and the angular isomers (XL)[70], (XLII)[74] and (XLIII)[84]. The linear dibenzobiphenylene differs markedly from, and is much more stable than the angular isomer (XL). Thus the melting points of the two compounds are, respectively, 372° and 137–9°; the linear hydrocarbon can be sublimed unchanged at 350° whereas the angular isomer decomposes at 160° *in vacuo*. This difference in stability is in agreement with the results of molecular orbital calculations[71]. Such differences are to be expected, for, if the central ring is to have a predominantly cyclobutane-like structure

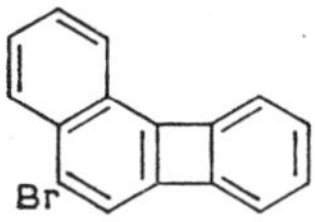

(XXII) 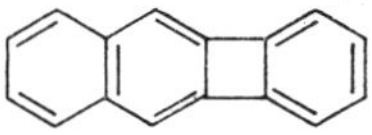   (XXXVIII) 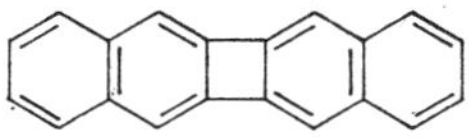   (XXXIX)

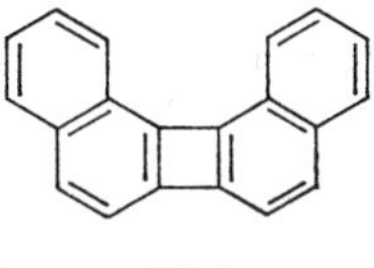

(XL) 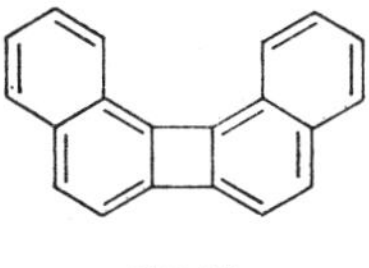   (XLI) 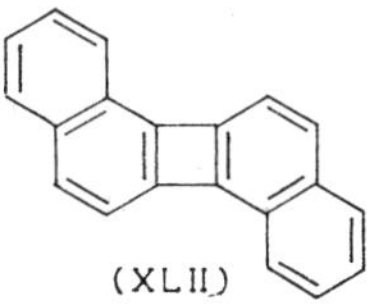   (XLII) 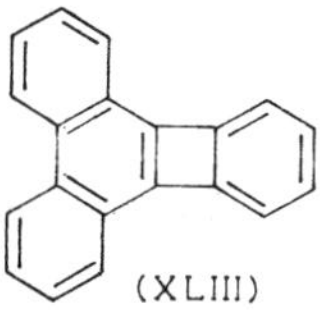   (XLIII)

(XXXIX), (XLI); (*cf.* formula XXVIII for biphenylene), then the two naphthalene rings in the angular isomer have to take up an unfavourable *ortho*-quinonoid structure, whereas in the linear isomer this is not the case[69,70,72]. The isomer (XLII) appears to resemble (XL) in its stability, as might be expected[74].

COMPARISON OF BENZOCYCLOBUTADIENES AND BIPHENYLENES

Molecular orbital calculations[73] suggest that benzocyclobutadienes, wherein the cyclobutadiene ring is stabilised by fusion with a benzene ring on one side only, differ markedly from biphenylenes wherein benzene rings are fused to both sides. The simplest description of the latter[72] is that there are two aromatic regions in the molecule with only a little cross-conjugation in the ground state. The corresponding description of the former is that there is one aromatic region and one olefinic double bond with only a little mutual conjugation. In neither case does it seem chemically appropriate to speak of a genuinely cyclobutadiene-like structure.

REFERENCES

1 J. D. ROBERTS, A. STREITWIESER AND C. M. REGAN, *J. Am. Chem. Soc.*, 74 (1952) 4579.
2 J. E. LENNARD-JONES AND J. TURKEVITCH, *Proc. Roy. Soc.*, A 158 (1937) 297.
3 W. H. PERKIN JR, *J. Chem. Soc.*, 65 (1894) 967.
4 R. WILLSTÄTTER AND W. VON SCHMAEDEL, *Ber.*, 38 (1905) 1892.
5 W. BAKER AND J. F. W. MCOMIE, *Chem. Soc. Special Publ.*, 12 (1958) 49.
6 A. E. FAVORSKII AND T. A. FAVORSKAYA, *Zhur. Russ. Fiz.-Khim. Obshch.*, 54 (1922) 310.
7 M. AVRAM, C. D. NENITZESCU AND E. MARICA, *Chem. Ber.*, 90 (1957) 1857.

8 E. R. BUCHMAN, M. J. SCHLATTER AND A. D. REIMS, *J. Am. Chem. Soc.*, 64 (1942) 2701.
9 R. CRIEGEE AND G. LOUIS, *Chem. Ber.*, 90 (1957) 417.
10 C. E. BERKOFF, R. C. COOKSON, J. HUDEC AND R. O. WILLIAMS, *Proc. Chem. Soc.*, (1961) 312.
11 A. T. BLOMQUIST AND Y. C. MEINWALD, *J. Am. Chem. Soc.*, 79 (1957) 5317.
12 G. W. GRIFFIN AND L. I. PETERSON, *J. Am. Chem. Soc.*, 85 (1963) 2268.
13 A. T. BLOMQUIST AND J. A. VERDOL, *J. Am. Chem. Soc.*, 77 (1955) 1806; 78 (1956) 109; Y. M. SLOBODIN AND A. P. KHITROV, *Zh. obshch. Khim.*, 33 (1963) 2819.
14 D. R. HOWTON AND E. R. BUCHMAN, *J. Am. Chem. Soc.*, 78 (1956) 4011.
15 D. E. APPLEQUIST AND J. D. ROBERTS, *J. Am. Chem. Soc.*, 78 (1956) 4012.
16 A. T. BLOMQUIST AND Y. C. MEINWALD, *J. Am. Chem. Soc.*, 81 (1959) 667.
17 F. F. CASERIO, S. H. PARKER, R. PICCOLINI AND J. D. ROBERTS, *J. Am. Chem. Soc.*, 80 (1958) 5507.
18 H. C. LONGUET-HIGGINS AND L. E. ORGEL, *J. Chem. Soc.*, (1956) 1969.
19 R. CRIEGEE AND G. SCHRÖDER, *Angew. Chem.*, 71 (1959) 70.
20 G. W. GRIFFIN AND D. F. VEBER, *Chem. and Ind.*, (1961) 1162.
21 M. AVRAM, E. MARICA AND C. D. NENITZESCU, *Chem. Ber.*, 92 (1959) 1088.
22 M. AVRAM, H. P. FRITZ, H. J. KELLER, C. G. KREITZER, G. MATEESCU, J. F. W. MCOMIE, N. SHEPPARD AND C. D. NENITZESCU, *Tetrahedron Letters*, (1963) 1611.
23 W. HÜBEL, et al., *J. Inorg. Nuclear Chem.*, 9 (1959) 204; 10 (1960) 250; L. MALATESTA et al., *Atti Accad. naz. Lincei, Rend. Classe Sci. fis. mat. nat.*, 27 (1959) 230; *Angew. Chem.*, 72 (1960) 34; R. P. DODGE AND V. SCHOMAKER, *Nature*, 186 (1960) 749; H. H. FREEDMAN, *J. Am. Chem. Soc.*, 83 (1961) 2194, 2195; A. NAKAMURA AND N. HAGIHARA, *Bull. Chem. Soc. Japan*, 34 (1961) 452; A. T. BLOMQUIST AND P. M. MAITLIS, *J. Am. Chem. Soc.*, 84 (1962) 2329; P. M. MAITLIS AND M. L. GAMES, *J. Am. Chem. Soc.*, 85 (1963) 1887; R. HÜTTEL AND H. J. NEUGEBAUER, *Tetrahedron Letters* (1964) 3541; R. C. COOKSON AND D. W. JONES, *J. Chem. Soc.*, (1965) 1881.
24 H. H. FREEDMAN AND A. M. FRANTZ, *J. Am. Chem. Soc.*, 84 (1962) 4165.
25 R. F. BRYAN, *J. Am. Chem. Soc.*, 86 (1964) 733.
26 T. J. KATZ, J. R. HALL AND W. C. NEIKAM, *J. Am. Chem. Soc.*, 84 (1962) 3199.
27 H. H. FREEDMAN AND A. E. YOUNG, *J. Am. Chem. Soc.*, 86 (1964) 734.
28 D. G. FARNUM AND B. WEBSTER, *J. Am. Chem. Soc.*, 85 (1963) 3502.
29 E. J. SMUTNY AND J. D. ROBERTS, *J. Am. Chem. Soc.*, 77 (1955) 3420.
30 E. J. SMUTNY, M. C. CASERIO AND J. D. ROBERTS, *J. Am. Chem. Soc.*, 82 (1960) 1793.
31 S. COHEN, J. R. LACHER AND J. D. PARK, *J. Am. Chem. Soc.*, 81 (1959) 3480; 84 (1962) 2919.
32 R. WEST, H. Y. NIU, D. L. POWELL AND M. V. EVANS, *J. Am. Chem. Soc.*, 82 (1960) 6204.
33 R. S. SKELL AND R. J. PETERSEN, *J. Am. Chem. Soc.*, 86 (1964) 2530.
34 R. WILLSTÄTTER AND H. VERAGUTH, *Ber.*, 40 (1907) 960.
35 H. FINKELSTEIN, *Ber.*, 43 (1910) 1528; *Chem. Ber.*, 92 (1959) XXXVII.
36 M. P. CAVA AND D. R. NAPIER, *J. Am. Chem. Soc.*, 78 (1956) 500.
37 M. P. CAVA AND D. R. NAPIER, *J. Am. Chem. Soc.*, 79 (1957) 1701; M. P. CAVA AND J. F. STUCKER, *J. Am. Chem. Soc.*, 79 (1957) 1706.
38 M. AVRAM, D. DINU AND C. D. NENITZESCU, *Chem. and Ind.*, (1959) 257.
39 C. D. NENITZESCU, M. AVRAM AND D. DINU, *Chem. Ber.*, 90 (1957) 2541.
40 M. P. CAVA AND M. J. MITCHELL, *J. Am. Chem. Soc.*, 81 (1959) 5409.
41 M. P. CAVA AND D. R. NAPIER, *J. Am. Chem. Soc.*, 80 (1958) 2255.
42 M. P. CAVA AND D. R. NAPIER, *J. Am. Chem. Soc.*, 79 (1957) 3606.
43 M. P. CAVA, B. HWANG AND J. P. VAN METER, *J. Am. Chem. Soc.*, 85 (1963) 4032.
44 M. P. CAVA AND D. MANGOLD, *Tetrahedron Letters*, (1964) 1751.

REFERENCES

45 W. BAKER, J. F. W. MCOMIE, D. R. PRESTON AND V. ROGERS, *J. Chem. Soc.*, (1960) 414.
46 H. H. BOSSHARD AND H. ZOLLINGER, *Helv. Chim. Acta*, 44 (1961) 1985.
47 T. C. W. MAK AND J. TROTTER, *J. Chem. Soc.*, (1962) 1.
48 W. C. LOTHROP, *J. Am. Chem. Soc.*, 63 (1941) 1187.
49 W. BAKER, M. P. V. BOARLAND AND J. F. W. MCOMIE, *J. Chem. Soc.*, (1954) 1476.
50 W. BAKER, J. W. BARTON AND J. F. W. MCOMIE, *J. Chem. Soc.*, (1958) 2658.
51 W. S. RAPSON AND R. G. SHUTTLEWORTH, *J. Chem. Soc.*, (1941) 487; W. S. RAPSON, R. G. SHUTTLEWORTH AND J. N. VAN NIEKERK, *J. Chem. Soc.*, (1943) 326.
52 G. WITTIG AND W. HERWIG, *Chem. Ber.*, 87 (1954) 1511.
53 G. WITTIG AND L. POHMER, *Chem. Ber.*, 89 (1956) 1334; G. WITTIG AND H. HÄRLE, *Ann.*, 623 (1959) 17; H. HEANEY, F. G. MANN AND I. T. MILLAR, *J. Chem. Soc.*, (1957) 3930.
54 J. WASER AND V. SCHOMAKER, *J. Am. Chem. Soc.*, 65 (1943) 1451.
55 J. WASER AND CHIA-SI LU, *J. Am. Chem. Soc.*, 66 (1944) 2035.
56 H. C. LONGUET-HIGGINS, *Proc. Chem. Soc.*, (1957) 157.
57 G. FRAENKEL, Y. ASAHI, M. J. MITCHELL AND M. P. CAVA, *Tetrahedron*, 20 (1964) 1179.
58 E. P. CARR, L. W. PICKETT AND D. VORIS, *J. Am. Chem. Soc.*, 63 (1941) 3231.
59 W. BAKER, J. W. BARTON AND J. F. W. MCOMIE, *J. Chem. Soc.*, (1958) 2666.
60 D. G. FARNUM, E. R. ATKINSON AND W. C. LOTHROP, *J. Org. Chem.*, 26 (1961) 3204.
61 R. D. BROWN, *Trans. Faraday Soc.*, 45 (1949) 300; 46 (1950) 146; J. I. F. ALONSO AND R. DOMINGO, *Anales real Soc. españ. Fis. Quim.*, 51 B (1955) 447; J. I. F. ALONSO AND F. PERADEJORDI, *Anales real Soc. españ. Fis. Quim.*, 50 B (1954) 253.
62 J. W. BARTON, D. E. HENN, K. A. MCLAUCHLAN AND J. F. W. MCOMIE, *J. Chem. Soc.*, (1964) 1622.
63 W. BAKER, J. F. W. MCOMIE AND V. ROGERS, *Chem. and Ind.*, (1958) 1236.
64 J. M. BLATCHLY, J. F. W. MCOMIE AND S. D. THATTE, *J. Chem. Soc.*, (1962) 5090.
65 E. R. ATKINSON, P. L. LEVINS AND T. E. DICKELMAN, *Chem. and Ind.*, (1964) 934.
66 M. P. CAVA, K. W. RATTS AND J. F. STUCKER, *J. Org. Chem.*, 25 (1960) 1101.
67 F. R. JENSEN AND W. E. COLEMAN, *Tetrahedron Letters*, No. 20 (1959) 7; W. BAKER, J. W. BARTON, J. F. W. MCOMIE AND R. J. G. SEARLE, *J. Chem. Soc.*, (1962) 2633.
68 R. F. CURTIS AND G. VISWANATH, *Chem. and Ind.*, (1954) 1174; *J. Chem. Soc.*, (1959) 1670; E. R. WARD AND B. D. PEARSON, *J. Chem. Soc.*, (1959) 1676; (1961) 515; B. D. PEARSON, *Chem. and Ind.*, (1960) 899.
69 R. F. CURTIS, *J. Chem. Soc.*, (1959) 3650.
70 M. P. CAVA AND J. F. STUCKER, *Chem. and Ind.*, (1955) 446; *J. Am. Chem. Soc.*, 77 (1955) 6022.
71 M. A. SILVA AND B. PULLMAN, *Compt. rend.*, 242 (1956) 1888.
72 M. A. ALI AND C. A. COULSON, *Tetrahedron*, 10 (1960) 41.
73 C. A. COULSON, *Chem. Soc. Special Publ.*, 12 (1958) 85.
74 J. W. BARTON, *J. Chem. Soc.*, (1964) 5161.
75 C. E. BERKHOFF, R. C. COOKSON, J. HUDEC, D. W. JONES AND R. O. WILLIAMS, *J. Chem. Soc.*, (1965) 194.
76 A. T. BLOMQUIST AND E. A. LALANCETTE, *J. Org. Chem.*, 29 (1964) 2331.
77 R. CRIEGEE AND W. FUNKE, *Chem. Ber.*, 97 (1964) 2934.
78 G. F. EMERSON, L. WATTS AND R. PETTIT, *J. Am. Chem. Soc.*, 87 (1965) 131.
79 A. R. KATRITZKY AND R. E. REAVILL, *Rec. Trav. chim.*, 83 (1964) 1230.
80 N. L. BAULD AND D. BANKS, *J. Am. Chem. Soc.*, 87 (1965) 128.
81 W. BAKER, N. J. MCLEAN AND J. F. W. MCOMIE, *J. Chem. Soc.*, (1964) 1067.
82 A. T. BLOMQUIST AND V. J. HRUBY, *J. Am. Chem. Soc.*, 86 (1964) 5041.
83 J. M. BLATCHLY, A. J. BOULTON AND J. F. W. MCOMIE, *J. Chem. Soc.*, (1965) 4930.
84 J. W. BARTON, *J. Chem. Soc.*, in press.

85 A. J. BOULTON AND J. F. W. MCOMIE, *J. Chem. Soc.*, (1965) 2549.
86 M. J. S. DEWAR AND G. J. GLEICHER, *J. Am. Chem. Soc.*, 87 (1965) 3255.
87 L. WATTS, J. D. FITZPATRICK AND R. PETTIT, *J. Am. Chem. Soc.*, 87 (1965) 3253.
88 J. D. FITZPATRICK, L. WATTS, G. F. EMERSON AND R. PETTIT, *J. Am. Chem. Soc.*, 87 (1965) 3254.
89 M. P. CAVA AND B. HWANG, *Tetrahedron Letters*, (1965) 2297.
90 C. D. NENITZESCU, M. AVRAM, I. G. DINULESCU AND G. MATEESCU, *Ann.*, 653 (1962) 79; M. AVRAM, I. G. DINULESCU, M. ELIAM, M. FARCASIN, E. MARICA, G. MATEESCU AND C. D. NENITZESCU, *Chem. Ber.*, 97 (1964) 372.
91 C. D. CAMPBELL AND C. W. REES, *Chem. Comm.*, (1965) 192.

CHAPTER IV

Derivatives of cyclopentadiene

INTRODUCTION

The *cyclopentadienide* anion was first prepared in 1901[1], by the action of potassium on cyclopentadiene in an inert solvent and atmosphere:

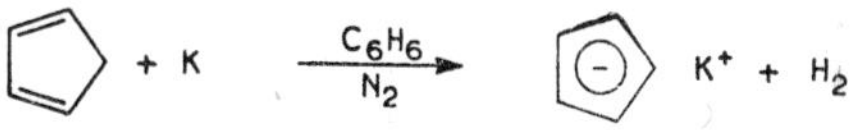

$$\text{C}_5\text{H}_6 + \text{K} \xrightarrow[\text{N}_2]{\text{C}_6\text{H}_6} \text{C}_5\text{H}_5^- \text{K}^+ + \text{H}_2$$

Although highly reactive the cyclopentadienide ion is none the less stable, deriving this stability from its delocalised sextet of π-electrons.

The negatively charged cyclopentadienide ring may occur not only as a discrete anion but also as part of a dipolar molecule, *viz.*

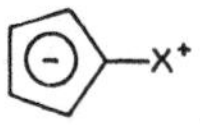

$$\text{C}_5\text{H}_4^- - \text{X}^+$$

Several types of compounds involve more or less contribution from such dipolar structures. In the cyclopentadienylides (I) the negatively charged

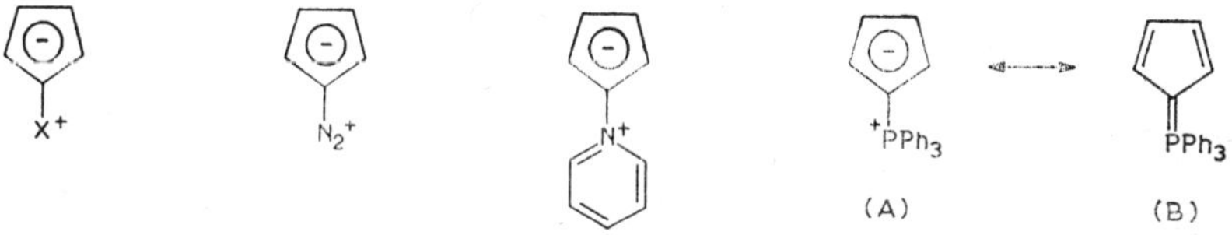

(I)   (II)   (III)   (A)   (B)
                           (IV)

cyclopentadienide ring is directly attached to a positively charged heteroatom. This hetero-atom may, for example, be part of a diazo group as in diazocyclopentadiene (II)[2] or of a pyridinium ring as in pyridinium cyclopentadienylide (III)[3,4]. If the exocyclic hetero-atom is phosphorus, as in triphenylphosphonium cyclopentadienylide (IV)[5] there may also be some contribution from a covalent form (B) as well as the zwitterionic form (A), since the phosphorus atom can expand its valency shell to accommodate ten electrons, utilising unfilled 3d orbitals to achieve this. This is not possible in the case of N-cyclopentadienylides.

There are also compounds closely related to the cyclopentadienylides in which the hetero-atom is separated from the cyclopentadiene ring by a con-

References p. 92

jugated chain of carbon atoms. In particular the cyclopentadiene ring may be attached to the carbon atom of a heterocyclic ring as in the *cyclopentadienylidene* derivatives (V)[6], (VI)[7] and (VII)[8]. In each of these compounds the molecule may be regarded as a hybrid of the dipolar structures (A) and purely covalent structures (B).

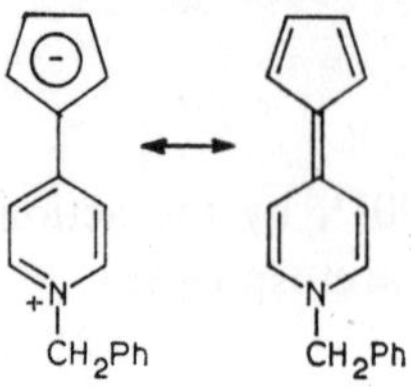

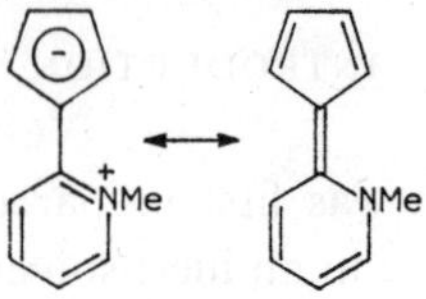

By condensation of cyclopentadiene with aldehydes or ketones in the presence of alkali, *fulvenes* (VIII) may be prepared[9]. Physical measurements show that even in the case of fulvenes there is a small contribution to the overall structure from charged forms such as (VIII B). This contribution is increased if the groups R or R' attached to the exocyclic carbon atom contain heteroatoms able to accommodate the positive charge as in the furylfulvene (IX)[10].

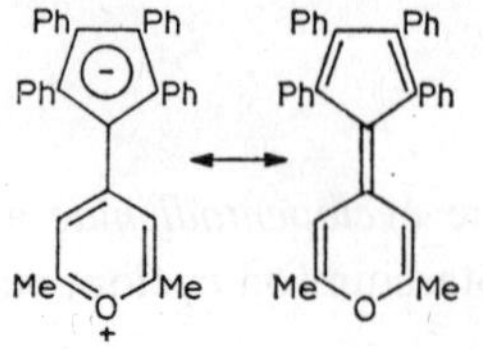

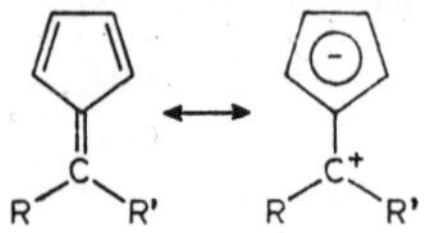

Another compound involving cyclopentadienide rings which has been prepared and intensively studied is the "sandwich" compound *ferrocene* (X).

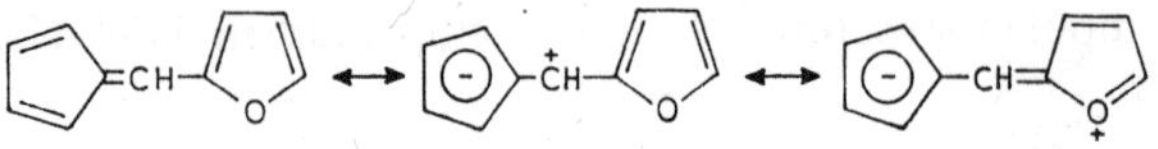

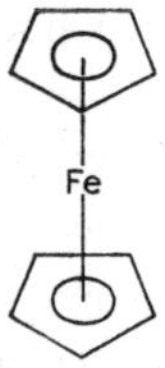

Since ferrocene may better be considered as an organo-metallic compound, deriving its characteristic properties from the presence of the metal atom, rather than as a straightforward carbocyclic non-benzenoid aromatic compound, it will be considered relatively briefly in the present text.

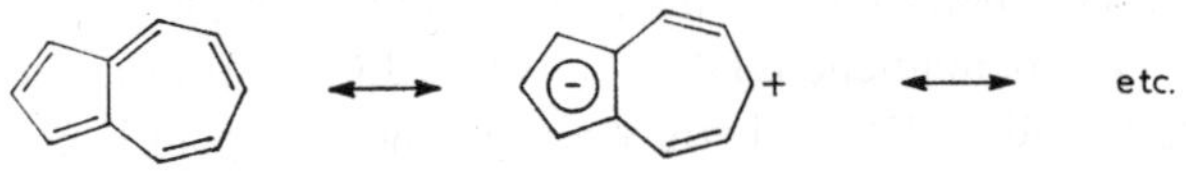

CYCLOPENTADIENIDE SALTS 57

Azulenes derived from the parent hydrocarbon bicyclo[5,3,0]decapentaene or *azulene* (XI) also involve charged structures having a negatively charged cyclopentadienide ring. They are considered in more detail in Chapter VIII. Cyclopentadienide salts, cyclopentadienylides, cyclopentadienylidene derivatives, fulvenes and ferrocenes will now be considered severally.

CYCLOPENTADIENIDE SALTS

Preparation

As mentioned above, the first cyclopentadienide salt to be prepared was the potassium salt, obtained by the action of potassium on cyclopentadiene dissolved in benzene and under an atmosphere of nitrogen[1]. It separated out as a yellow precipitate, insoluble in benzene. Since then many others salts have been made including a variety of alkali metal, alkaline earth metal and rare earth metal salts. Calcium cyclopentadienide has been prepared by the action of calcium carbide on cyclopentadiene in liquid ammonia; acetylene is also evolved[11].

The cyclopentadienide salts are stable in an inert atmosphere, and sodium cyclopentadienide remains unchanged on heating to 300° in nitrogen for a long period[11]. They rapidly resinify and may inflame in air. They are insoluble in most organic solvents but sodium cyclopentadienide is readily soluble in 1,2-dimethoxyethane and in tetrahydrofuran, giving stable solutions. It is readily soluble in liquid ammonia. The cyclopentadienide anion reacts rapidly with water and with halogen compounds[1].

The first clear correlation between the formation of the stable cyclopentadienide anion and the fact that this ion had a sextet of π-electrons was made in 1928; "in regard to its ability to provide the electrons for the stable sextet – cyclopentadiene can do so only by the appropriation of the electrons of one of its hydrogen atoms, it is this circumstance which gives to the hydrocarbon and its derivatives properties analogous to those of an acid, and confers stability on the corresponding anion[12]".

Properties of cyclopentadienide salts

The pK_a of cyclopentadiene<19; it is thus more acidic than other hydrocarbons, except the acetylenes[13]. For example in the case of triphenyl-

References p. 92

methane, $pK_a = 33$. In keeping with this cyclopentadiene is converted into its sodium salt by the action of sodium tertiary butoxide but not by sodium ethoxide[11]. The ready formation of the anion is illustrated by the ease with which cyclopentadiene undergoes deuterium exchange with deuterium oxide in the presence of strong base[14].

The ionic nature of the cyclopentadienide salts has been confirmed in a variety of ways. When they were first prepared it was noted that it was possible to prepare other metal salts from potassium cyclopentadienide by double decomposition[1]. The ionic character has been demonstrated by conductivity measurements in liquid ammonia[15] and from spectra determinations[16].

The n.m.r. spectrum of sodium cyclopentadienide shows one sharp peak at 4.5 τ, indicating the symmetry of the anion[17,18]. The difference between the position of this peak and that shown by benzene (2.73 τ) can be mainly accounted for by electrostatic effects due to the charge on the cyclopentadienide ring. Since benzene and the cyclopentadienide anion each have six π-electrons and the radii of the two rings is similar the resultant screening is almost the same in each case; hence the relative hydrogen shifts in a common solvent are due almost entirely to different electron densities[18].

The symmetry of the cyclopentadienide anion also follows from work on ^{14}C labelled cyclopentadiene[19]. Attempts to convert 1,2-dibromocyclopentane labelled at the 4-position with ^{14}C into labelled cyclopentadiene by three different methods in each case produced cyclopentadiene with a random distribution of [14]C. This leads to the conclusion that the preparative routes involved formation of the cyclopentadienide anion, and that all five positions in this ion are equivalent, thus leading to the random distribution in the resulting cyclopentadiene.

The resonance energy of the cyclopentadienide anion has been calculated to be 42 kcal/mole[20].

Chemical reactions of cyclopentadienide salts

Not surprisingly the cyclopentadienide anion reacts readily with electrophilic reagents. With carbon dioxide it gives a dicarboxylic acid which dimerises[1,11]. Similarly it is alkylated by alkyl halides but the products are dimeric[21]. It is aroylated by acid chlorides, no catalyst being necessary, to give diaroyl derivatives, which are monomeric[22,23]. On acidification these diaroylcyclopentadienide salts give rise to diaroylcyclopentadienes which exist mainly in a mono-enol form[22,23]. Reaction of potassium cyclopentadienide with chloroformic ester results in the formation of mono- and di-carbomethoxycyclopentadienide salts[27]. It is interesting to note that in the presence of a quaternary ammonium hydroxide, acrylonitrile substitutes in the cyclopentadiene ring; in the absence of this base the two compounds undergo a normal Diels–Alder addition reaction[24]. The cyclopentadienide anion undergoes electrophilic substitution with methyl iodide in liquid ammonia, with or without the presence of sodamide, to give methylcyclopentadienes[203].

Substituted cyclopentadienide salts

Introduction of electron-withdrawing groups attached to the cyclopentadiene ring greatly lowers the reactivity of the related cyclopentadienide anion, presumably owing to enhanced delocalisation of the negative charge. An example of such an unreactive anion was first prepared in 1912 by the condensation of nitromalondialdehyde with hexan-2,5-dione in the presence of alkali[25]:

$$NO_2\text{—}CH(CHO)_2 \ + \ \begin{array}{c} CH_2COCH_3 \\ | \\ CH_2COCH_3 \end{array} \xrightarrow{NaOEt} O_2N{\displaystyle\bigcirc^{\ominus}}{\begin{array}{c} COCH_3 \\ COCH_3 \end{array}} \ Na^+$$

This salt appears to be stable indefinitely in air; it is unattacked by bromine, hydrogen bromide or hydrogen iodide. Other examples of substituted cyclopentadienide anions which can be handled in air include aroyl[22,23], formyl[26], cyano[200] and carbomethoxycyclopentadienides[27–30].

Sodium dicarbomethoxycyclopentadienide, which is a red solid stable in air or in neutral or alkaline aqueous solution, has been shown from conductivity measurements and other data to be an ionic compound[27].

1,2,3,4,5-Pentacarbomethoxycyclopentadiene is an extremely strong acid,

$$HOHC{=}\!\!\begin{array}{c}CHO\\CHO\end{array} \rightleftharpoons \left[{}^-OHC{=}\!\!\begin{array}{c}\\ \end{array} \leftrightarrow OHC{\bigcirc^{\ominus}}\!\!\begin{array}{c}CHO\\CHO\end{array} \leftrightarrow OHC\!\!\begin{array}{c}CHO^-\\CHO\end{array} \atop \updownarrow \atop OHC\!\!\begin{array}{c}CHO\\CHO^-\end{array} \right]$$

(XII)   (XIII)

stronger than sulphuric acid (as a dibasic acid) and probably stronger than hydrochloric acid[29]. Its potassium salt couples with benzene diazonium chloride[30].

The pentacyanocyclopentadienide anion is not protonated by perchloric acid in acetonitrile. The potassium salt of this anion can be heated to 400° in air without decomposition[200].

The cyclopentadiene derivative (XII)[26] is also a strong acid (pK = 1.8) no doubt owing to the high delocalisation of negative charge in the resultant anion (XIII).

Salts derived from indene, fluorene, etc.

Indene and fluorene (benzo- and dibenzo-cyclopentadiene) also have acidic methylene groups and alkali metal salts have been made by the action of, for example, sodamide on the hydrocarbons[31]. A tetramethylammonium salt of fluorene has also been prepared; it is an orange compound which remains unchanged when kept under nitrogen[32]. It reacts with benzophenone to give 9-(hydroxydiphenyl)methylfluorene and decomposes to fluorene and 9-methylfluorene on heating in ether. Fulvenes also form cyclopentadienide salts on treatment with lithium alkyls[33,34] or with sodamide[33,146] in liquid ammonia, *e.g.*

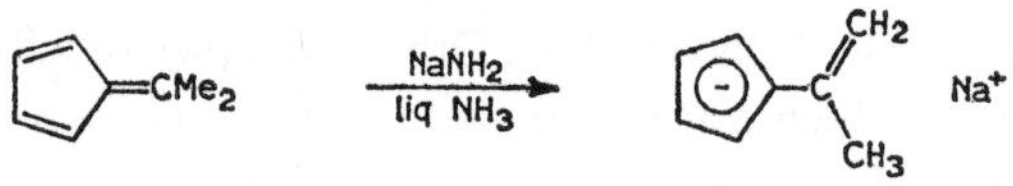

Among more sophisticated cyclopentadienide derivatives which have been prepared are the dianion (XIV)[35], made by treating 3-methylcyclopent-2-enone with sodamide in liquid ammonia, the pentalene dianion (XV)[36] (see also Chapter VIII) and the strongly coloured cyclopentadienylidene methine anions (XVI)[37] and (XVII)[38].

The tribenzopentalene derivative *fluoradene* is so strongly acidic that it dissolves in aqueous sodium hydroxide[201]; its $pK_a = 13.6$:

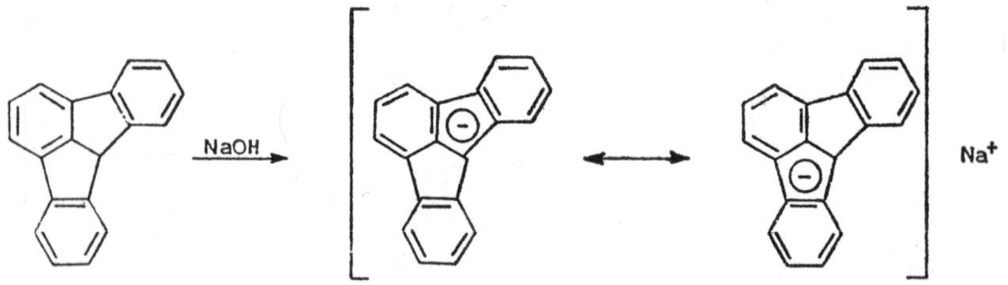

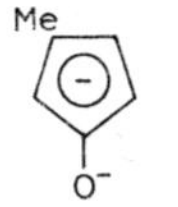

CYCLOPENTADIENIDE SALTS

(XIV)

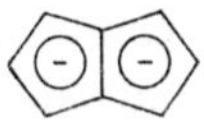

(XV)

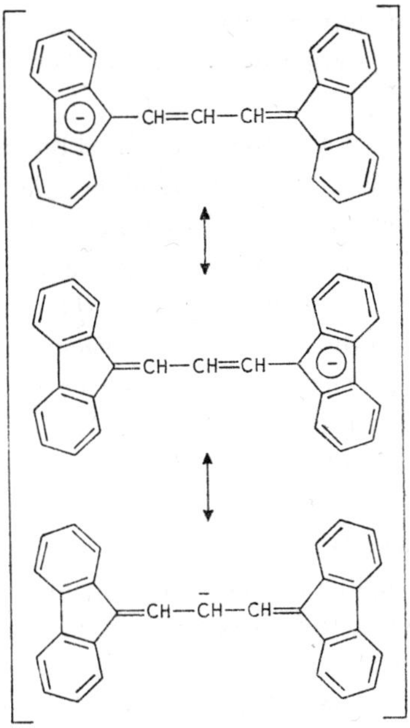

(XVI)

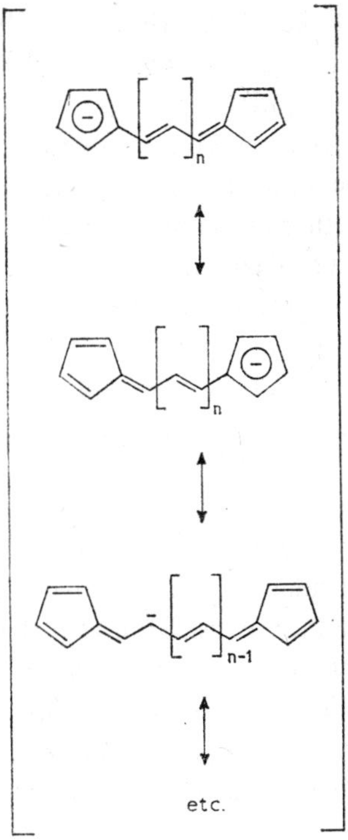

(XVII)

DIAZOCYCLOPENTADIENES

Preparation

Diazocyclopentadiene was first prepared in 1953[2], although diazofluorene had been obtained more than forty years before[39] (see below, p. 69). Reaction between lithium cyclopentadienide and p-toluenesulphonazide led to the formation of diazocyclopentadiene, the reaction path being postulated as follows[2]:

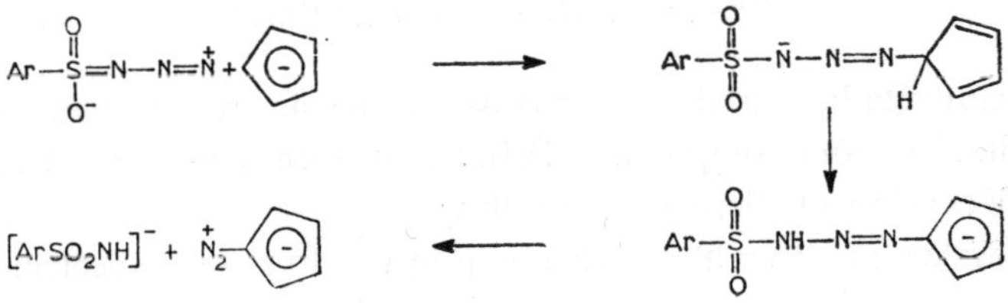

References p. 92

Diphenyl-[40], triphenyl-[41-44] and tetraphenyl-diazocyclopentadienes[43,44,202] have been prepared from the corresponding cyclopentadienide salts by the same method.

A variant on this method dispenses with the need of first preparing a cyclopentadienide salt. Cyclopentadiene and *p*-toluenesulphonylazide react together in the presence of an amine such as diethylamine or ethanolamine to give diazocyclopentadiene[45]. Phenylated diazocyclopentadienes have also been prepared in this way the yields being notably better than those obtained by the original method[44]. In view of the relatively low basicity of the catalysts involved it seems unlikely that this reaction proceeds *via* a cyclopentadienide anion as intermediate; the function of the base is more likely to be to remove a proton from an intermediate formed by attack of the azide on cyclopentadiene itself[44].

Tetraphenyldiazocyclopentadiene has also been prepared by the action of alkali on the *p*-toluenesulphonylhydrazone of cyclopentadienone[8]. This method is useful in that the latter ketone is very readily accessible and the rather tedious preparation of tetraphenylcyclopentadiene is avoided.

Tetrachlorodiazocyclopentadiene has been prepared by reacting hexachlorocyclopentadiene with hydrazine and oxidising the hydrazone which is obtained with aqueous hypochlorite[46] or mercuric oxide[202].

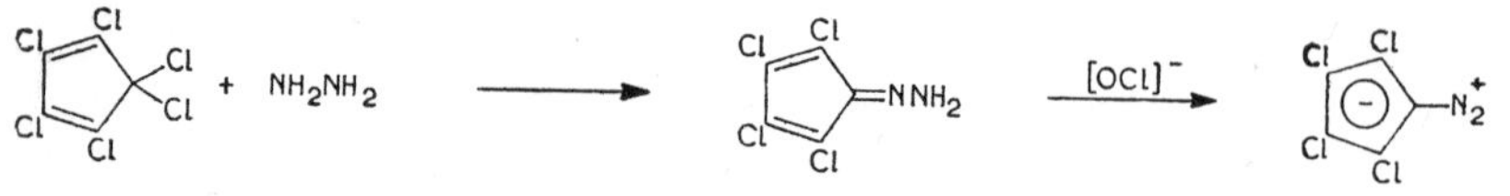

The substituted diazocyclopentadiene (XVIII) results from the action of nitrous acid on the corresponding amino-compound[47].

(XVIII)

Tetracyanodiazocyclopentadiene has also been prepared by this method[200].

Spectra of diazocyclopentadienes

Diazocyclopentadiene and its phenyl derivatives are red; tetrachlorodiazocyclopentadiene and compound (XVIII) both form yellow needles. Tetracyanodiazocyclopentadiene is pale yellow.

Diazocyclopentadiene has a maximum in its ultra-violet spectrum at 298 mμ (log ε = 4.17), the red colour being due to long tailing[2]. As in the case of other aromatic compounds, substitution of conjugating substituents in the ring results in shifts of the spectra to longer wavelengths[48]. 2,3,4-Triphenyldiazocyclopentadiene has absorption maxima at 239 and 330 mμ; on addition of perchloric acid the peak at 330 mμ disappears, showing that it arises from the aromatic five-membered ring[44].

Infra-red spectra are most useful for characterising diazocyclopentadienes. Diazocyclopentadiene itself shows strong absorption at 4.8 μ due to the diazogroup[2], and a similar band is found in the infra-red spectra of all its derivatives.

The n.m.r. spectrum of diazocyclopentadiene[48,49] consists of two equivalent unsymmetrical quartets corresponding to an A_2X_2 system. The chemical shifts have been reported as 3.3 and 4.2 τ[48] or 3.38 and 4.14 τ[49]. Since the electron-withdrawing diazo-group exerts a stronger deshielding influence on the 2,5-positions than the 3,4-positions the absorption at 4.2 τ has been assigned to the 3,4-hydrogens and that at 3.3 τ to those at the 2,5-positions. The τ-value for the 3,4-hydrogen atoms corresponds closely to that for the hydrogen atoms in a cyclopentadienide anion (4.5 τ) and indicates that the compound exists almost completely in a dipolar form.

The n.m.r. spectrum of 2,3,4-triphenyldiazocyclopentadiene suggests that the phenyl groups share the negative charge on the five-membered ring[44]. In deuterotrifluoroacetic acid the signal due to the hydrogen atom in the 5-position disappears, showing that rapid deuterium exchange takes place[44].

Chemical reactions of diazocyclopentadiene

Diazocyclopentadiene has been reported to explode on heating[52]. Catalytic reduction of diazocyclopentadiene converts it into cyclopentanone hydrazone[2]. Lithium aluminium hydride reduces it to the hydrazone of the as yet unknown cyclopentadienone; this hydrazone is probably a hybrid of purely covalent and dipolar forms, viz.[50].

Not surprisingly diazocyclopentadiene readily undergoes electrophilic substitution reactions. Electrophilic attack takes place preferentially at the 2- and 5-positions, presumably because this involves a more stable transition state than does attack at the 3- or 4-positions[48]. Thus nitration by means of benzoyl nitrate in acetonitrile produces both 2- and 3-nitrodiazocyclopentadiene, but the two products are formed in the ratio 2:1[48]. Reaction with mercuric acetate followed by sodium iodide gives rise to 2,5-diiodomercuridiazocyclopentadiene[48]. This is an unstable compound which explodes on grinding; it reacts with iodine to give the orange, light-sensitive 2,5-diiododiazocyclopentadiene[48]. N-Bromosuccinimide brominates diazocyclopentadiene readily, giving the tetrabromo-derivative[48]. This compound is also light-sensitive. Benzenediazonium fluoroborate couples with diazocyclopentadiene giving a deep violet 2-phenylazo-derivative[48].

2,3,4-Triphenyldiazocyclopentadiene may similarly be mercurated and brominated, reaction taking place at the 5-position but it does not appear to couple with diazonium compounds[44]. It is protonated by strong acid; the n.m.r. spectrum shows that protonation has taken place at the 5-position[44].

Tetracyanoethylene reacts with diazocyclopentadiene by substitution and not by a Diels–Alder reaction, but dimethyl acetylene dicarboxylate reacts by addition to form a pyridazine derivative[48]:

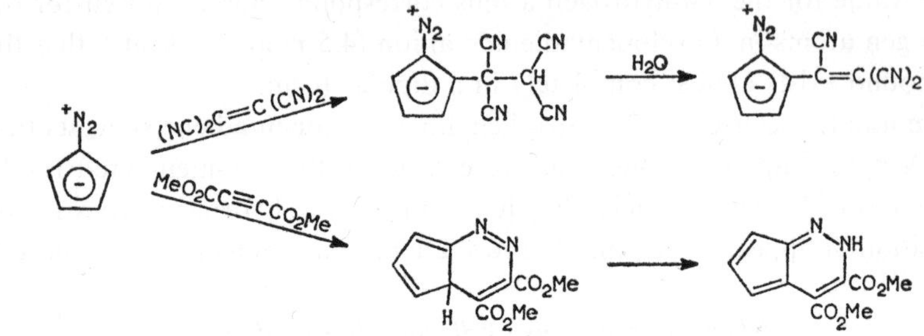

Diazocyclopentadiene reacts with methyl lithium or phenyl lithium to form azo-substituted cyclopentadienide salts[51]. Phenylsubstituted diazocyclopentadienes react similarly[43,44]. Protonation of the phenylazocyclopentadienide salts gives rise to cyclopentadienone phenylhydrazones. The u.v. and n.m.r. spectra of the products confirm that protonation has taken place at the β-nitrogen atom and not in the cyclopentadiene ring, e.g.

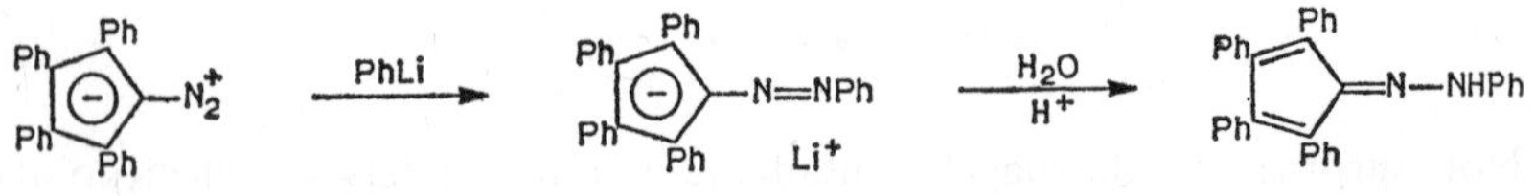

Diazocyclopentadiene couples with triphenylphosphine to give a phosphazine (XIX)[45,52]. Tetrachlorodiazocyclopentadiene couples similarly[46] but tetraphenyldiazocyclopentadiene only couples with triphenylphosphine at

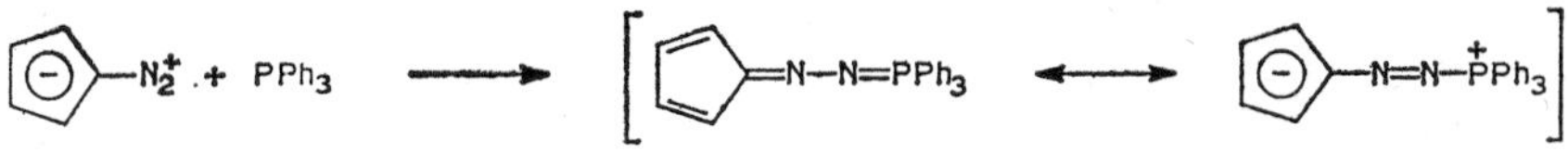

$$\text{Cp-N}_2^+ + \text{PPh}_3 \longrightarrow \left[\text{Cp=N-N=PPh}_3 \longleftrightarrow \text{Cp-N=N-}\overset{+}{\text{P}}\text{Ph}_3 \right]$$

(XIX)

high temperatures[44]. The phosphazine (XIX) is thermally stable[52]; it is converted into cyclopentadienone hydrazone by acid hydrolysis[50]. Diazocyclopentadiene also couples with dialkyl phosphites[45].

Diazocyclopentadienes tend to lose nitrogen on heating, thereby forming carbenes. A similar decomposition may be induced by light, for example irradiation of a solution of diazocyclopentadiene in either cyclopentane or cyclohexane gives rise to the formation of cyclopentyl- and cyclohexylcyclopentadiene respectively[53]. Photolysis of diazocyclopentadiene in a rigid matrix at very low temperatures produces a hydrocarbon which decomposes on heating and which is believed from its u.v. spectrum to be fulvalene (XX)[54].

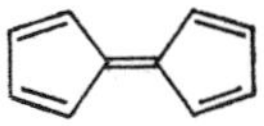

(XX)

2,3,4-Triphenyldiazocyclopentadiene decomposes partially on recrystallisation from organic solvents[41,44]; on heating in boiling ethanol the corresponding hexaphenylfulvalene is formed[41]. In contrast the 2,3,5-triphenyl isomer could not be converted into a fulvalene by thermal decomposition, presumably owing to steric hindrance[42].

Other reactions presumably involving carbenic decomposition of diazocyclopentadienes include the reaction of the tetrachloroderivative with diazodiphenylmethane to yield the azine (XXI)[55], and the reactions of tetraphenyldiazocyclopentadiene with pyridine[56] and with hydrazine[44] to give, respectively, pyridinium tetraphenylcyclopentadienylide (see p. 66) and a product which appears to be aminotetraphenylcyclopentadiene.

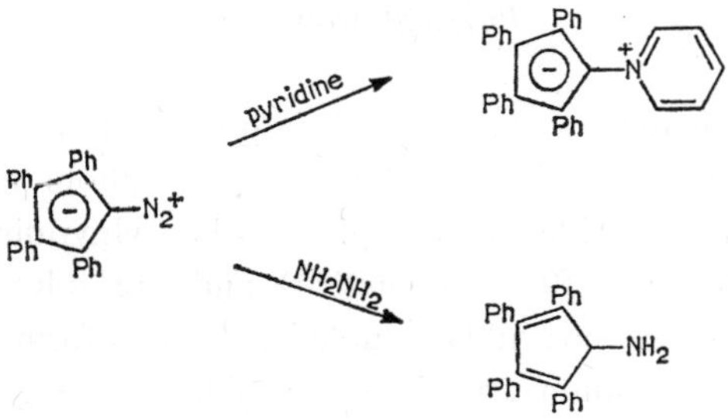

References p. 92

(XXI)

Tetracyanodiazocyclopentadiene resembles benzene diazonium salts in that the diazo group can be replaced by a cyano group by reaction with cuprous cyanide and can be reduced by ethanol in the presence of copper powder[200]. The product in the latter case is the tetracyanocyclopentadienide anion.

OTHER CYCLOPENTADIENYLIDES

Preparation

Pyridinium cyclopentadienylide was prepared in 1955 by the following route[3]:

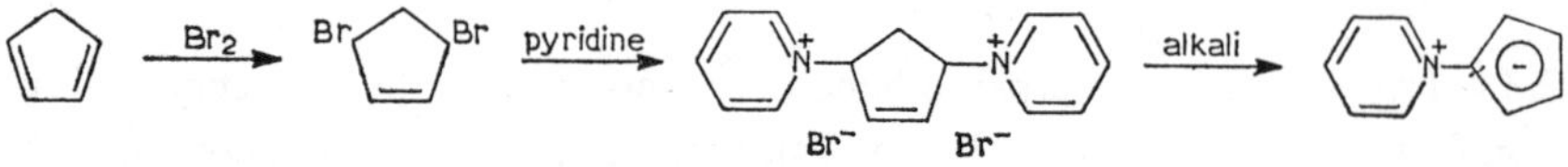

A similar sequence of reactions has been used to prepare substituted pyridinium cyclopentadienylides[4,57,58], and trimethylammonium[59], triphenylphosphonium[55,60] and dimethylsulphonium[204] cyclopentadienylides.

Pyridinium tetraphenylcyclopentadienylide and methyl homologues have been prepared by reacting 5-bromo-1,2,3,4-tetraphenylcyclopentadiene with pyridine (or its homologues) and treating the product with alkali[4]. α-Substituted pyridines could not be used satisfactorily, presumably owing to steric hindrance[4]. The pyridinium compound has also been prepared by heating tetraphenyl diazocyclopentadiene with pyridine[56].

Physical properties

Since cyclopentadienylides are dipolar compounds it is not surprising that they have high melting-points (*e.g.* triphenylphosphonium 229–231°[5]; trimethylammonium *ca.* 200° (decomp.)[59]; 4-benzylpyridinium>250° (decomp.)[57]; pyridinium>350° (decomp.)[3,4]) and are soluble only in polar solvents[3–5,59]. They are reversibly soluble in acid and form crystalline picrates[4,5]. Trimethylammonium cyclopentadienylide has $pK_b = 3.94$[59].

Pyridinium cyclopentadienylide has a dipole moment of 13.5 D[61,62] but the values for triphenylphosphonium and dimethylsulphonium cyclopentadienylides are much lower (7.0, 5.7 D)[60,61,204] since in the latter cases covalent as well as dipolar forms contribute to the structure of the molecules (see above, p. 55).

The polarography of pyridinium cyclopentadienylide has been studied and it has been shown that in neutral solution an ylide molecule reversibly adds an electron to give a radical which readily dimerises[63].

Spectra

The pyridinium cyclopentadienylides are intensely coloured, unphenylated ones being copper coloured and tetraphenyl derivatives very dark blue. The colours of their solutions depend on the polarity of the solvents; thus pyridinium cyclopentadienylide gives an orange solution in alcohol and a red solution in acetone or chloroform[3,4]. Similarly its tetraphenyl derivative dissolves in ethanol giving a red solution but in benzene giving a blue solution[4].

It was suggested[64] that the colours were consistent with the compounds behaving as "internal" charge-transfer complexes. In less polar solvents the ionic ground state becomes progressively destabilised through loss of solvation energy with respect to an uncharged excited state. The ultra-violet and visible spectrum of pyridinium cyclopentadienylide has been recorded in a variety of solvents, and the band at *ca.* 500 mμ which has been attributed to intramolecular charge transfer has been shown to vary with the polarity of the solvent[58], *viz.*

Solvent	λ_{max}(mμ)	Solvent	λ_{max}(mμ)
water	453	acetonitrile	495
methanol	486	acetone	509
ethanol	496	chloroform	511
isopropanol	498	n-heptane	524, 546
tert. butanol	501		

Similarly in the case of the tetraphenyl derivative the maximum shifts from 532 mμ in methanol to 595 mμ in benzene[56]. In aqueous solution pyridinium[58] and trimethylammonium[59] cyclopentadienylides have the following maxima in their absorption spectra:

pyridinium: $\lambda_{max}(m\mu)$ 224,245,333,453 (log ε 4.08, 4.04, 3.84, 3.66)
trimethylammonium: $\lambda_{max}(m\mu)$ 234,286,313,395,515(log ε 3.51, 2.70, 2.74, 2.32, 1.85)

The longer wavelength peaks disappear in acid solution.

In contrast triphenylphosphonium cyclopentadienylide is yellow. It shows the following maxima in solution in acetonitrile: λ_{max} (mμ) 222, 250, *ca.* 295 (log ε 4.58, 4.33, 3.77)[5,60].

Stability

Pyridinium and trimethylammonium cyclopentadienylides are destroyed by prolonged contact with air, but may be kept under nitrogen[4,59]. Substitution of an electron-withdrawing group into the pyridine ring appears to lower their stability[58]. Both triphenylphosphonium cyclopentadienylide and pyridinium tetraphenylcyclopentadienylide appear to be indefinitely stable in air in the solid state, although solutions of the latter rapidly decompose[4].

Chemical reactions

Pyridinium cyclopentadienylide has been catalytically reduced to cyclopentylpiperidine using platinum[3,4] or rhodium[65] as catalyst. Substitution of a methyl group in the pyridine ring increases the difficulty of reducing the ylide[4]. Triphenylphosphonium cyclopentadienylide could not be reduced using platinum oxide as catalyst[5]. The trimethylammonium ylide could be catalytically reduced in acetic acid solution to cyclopentane and trimethylamine; it seems likely that the species actually reduced in this case was the conjugate acid of the ylide[59].

A number of electrophilic substitution reactions have been carried out on cyclopentadienylides. Thus the pyridinium, trimethylammonium and triphenylphosphonium ylides couple with diazonium salts[4,59,69] and can be brominated[59,66–68]; the trimethylammonium ylide has been mercurated[59] and the triphenylphosphonium compound has been formylated and acetylated[68]. Pyridinium cyclopentadienylide readily undergoes deuterium exchange with deuteroethanol in alkaline, neutral or feebly acidic solution[61,62,66]; exchange takes place less readily in the case of triphenylphosphonium cyclopentadienylide[62]. Cyclopentadienylides do not appear to react with aldehydes or ketones[4,5,60].

Triphenylphosphonium cyclopentadienylide reacts with certain metal carbonyls to form organo-metallic derivatives[70]:

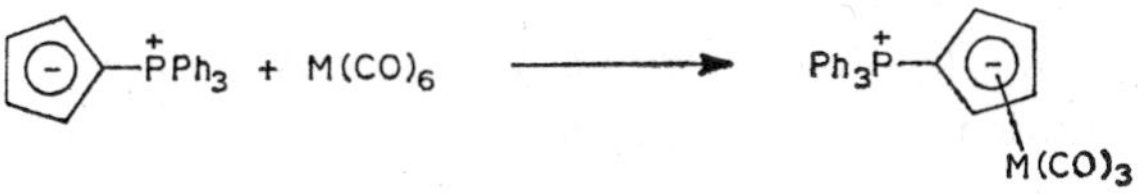

$$\text{Cp-}\overset{+}{\text{P}}\text{Ph}_3 + M(CO)_6 \longrightarrow \text{Ph}_3\overset{+}{\text{P}}\text{-Cp-}M(CO)_3 \quad (M = Cr, Mo, W)$$

INDENYLIDES AND FLUORENYLIDES

Indenylides

When 1-bromoindene is treated with pyridine and then alkali a transient blue colour is observed[4,71]. This is presumably due to the formation of pyridinium indenylide; it would appear that this ylide has a very fugitive existence. 1-Diazoindene has been prepared by reacting indene with p-toluenesulphonylazide in the presence of diethylamine as a catalyst[45].

9-Diazofluorene

9-Diazofluorene could not be prepared by reacting fluorene with p-toluenesulphonylazide in the presence of an amine catalyst[45].

It was first prepared in 1911 by the oxidation of fluorenone hydrazone by means of mercuric oxide[39]. This method has been refined to give an almost quantitative conversion of the hydrazone into the diazo-compound[72]. Various substituted diazofluorenes have been made by the same method[73–75,205]. Autoxidation of fluorenone hydrazone in alkali also leads to the formation of diazofluorene[76].

Diazofluorene is a stable red compound which melts without decomposi-

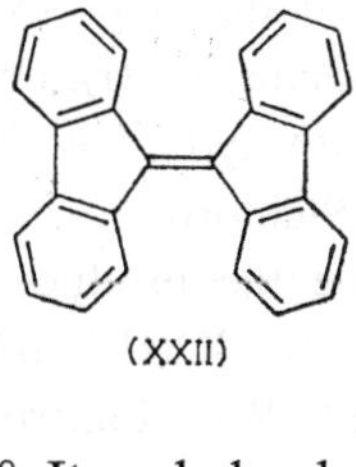

(XXII) 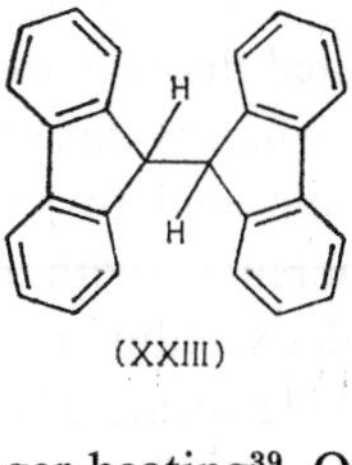   (XXIII)

tion at 94–95°. It explodes, however, on stronger heating[39]. On heating in a variety of solvents such as water, ethanol, benzene and toluene it decomposes

References p. 92

with loss of nitrogen; the carbene so formed dimerises to give 9,9'-bifluorenylidene (XXII). Photolysis of diazofluorene in solution in cyclohexane gives rise to a mixture of bifluorenyl (XXIII) and 9-cyclohexylfluorene[53].

Diazofluorene is unchanged after shaking with oxygen for several days[39]; it is however oxidised to fluorenone by perbenzoic acid[72].

It is not reduced by aluminium amalgam[39] but is reduced to fluorene either by zinc dust and alkali[39] or by prolonged shaking with palladium in aqueous ethanol[77]. When it is shaken with powdered sodium in ether under nitrogen, diazofluorene gives rise to a radical which is deep blue and very sensitive to oxygen[78]. Electron spin resonance measurements show that this product exists to about 60% in radical form and that there is an equilibrium between this radical and a non-radical dimer[78].

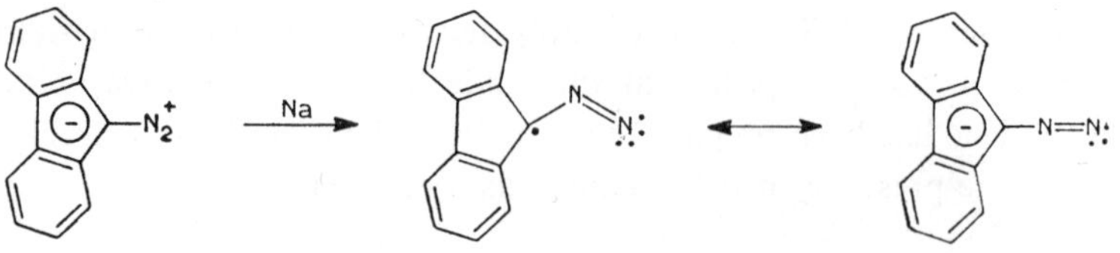

The majority of the reactions of diazofluorene involve its decomposition to nitrogen and a carbene, which then makes an electrophilic attack on the reagent present. An exception to this mode of reaction takes place with phosphine derivatives which react additively to form phosphazines[79]:

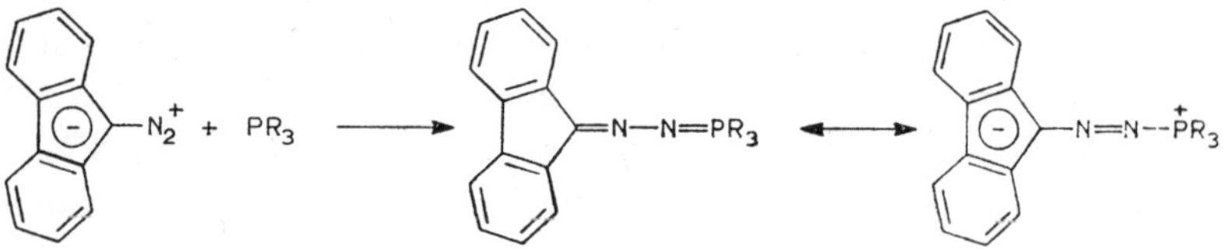

PR_3 = PPh_3, PEt_3 or PEt_2Ph

In reactions proceeding *via* a carbene intermediate the product may either be a 9-substituted fluorene or 9,9'-bifluorenylidene (XXII), or a mixture of the two. Thus whereas diazofluorene reacts with hydrogen chloride, acetic acid and benzoic acid to give, respectively, 9-chloro-, 9-acetoxy- and 9-benzoyloxy-fluorene[76], with hydrogen bromide the product is the dimer (XXII)[39]. Similarly iodine in ethanol gives rise to (XXII)[39] but bromine in carbon disulphide forms 9,9-dibromofluorene[76]. With dinitrogen tetroxide in benzene 9,9-dinitrofluorene is formed[80], and with aniline 9-phenylaminofluorene[76].

Like other carbenes, that arising from diazofluorene adds to olefins to give cyclopropane derivatives[76,81]. The heat or light induced reaction with either dimethyl fumarate or diethyl maleate in each case produces a preponderance of the *trans*-dicarbethoxy-compound[81]. This non-stereospecific addition suggests a triplet state for the biphenylenecarbene which is formed[81,206]. This is also supported by the fact that the light induced decomposition of diazofluorene in either *cis*- or *trans*-1,2-dichloroethylene results in the formation of 9-chlorofluorene, and that in carbon tetrachloride 9,9'-dichlorobifluorenyl is formed. Both of these products probably result from radical abstraction reactions[81], as do the reactions with carbonyl and oxalyl chlorides in which the products are as follows[76]:

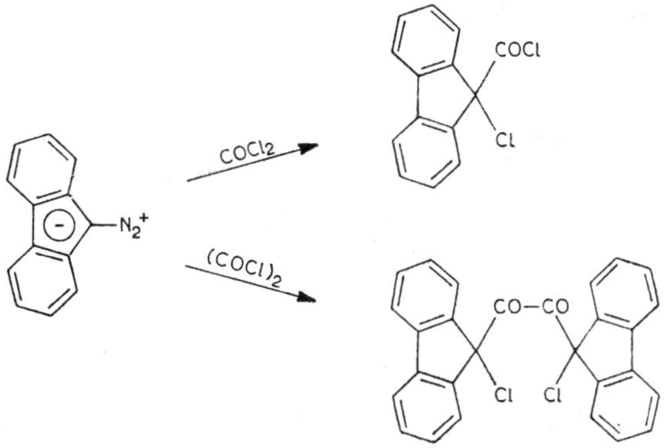

Stereospecific addition of the carbene to olefins is achieved by also having present oxygen or a conjugated diene. The latter compounds act as scavengers for triplet species present and addition to the mono-olefin presumably involves some residual singlet species[206].

Diazofluorene reacts with thiocarbonyl compounds to give thiiranes which decompose on heating to form fluorenylidene derivatives[82–85] (see also below, p. 80); *e.g.*

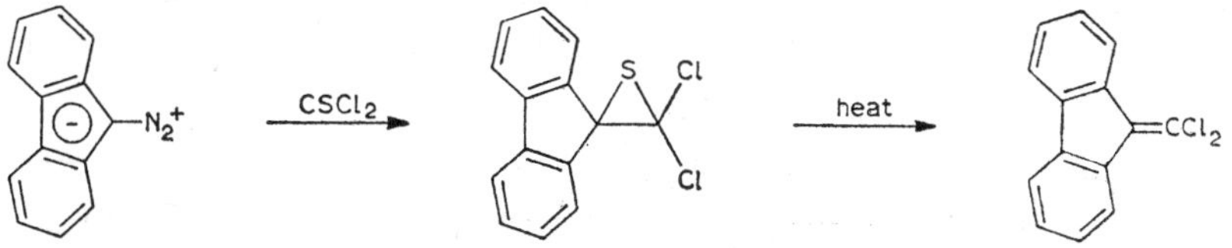

With nitrosobenzene a nitrone is obtained[86,87], and with pyridine oxide an azine; the following reaction path has been suggested[88]:

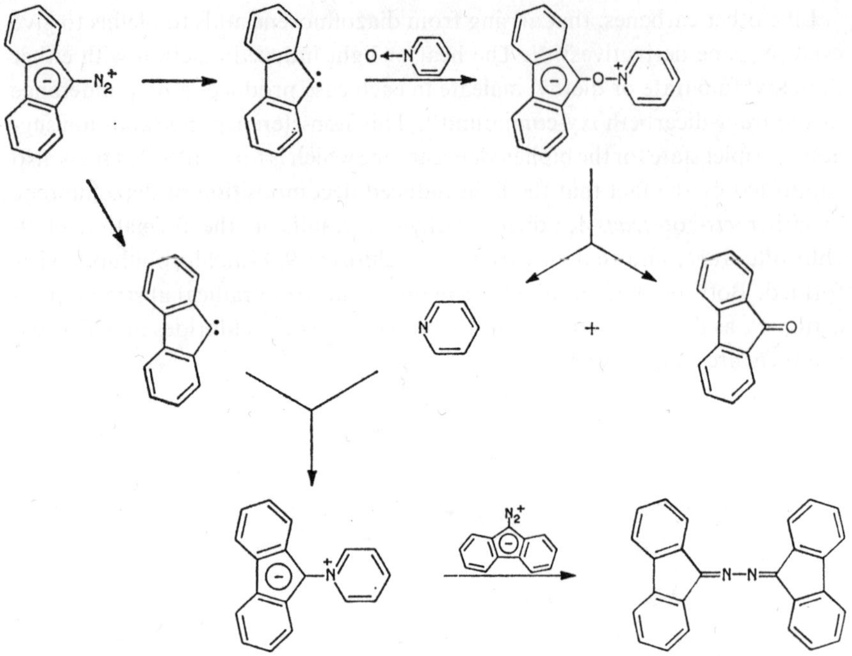

The acid catalysed solvolysis of diazofluorenes in water/ethanol mixtures has been investigated[73,74,205]. This appears to proceed *via* a carbonium ion intermediate. As expected solvolysis is facilitated by electron donating groups and retarded by electron withdrawing groups[74,205].

Other fluorenylides

On treating fluorenone with hydrogen peroxide in the presence of phosphorus pentoxide in ether fluorene peroxide (XXIV) is formed[89]. This yellow compound has some contribution to its structure from a fluorenylide form. As a dipolar molecule it is readily soluble in polar solvents such as methanol but insoluble in hydrocarbon solvents such as benzene. Molecular weight deter-

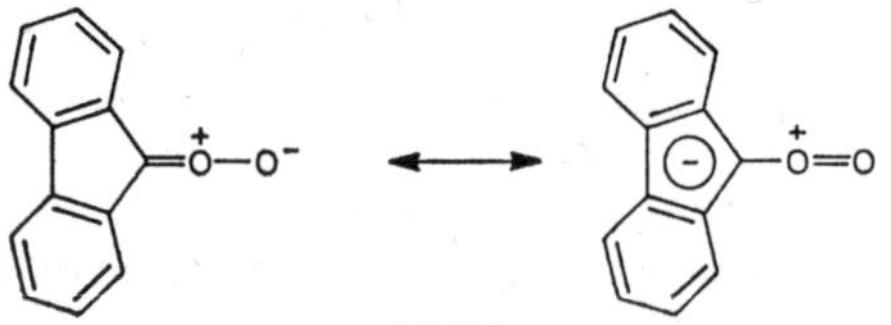

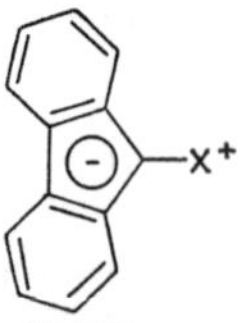

FLUORENYLIDES

minations show the molecule to be monomeric. It melts without decomposition at 108° but above 115° decomposes to fluorenone and oxygen. On sudden heating or on treatment with concentrated sulphuric acid it explodes[89].

Fluorenylides can in general be represented by formula (XXV), where $X = NR_3$, PR_3, SR_2, etc. They are normally prepared by reacting 9-bromofluorene with a suitable amine, phosphine or sulphide and treating the resultant salt with alkali, e.g.[90].

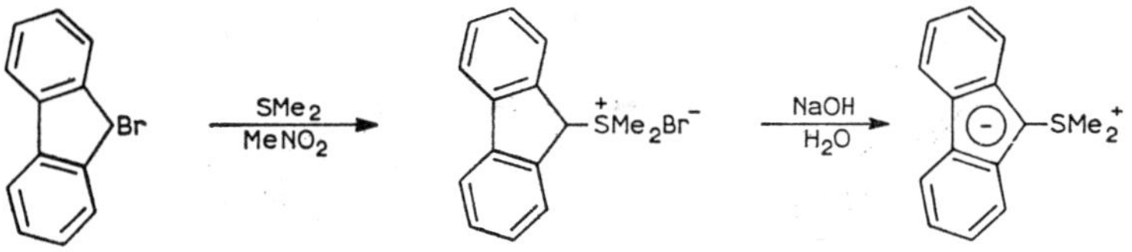

The first attempted preparation of a fluorenylide was by treating fluorenyltrimethylammonium bromide with silver oxide[91]. A purple solution was obtained but no ylide could be isolated and on heating the solution 9,9'-bifluorenylidene (XXII) and trimethylamine were formed. Some years later the desired ylide was isolated by treating the bromide with phenyl lithium in ether and in an atmosphere of nitrogen[92]. It proved to be stable under nitrogen but decomposed in air or on heating as previously described. It reacted readily with electrophilic reagents, for example, giving 9-alkylfluorenyl-9-trimethylammonium salts with alkyl halides. It did not however react with carbonyl compounds[92]. With water fluorenyl-9-trimethylammonium hydroxide was formed. When this ylide is heated with benzyldimethylamine, 9-benzyl-9-dimethylaminofluorene is obtained; reaction presumably proceeds via carbenic decomposition of the ylide[198]:

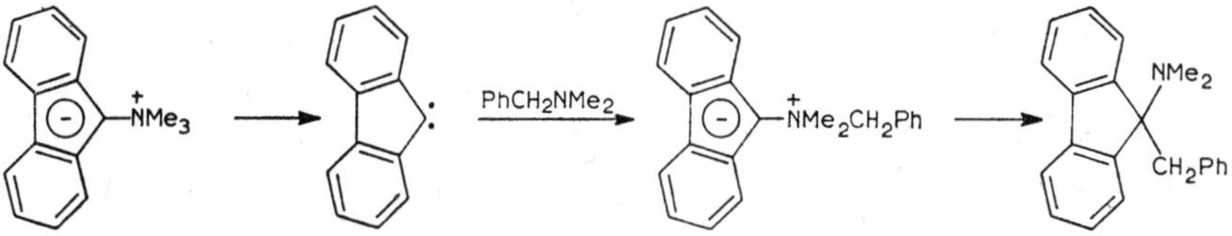

On treatment of N-fluorenylpyridinium bromide (and also the picolinium and quinolinium analogues) with alkali a deep blue ylide is similarly obtained but it decomposes within a few hours[71,93,94]. Despite the fugacity of this ylide its dipole moment has been measured[95]; the value of 4.13 D seems surprisingly low, however, and may reflect the non-homogeneity of the sample.

Introduction of an electron-withdrawing group into the fluorene portion of the molecule stabilises the resulting ylide; for example pyridinium 2-nitro-

References p. 92

fluorenylide may be isolated as a blue-green solid which remains unchanged away from air and light although it decomposes slowly in their presence[96].

Dimethylsulphonium fluorenylide was the first fluorenylide to be isolated other than diazofluorene[90]. It is prepared by the action of aqueous sodium hydroxide on fluorenyldimethylsulphonium bromide[90]. Its greater stability relative to the N-fluorenylides is probably due to the fact that the sulphur atom can expand its valency shell by utilising $3d$-orbitals, and that in consequence the ylide is a hybrid of dipolar and covalent forms:

This is demonstrated by the dipole moment [97], which = 6.2 D. None the less it decomposes in a few hours in air, with evolution of dimethyl sulphide[90,98]. Once again introduction of electron-withdrawing groups into the fluorene rings stabilises the resulting ylide; the 2-nitro-compound is stable for some days while the 2,7-dinitro-compound appears to be quite stable at room temperature[99]. Both decompose on heating in nitromethane to give bifluorenylidene derivatives[99].

Dimethylsulphonium fluorenylide is also decomposed by chromatography on neutral or slightly basic alumina to give a mixture of 9,9'-bifluorenylidene and fluorenone[98]. Decomposition probably proceeds by a carbenic mechanism. Dimethylsulphonium fluorenylide is less basic than phosphorus and arsenic fluorenylides (see below), suggesting that there is a greater contribution from the covalent form in this case[98]. It is also less reactive than phosphorus and arsenic fluorenylides towards carbonyl compounds and reacts only with aromatic aldehydes having powerful electron-withdrawing substituents[98,100]. It reacts with nitrosobenzene to form a nitrone[87,98]. On prolonged treatment with alcoholic sodium hydroxide or with liquid ammonia an interesting rearrangement reaction ensues, probably of the Sommelet type, producing 1-methylthiomethylfluorene[98,101].

A number of phosphonium fluorenylides have been prepared by the standard method[102-105]. The dipole moment of the triphenylphosphonium fluorenylide is 7.09 D, which suggests that there is a nearly equal contribution from the dipolar and covalent forms[106]. These phosphorus ylides are stable to heat[103]. The trialkylphosphonium fluorenylides are very rapidly hydrolysed by atmospheric moisture but the triaryl ones react more slowly[103-105]. Trialkylphosphonium fluorenylides are cleaved on passing through a column of nearly neutral alumina[105]. Cleavage takes place in two ways to give fluorene and trialkylphosphine oxide or fluorenol and trialkylphosphine. Triphenylphosphonium fluorenylide reacts with aromatic aldehydes but not normally with ketones, but the trialkyl ylides react with activated aromatic ketones as well[104-106]. An increase in the electron-withdrawing character of the groups attached to the phosphorus atom should increase the possibility of d-orbital participation in the bonding and therefore of the contribution of the covalent form to the overall structure. This greater contribution of the covalent form in the case of the triaryl compounds as compared with the trialkyl compounds provides an explanation of the greater reactivity of the latter compounds towards ketones and also their more facile hydrolysis[105].

Trimethylarsonium[103] and triphenylarsonium[107] fluorenylides have also been prepared, but an attempted preparation of trimethylstibonium fluorenylide led instead to bifluorenyl[103]. The trimethylarsonium ylide is readily hydrolysed to fluorene[103] but the triphenyl compound is only hydrolysed on prolonged boiling with ethanolic sodium hydroxide[107]. The latter compound reacts with aromatic aldehydes but not with ketones[107].

In concluding this discussion of benzocyclopentadienylides mention must be made of the anhydronium base (XXVI).

(XXVI)

One contributing form has an indenylide structure and the first postulate that such structures might represent stable aromatic compounds was made in connexion with this compound in 1925[108].

CYCLOPENTADIENYLIDENE DERIVATIVES

Preparation

A variety of heterocyclic cyclopentadienylidene derivatives of type (XXVII) or (XXVIII) (X = NR or O) have been prepared during the last decade.

(XXVII)    (XXVIII)

Thus in 1956 compound (XXVII), (X = N-(o,o'-dichlorobenzyl)) was prepared by treating N-o,o'-dichlorobenzylpyridinium bromide with cyclopentadiene in the presence of sodium methoxide[109]. It has since been suggested[7] that the red product might in fact be a mixture of isomers, corresponding to (XXVII) and (XXVIII).

A year later N-benzyl-4-cyclopentadienylidene-1,4-dihydropyridine (XXVII), (X = N-benzyl) was prepared by the reaction between N-benzylpyridinium chloride and lithium cyclopentadienide[6]:

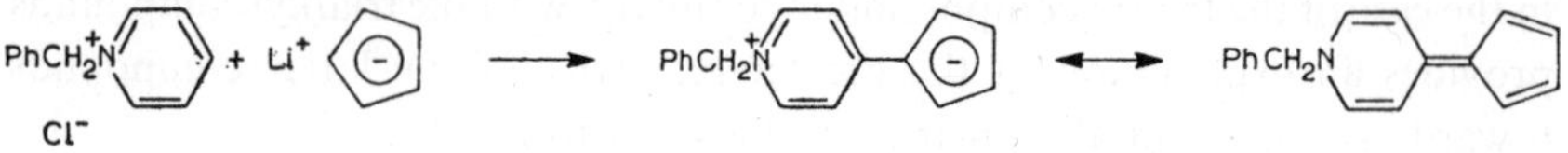

That this compound is indeed the 4-isomer has been confirmed by an alternative method of preparation form N-benzyl-4-bromopyridinium bromide and sodium cyclopentadienide[110,207], and by oxidising it by means of potassium permanganate in acid solution to isonicotinic acid[111]. The N-methyl analogue has also been prepared by both of the methods used to obtain the benzyl compound[110,207].

An alternative method which has been used to prepare a number of N-alkyl-4-cyclopentadienylidene-dihydropyridines starts from N-alkyl-4-methoxypyridinium salts, which are treated with cyclopentadiene in the presence of potassium tertiary butoxide[112,113], e.g.

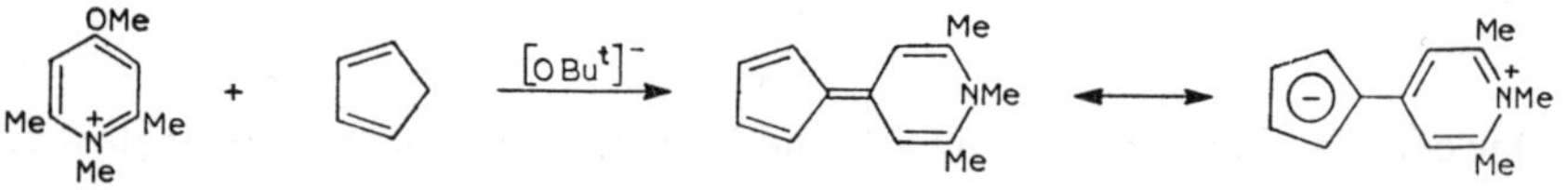

N-Methyl- and N-benzyl-2-cyclopentadienylidene-dihydropyridines have been prepared by treating the appropriate N-alkyl-2-halopyridinium salts with sodium cyclopentadienide[7,110,207]:

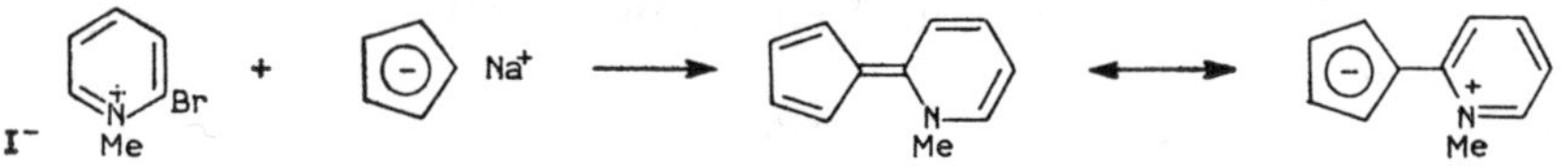

The N-methyl compound has also been obtained but only in spectroscopically detectable amounts by the following reaction sequence[7,207]:

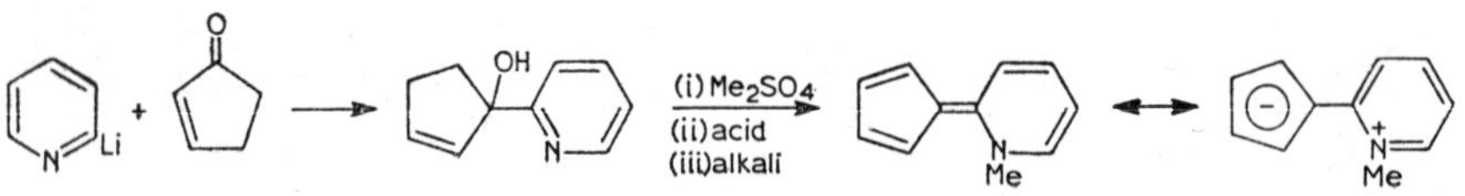

A 2-cyclopentadienylidenequinoline derivative has been prepared as follows[114]:

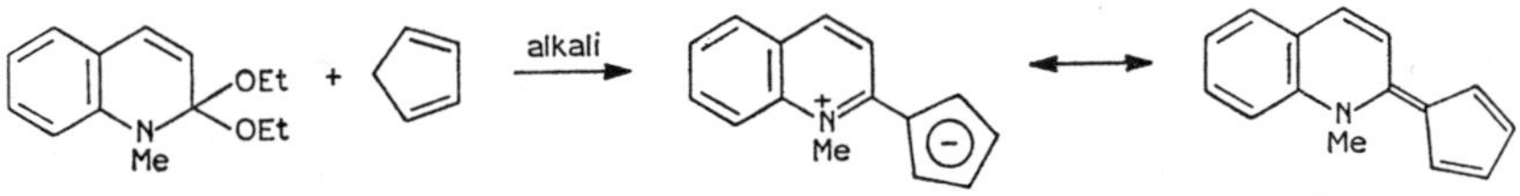

Whereas 4-methoxypyridinium salts react with cyclopentadienide anions to give cyclopentadienylidene-dihydropyridine derivatives, an attempt to carry out an analogous reaction using 4-methoxypyrylium salts led to formation of an azulene instead[113].

The difference in the mode of reaction is a consequence of the greater ease with which the pyrylium ring is opened by alkaline reagents.

Cyclopentadienylidenepyrans have been prepared by reaction of diazo-cyclopentadienes with 4-thiopyrones[8,115], *e.g.*

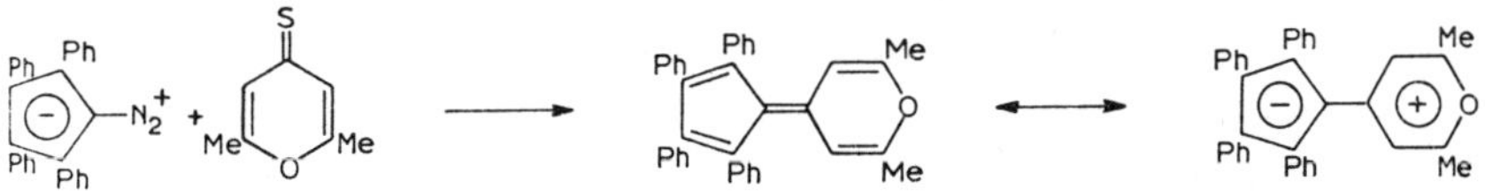

This reaction has been used with tetraphenyl- and 2,3,4-triphenyldiazocyclopentadiene but was unsuccessful when applied to unsubstituted diazocyclopentadiene[115]. Reaction apparently proceeds *via* carbene formation. In the case of the unsubstituted diazo-compound competing reactions appear to take place in preference.

The same method has also been used to prepare cyclopentadienylidene-thiapyrans[8].

It has proved possible to convert cyclopentadienylidenepyrans into the corresponding dihydropyridinium compounds by reaction with suitable amines[115] (see also below, p. 79).

Dithiafulvalenes (dithiacyclopentadienylidene-cyclopentadienes) have been prepared by reacting the sodium salts of cyclopentadiene and its derivatives with dithiolium salts, dithiolethiones or alkylmercaptodithiolium salts[209].

Properties of cyclopentadienylidenedihydropyridines

The cyclopentadienylidenedihydropyridines are yellow or orange solids. The 4-cyclopentadienylidene compounds have λ_{max} (in dioxan) at about 430 mμ, which is independent of the polarity of the solvent but the 2-isomers have peaks at $\lambda_{max} \sim$370 and 430 (dioxan) which show a hypsochromic shift with increasing polarity of solvent[110,208]. The dipole moment of the 4-benzyl compound = 9.7 D, showing a considerable contribution from the dipolar form[6,62,208]. There is disagreement in the literature concerning the stability of these pyridinium compounds, some workers describing them as stable, others as unstable[6,110,112,113,207]. They are reversibly soluble in acid to give colourless solutions[6,7,113,207]. Since the protonated compounds are colourless, protonation cannot take place at the nitrogen atom, for the resulting products would be fulvenes, which are coloured compounds[113]. The u.v. and n.m.r. spectra of the protonated 4-cyclopentadienylidene compounds suggest that two distinct products are formed and these are probably protonated at the 2 and 3 positions of the five-membered ring respectively[113]. The spectra of the protonated 2-cyclopentadienylidene compounds suggest that two different cations are formed in this case also, protonation again taking place at the 2 and 3 positions of the cyclopentadiene ring[7,110,207]. N-Methyl-2-cyclopentadienylidenedihydropyridine[7,207,208] has a pK_a value of 9.1. (in water).

N-Benzyl-2-cyclopentadienylidenedihydropyridine has been reduced catalytically; it absorbs five moles of hydrogen[6,65,66]. In ethanol, with platinum black as catalyst, reduction proceeds very slowly, but in acetic acid, two moles of hydrogen are taken up rapidly, the rate then falling off. The difference is presumably due to the fact that in acetic acid the protonated species is present and that this is much more readily reduced[66]. This N-benzyl compound undergoes deuterium exchange with deuteroethanol in alkaline, neutral or feebly acidic media[66].

Properties of cyclopentadienylidenepyrans

Triphenyl- and tetraphenyl-cyclopentadienylidenepyrans are red solids which are stable in air[115]. The ultra-violet spectrum of the latter compound resembles that of tetraphenylcyclopentadienylidenecycloheptatriene, with which it is isoelectronic[115]. Both cyclopentadienylidenepyrans give coloured salts with perchloric acid; n.m.r. determinations on the 2,3,4-triphenyl compound indicate that protonation occurs at the 5-position[115]. On solution of triphenylcyclopentadienylidenepyran in deuterotrifluoracetic acid, deuterium exchange does not take place between the 5-hydrogen atom and the solvent; this contrasts with the behaviour of 2,3,4-triphenyldiazocyclopentadiene[115]. When the perchlorate salts are treated with base, different reactions ensue depending on the strength of the base which is used. Amines having pK_b between 5 and 10 react to form the corresponding cyclopentadienylidenedihydropyridines; bases (including amines) having $pK_b < 5$ or > 10 regenerate the original cyclopentadienylidenepyran, which is not further attacked by these bases[115].

These cyclopentadienylidenepyrans do not react with maleic anhydride[115]. They are oxidised to pyrones by means of chromic acid/acetic acid[115].

Attempts to carry out electrophilic substitution at the 5-position of the 2,3,4-triphenyl compound indicate that its reactivity towards such reagents is low[115]. Bromination and mercuration proceeded only incompletely, diazonium salts did not couple at all, and chlorine appears to add to the double bond joining the rings[115].

Other cyclopentadienylidene derivatives

Among other cyclopentadienylidene derivatives which have some contribution from a dipolar form may be mentioned the deep violet compound (XXIX)[116–118], obtained by condensation of p-nitrosodimethylaniline with tetraphenylcyclopentadiene, and the azo compound (XXX) whose dipole moment of 3.3 D shows that there is a small contribution from the dipolar form[119].

Indenylidene and fluorenylidene derivatives

A number of indenylidene- and fluorenylidene-dihydropyridines have been prepared by methods identical to those used to obtain the cyclopentadienylidene analogues[110,112,113,120,207]. Annellation causes a bathochromic shift of the long wavelength absorption maxima of cyclopentadienylidene compounds[110].

These compounds are reversibly soluble in mineral acid, protonation of the fluorenylidene compounds taking place at the 9-position of the fluorene moiety and of the indenylidene compounds probably at the 3-position of the indene moiety[113].

Fluorenylidene- and indenylidene-pyrans have been prepared either by reacting diazofluorene with thiopyrones[83] or as shown in the following chart[113,121]:

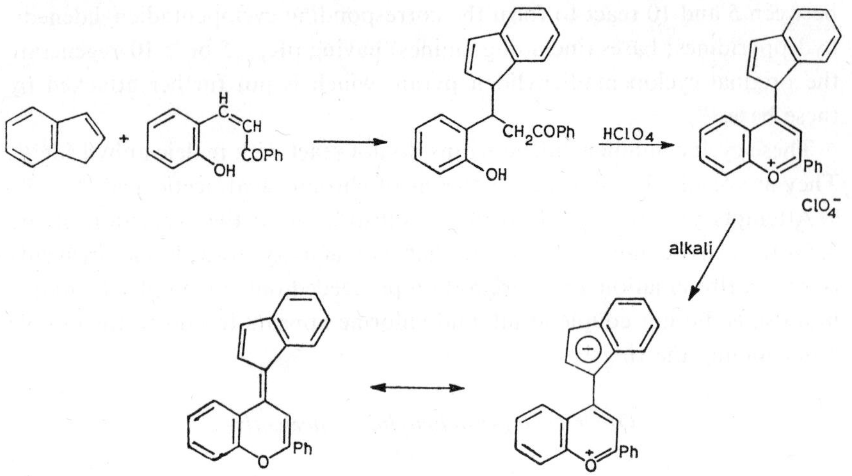

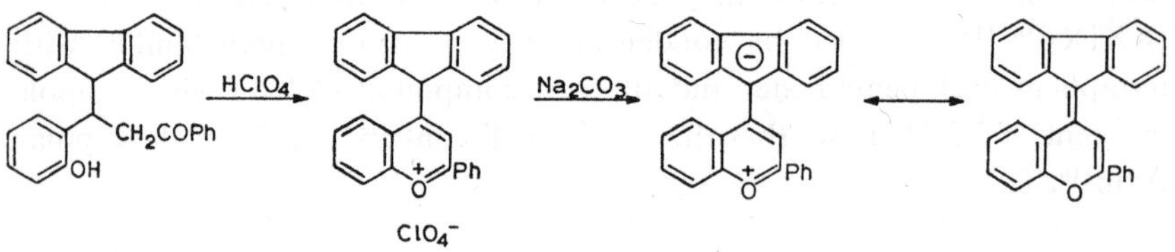

These compounds are stable in air, are reversibly soluble in mineral acids, and resist alkaline hydrolysis[113].

SESQUIFULVALENES

Isoelectronic with the cyclopentadienylidenepyrans and cyclopentadienyl-idenedihydropyridines are the cyclopentadienylidenecycloheptatrienes or *sesquifulvalenes*:

The parent compound has not yet been isolated but some derivatives of sesquifulvalene are known. They are discussed further in Chapter VIII, p. 203.

FULVENES

Fulvenes have the general formula (XXXI).

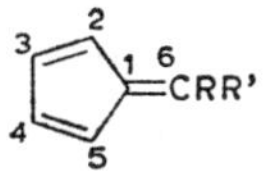

(XXXI) 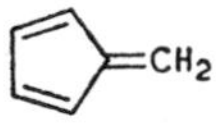   (XXXII)

Thus the cyclopentadienylidene derivatives which have been discussed in the previous sections are really special examples of substituted fulvenes. The parent compound, *fulvene*, (XXXII) is thermally unstable and very susceptible to autoxidation[122—124,128,172,210]. It is a yellow oil which decomposes in minutes in air but *in vacuo* or under nitrogen it lasts about a week. Fulvenes wherein R = alkyl and R' = H (XXXI) polymerise and decompose to brown tars within minutes of their formation, but increasing substitution at the 6-position and especially substitution by phenyl groups increase the longevity. Fulvenes are usually orange or red compounds.

Preparation

Fulvenes are usually prepared by condensing aldehydes or ketones with cyclopentadiene in the presence of a strong base[9,125]. Primary and secondary amines have also been used as condensing agents[126]. Other fulvenes have been prepared by the reaction between cyclopentadienyl magnesium bromide and appropriate ketones[127]. It is remarkable that benzene may be made to undergo partial isomerisation to fulvene on irradiation with ultra-violet light[199]. The change is apparently irreversible although fulvene itself is destroyed by ultra-violet light. Analogous isomerisations have been shown to take place with toluene, isopropylbenzene and anisole.

Structure of fulvenes

Fulvenes have small dipole moments, *e.g.* for fulvene itself, 1.1 D[128], for 6,6-dialkylfulvenes *ca.* 1.5 D, and for 6,6-diphenylfulvene, 1.9 D [129,130]; the dipole moment in each case being directed towards the five-membered ring. This indicates that there is some contribution to the structure of fulvenes from a dipolar form, *viz.*

$$\text{[cyclopentadiene]}=CRR' \longleftrightarrow \text{[cyclopentadienide]}^{(-)}-\overset{+}{C}RR'$$

Measurements of the dipole moments of a series of fulvenes wherein R = Ph, and R' varied, showed that the greater the electron donating character of R', the larger the dipole moment of the fulvene[131].

From these dipole moments it appears that the contribution of the dipolar form in simple fulvenes is about 5–10%, but with increasing electron-donating character of substituents at the 6-position the contribution becomes more significant[131,132]. This contribution is also larger in the excited state of the molecule and is responsible for a decrease in the energy difference between the ground and excited states. This in turn accounts for the fact that fulvenes absorb light at longer wavelengths compared with isomeric benzenoid compounds[132].

The resonance energy of the simple fulvene system has been calculated (from heats of combustion) to be about 12 kcal/mole[133].

Spectra of fulvenes

The ultra-violet spectra of fulvenes are characterised by a strong band at ca. 270 mμ and a weaker band ca. 365 mμ. Fulvene itself has absorption maxima at $\sim$ 241 mμ and $\sim$360 mμ (log $\varepsilon \sim$4.1, 2.35)[123]. Phenyl groups at the 6-position shift the shorter wavelength peak to longer wavelengths, e.g. λ_{max} 6-phenylfulvene = 300 mμ, 6,6-diphenylfulvene = 328 mμ. Electron donating groups substituted into these phenyl groups cause even greater shifts. The ultraviolet spectra of 6-furyl-[10] and 6-thienyl-fulvenes[134] are significantly different from those of simple fulvenes, presumably owing to much greater contributions from the dipolar forms (see above, p. 56). For further discussion of the ultra-violet and infra-red spectra of fulvenes see refs. 131, 135, 136.

The n.m.r. spectrum of fulvene has been recorded[123,124]. It shows an $A_2B_2X_2$ system, the ring protons absorbing at 3.56 τ and 3.89 τ and the exocyclic protons at 4.22 τ. This represents a slight shift from the values for cyclopentadiene and is in accord with the presence of a small but none the less real ring current[123,211]. The n.m.r. spectra of 6,6-dimethyl-, 6,6-dibenzyl- and 6,6-diphenyl-fulvenes have also been measured and discussed[137]. It is suggested that the coupling constants indicate that it is better to treat fulvenes as olefins rather than as aromatic systems[137]; this is not unreasonable in view of the much greater contribution from the covalent form.

Chemical reactions

In general fulvenes behave as olefinic compounds, for example, they readily undergo Diels–Alder reactions with maleic anhydride (*inter alia*, ref. 136, 138). It has been claimed that fulvenes react with halogens both by addition[117,122,139] and by substitution[139,140]. The polar character of the exocyclic double bond is shown, however, by the fact that fulvenes may be reduced with lithium aluminium hydride[124,141–143].

Fulvenes react with organo-lithium compounds by addition; aqueous work-up yields substituted cyclopentadienes: (*inter alia* ref. 142, 144, 145).

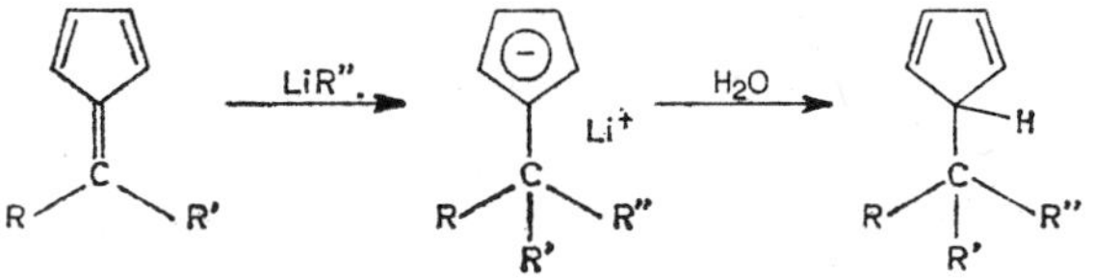

The driving force for this reaction is undoubtedly the gain in delocalisation energy resulting from the formation of the cyclopentadienide anion. In this reaction the nucleophile adds to the exocyclic carbon atom as would be expected. Fulvenes form sodium salts with sodamide[33,146], *e.g.*

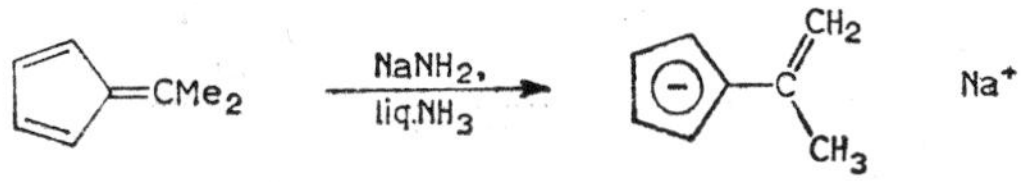

One example of electrophilic substitution of a simple fulvene is known, however, for 6,6-diphenylfulvene has been formylated by means of dimethylformamide and phosphorus oxychloride to give the 2-formyl derivative[26].

6,6-Dimethyl- and 6,6-diphenyl-fulvenes have also been shown to undergo protonation, alkylation and nitrosation at temperatures as low as $-80°$ [132]:

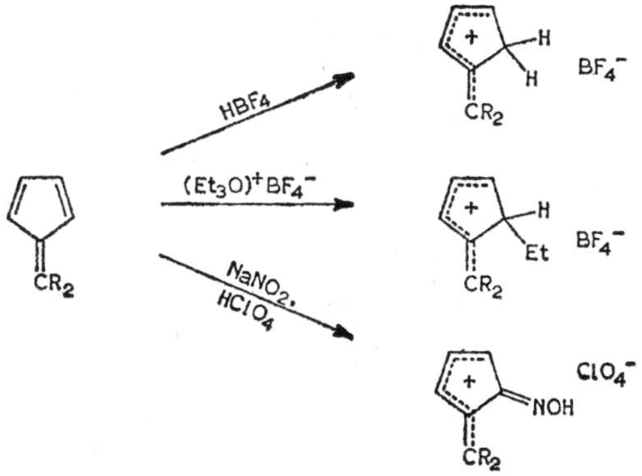

References p. 92

These salts, which are colourless, are however only stable at low temperatures. The conversion of these intermediates into substituted fulvenes has not been achieved, polymeric products being formed instead[132]. Similarly attempts to condense fulvenes with benzaldehyde in the presence of fluoroboric acid gave a product which rapidly polymerised[132].

Fulvenes with nitrogen or oxygen atoms attached to the 6-carbon atom

Substitution of electron-donating groups at the 6-carbon atom of fulvenes greatly modifies their properties; in particular the dipolar character of the fulvene is enhanced. The 2-cyclopentadienylidenedihydropyridines discussed above (p. 76) are in fact special cases of fulvenes of this type, while their 4-isomers and the cyclopentadienylidenepyrans are vinylogues.

A simple example of a 6-hetero-substituted fulvene is 6-dimethylaminofulvene (XXXIII)[114,147].

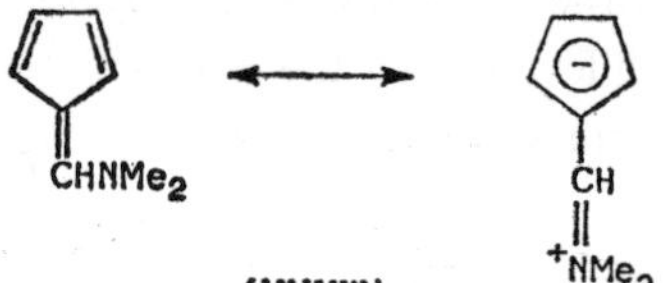

The dipole moment of this fulvene is 4.5 D, indicating about 25% participation of the dipolar form[147]. It undergoes electrophilic substitution readily, giving for example 2-tricyanovinyl-6-dimethylaminofulvene with tetracyanoethylene[148] and mono- and di-formyl derivatives by means of the Vilsmeier reaction[147].

These mono- and di-formyl derivatives may be obtained more simply by carrying out a Vilsmeier reaction on cyclopentadiene[26,147,149,150].

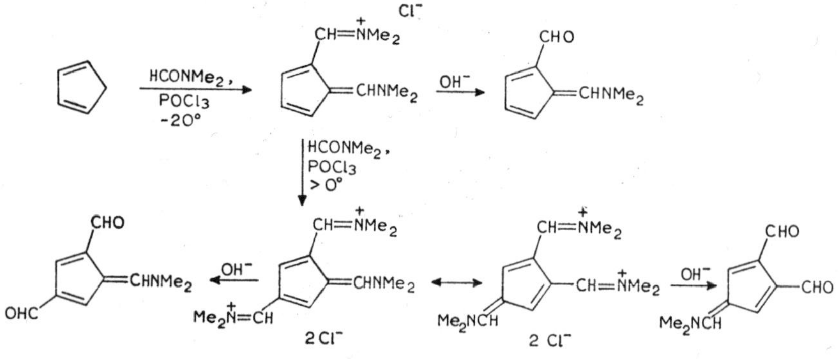

Their infra-red spectra suggest strong dipole interactions between the amine and aldehyde groups, *viz.*

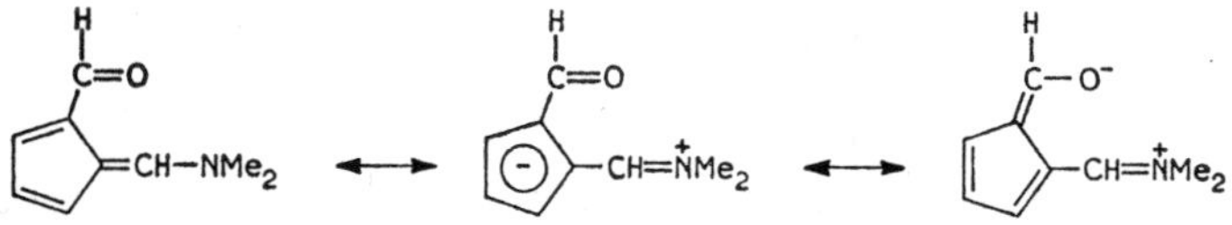

The aldehyde groups only faintly resemble aromatic aldehyde groups in their properties and these compounds may be regarded rather as vinylogous amides. Thus attempts to oxidise the aldehyde groups to carboxyl groups fail, while reduction of the diformyl compound with lithium aluminium hydride reduces the carbonyl groups to methyl groups and also reduces the 1-6 double bond[147].

Alkaline hydrolysis of the dimethylaminofulvene gives a 6-hydroxyfulvene, whose i.r. and n.m.r. spectra suggest that it exists as a tautomeric mixture consisting largely, if not entirely, of formylcyclopentadienes[50,151]:

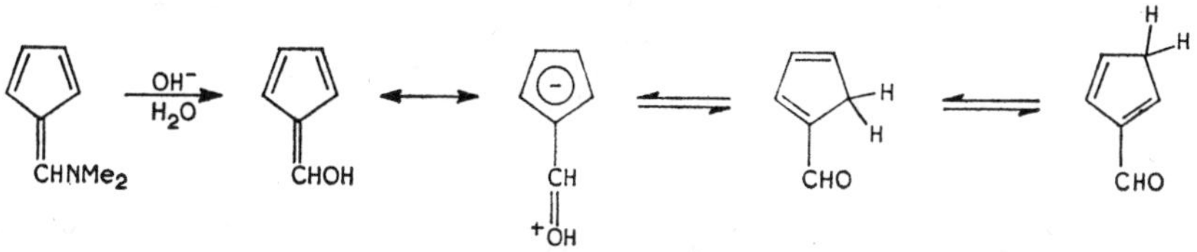

This hydroxyfulvene is not very stable thermally and is susceptible to autoxidation. Its sodium salt, wherein the negative charge is delocalised over the ring and the oxygen atom, is however more stable.

Hydrolysis of the monoformyl- and diformyl-dimethylaminofulvenes gives rise to 2-formyl- and 2,4-diformyl-6-hydroxyfulvenes respectively[147]. In both of these compounds the hydroxyfulvene structure is stabilised by strong intramolecular hydrogen bonding between the hydroxyfulvene group and an adjacent formyl group and is thus preferred to the tautomeric diformylcyclopentadiene (or triformylcyclopentadiene) structure.

Infra-red and n.m.r. spectra show that the molecules have an almost symmetrical structure, probably due to a time-averaging effect[147,151].

References p. 92

Both the formyl- and diformyl-hydroxyfulvenes are strong acids (pK_a = 4.5, 1.8 respectively)[147]. In the resulting anions the negative charge is delocalised over the five-membered ring and the formyl groups *e.g.*

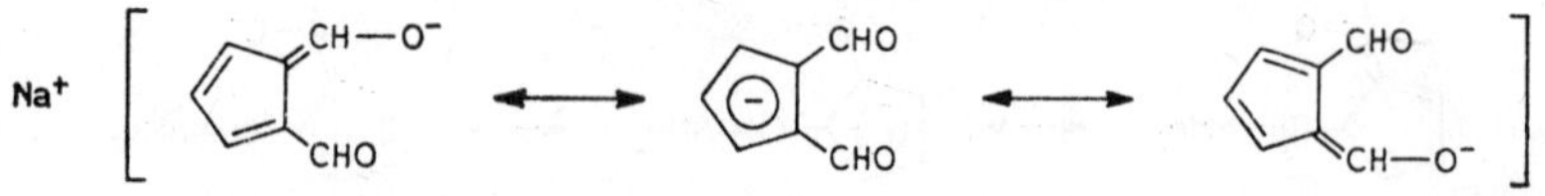

It is possible to carry out electrophilic substitution reactions in the ring of the anion derived from the diformylhydroxyfulvene; for example it couples with diazonium salts[147].

The 6-substituted fulvenes (XXXIV A-C) have also been prepared; their dipole moments are 3.6 D, 4.5 D, and 5.4 D respectively, and show that in these compounds also the dipolar form plays a not inconsiderable role[50].

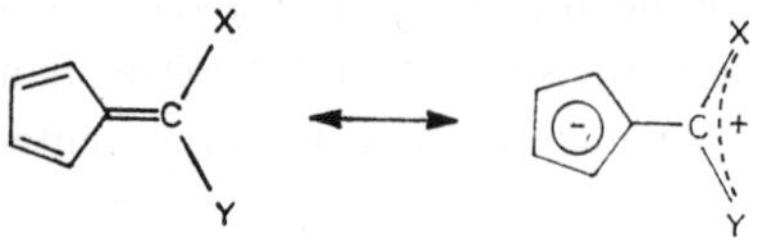

(XXXIV)

(XXXIV) (A, X = Y = OEt; B, X = OEt, Y = NMe_2; C, X = Y = NMe_2)

FERROCENE

Dicyclopentadienyl iron or *ferrocene* (XXXV) has been studied in great

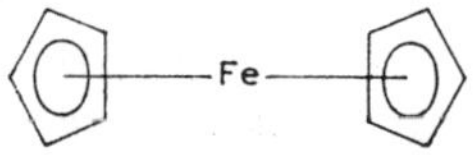

(XXXV)

detail owing to its initial novelty and to the subsequent interest in its detailed electronic structure. Originally regarded roughly as a ferrous ion sandwiched between two negatively charged cyclopentadienide rings it is now realised that the situation is much more complicated than this. Although all the π-electrons of the rings are involved in bonding with the central iron atom, the properties of ferrocene make it evident that the overall electron distribution in the molecule is such as to leave the rings nearly neutral. Ferrocene is thus far from being a simple derivative of the cyclopentadienide anion, and it is better classified as an organo-metallic compound rather than as a simple carbocyclic non-benzenoid aromatic compound. For this reason and since in any case its chemistry has been the subject of extensive reviews[152] it will only be considered briefly here, with most attention being paid to the more straightforward organic chemistry rather than to details of its electronic structure or the nature of the bonding.

Preparation

The discovery of ferrocene was serendipitous. An attempted preparation (1951) of fulvalene (cyclopentadienylidenecyclopentadiene) by allowing cyclopentadienyl magnesium iodide to react with ferric chloride led instead to the formation of ferrocene[153]:

$$2\ C_5H_5MgI + FeCl_3 \rightarrow (C_5H_5)_2Fe$$

A year later ferrocene was also prepared[154] by the reaction of cyclopentadiene with iron in an atmosphere of nitrogen at 300°:

$$Fe + 2\ C_5H_6 \xrightarrow{300°,\ N_2} (C_5H_5)_2Fe$$

Two later preparative methods which have been developed involve in one case the reaction of ferrous chloride with sodium cyclopentadienide[155] and in the other the reaction of cyclopentadiene with ferrous chloride in the presence of diethylamine[15,156,157],

$$(C_5H_5)^- Na^+ + FeCl_2 \rightarrow (C_5H_5)_2Fe$$

$$C_5H_6 + FeCl_2 \xrightarrow{Et_2NH} (C_5H_5)_2Fe$$

The latter method produces ferrocene in *ca.* 80% yield[157]. Substituted ferrocenes have been prepared from fulvenes by reaction with lithium aluminium hydride and then ferrous chloride[146].

Unsymmetrical ferrocenes have been prepared by reacting cyclopentadiene with iron pentacarbonyl at *ca.* 135° and treating the product (XXXVI) (for structure see ref. 158) with a suitably substituted cyclopentadiene[159]:

$$Fe(CO)_5 + 2\ C_5H_6 \rightarrow (C_5H_5)_2Fe_2(CO)_4 \xrightarrow{C_5H_5R} (C_5H_5)Fe(C_5H_4R)$$

(XXXVI)

If the first step of this reaction is carried out at 200–250°, ferrocene itself results instead of the intermediate (XXXVI).

Structure

The anti-prism structure (XXXVII) was first proposed for ferrocene in 1952[160].

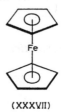

(XXXVII)

This has been confirmed by a variety of physical methods.

The geometry of the ferrocene molecule has been established by means of X-ray crystallographic examination[161,171]. The molecule is centrosymmetrical with the two rings lying in parallel planes about 3.4 Å apart. All the carbon atoms are equidistant from the iron atom. All the carbon–carbon bonds in the rings are of the same length, 1.40 Å, *i.e.* comparable to the bonds in benzene.

The infra-red spectrum of ferrocene is of striking simplicity due to the symmetry of the molecule[162]. There is a single carbon–hydrogen stretching band at 3075 cm^{-1} and only four other strong bands, at 811, 1002, 1108 and 1141 cm^{-1}. Furthermore the position of this carbon–hydrogen stretching band is in the range associated with carbon–hydrogen bands in benzenoid compounds. Nuclear magnetic resonance studies on solid ferrocene suggest that the rings rotate about their axis and that such rotation is slowed down by the presence of substituent groups in the rings[163].

In this sandwich structure all of the π-electrons of the rings may be considered to be involved in bonding with the central iron atom, but the reactivity of ferrocene leads to the conclusion that the overall electron distribution is such as to leave the rings nearly neutral.

Stereochemistry of ferrocene derivatives

No isomers of heteroannular disubstituted (1,1′) ferrocenes have ever been isolated, in accord with the concept that there is free rotation of the rings about their axis.

If, on the other hand, in a homoannular disubstituted ferrocene, $(C_5H_5)Fe(C_5H_3XY)$, the two substituent groups are different, then optical isomers are possible. (See XXXVIII and XXXIX.) Resolution of compound (XL) has indeed been achieved[164].

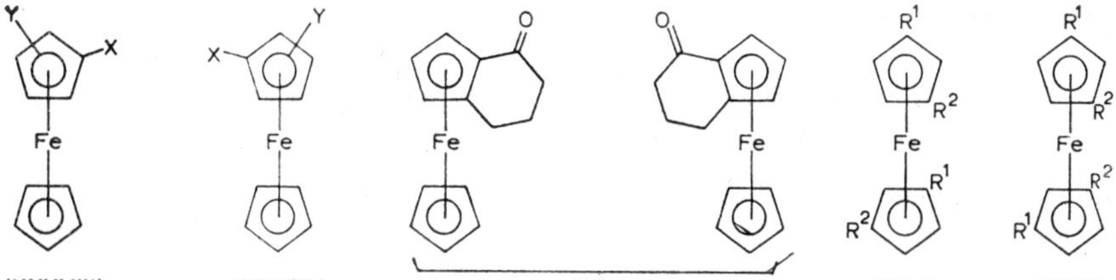

(XXXVIII)   (XXXIX)   (XL)   (XLI)   (XLII)

Geometrical isomers of ferrocenes having two different substituent groups in each ring are possible, *e.g.* (XLI) and (XLII) and such isomers have been obtained[35,152].

Chemical properties of ferrocene

Ferrocene is an orange crystalline compound, m.p. 173°, which can be recrystallised from organic solvents. It is remarkably stable. For example it is insoluble in, and unaffected by, water, caustic soda, or concentrated hydrochloric acid, even at their boiling points. It sublimes at 100° and is volatile in steam and in ethanol. It resists pyrolysis up to at least 470° (ref. 165). It does not take part in a Diels–Alder reaction with maleic anhydride[166], and does not undergo deuterium exchange with deuterium oxide in the presence of a basic catalyst[14].

It is not reduced catalytically except under extreme conditions, and even then reduction may be incomplete[166,167]. At near 350° and in the presence of Raney nickel it is converted into cyclopentane and iron. It is notable that benzene rings may be reduced preferentially whether they are fused to the five-membered rings or otherwise attached to them[168,169]. It is unaffected by lithium aluminium hydride but is reduced by lithium in ethylamine to iron and cyclopentadiene[170]. This reductive method has proved useful for the orientation of substituent groups in substituted ferrocenes.

On oxidation a blue *ferricinium* cation $[(C_5H_5)_2Fe]^+$ is obtained. The formation of this cation foils attempts to carry out bromination, nitration or sulphonation under the conditions normally used for benzenoid compounds, for the cation is at once formed and then not surprisingly resists electrophilic attack. Oxidation is facilitated by acid conditions and at high hydrogen ion concentration even air oxidises ferrocene. Ferricinium salts have been isolated[160,171].

Electrophilic substitution reactions

Despite the difficulties imposed by oxidation to the ferrocinium cation, it is possible to achieve electrophilic substitution in the carbocyclic rings of ferrocene. It has been suggested that the electrophilic reagents are first of all attached to the iron atom[173,174]; n.m.r. studies of ferrocene in boron trifluoride hydrate show that protonation occurs exclusively on the metal atom[174]. Substituent groups in one ring affect the properties of the other ring; for example the presence of electron withdrawing groups in one ring deactivates the other ring.

Haloferrocenes cannot be obtained by direct electrophilic substitution and are obtained *via* mercuration, the latter reaction being carried out under the mildest conditions, since polymers are otherwise obtained[175]. Ferrocene is broken down by bromine, probably through intermediate formation of the ferricinium ion, giving pentabromocyclopentane[175]. Halogen atoms directly attached to a ferrocene ring are very inert to substitution reactions.

Ferrocene may be sulphonated using acetic anhydride as solvent and mono- and di-sulphonic acids can be obtained by using the appropriate conditions[176]

To obtain nitroferrocene, ferrocene is first metalated by means of butyl lithium. Treatment of the lithium derivative with either propyl nitrate or with dinitrogen tetroxide in ether at $-70°$ gives nitroferrocene[177,178]. The latter compound may be reduced to aminoferrocene by means of either iron and hydrochloric acid[177] or hydrogen in the presence of Raney nickel[178]. Aminoferrocene cannot be diazotised owing to its susceptibility to oxidation; it is very sensitive to heat and light.

The lithium derivative of ferrocene also reacts with carbon dioxide to give ferrocene carboxylic acid[179-181].

Ferrocene can be acylated by means of a Friedel–Crafts reaction. Very mild conditions suffice; thus monoacetylferrocene is formed by the action of acetic anhydride in the presence of phosphoric acid[166,182]. The acetyl group deactivates the whole molecule, and further substitution, even in the ring to which the acetyl group is not directly attached, does not occur under these mild conditions. Acetyl chloride and aluminium chloride react with ferrocene to give a diacetyl derivative[166,183]. The acetyl group which enters first would be expected to deactivate the ring in which it is substituted more than the other ring and not surprisingly the diacetyl ferrocene formed consists almost entirely of the 1,1'-isomer, *i.e.* the isomer having one acetyl group substituted in each ring. A very small amount of 1,2-diacetylferrocene is also obtained but none of the 1,3-isomer[184]. 1,1'-Diacetylferrocene can be oxidised to ferrocene-1,1'-dicarboxylic acid[166].

Alkylation by means of the Friedel–Crafts reaction has been only moderately successful and alkylferrocenes have usually been obtained by Clemmensen or catalytic reduction of the appropriate acyl derivative[185].

Diazonium salts react readily and smoothly with ferrocene to give monoaryl and 1,1'-diaryl derivatives[180,186-188].

Ferrocene can be formylated by reaction with N-methylformanilide and phosphorus oxychloride[182,189,190]. The ferrocenealdehyde so formed is reduced by sodium borohydride[190] or lithium aluminium hydride[182,192] to the corresponding carbinol. The latter compound is also obtained by lithium aluminium hydride reduction of methyl ferrocenecarboxylate[191,192]. The carbinol may be oxidised back to ferrocenealdehyde by using manganese dioxide[193,194].

Formaldehyde reacts with ferrocene in the presence of dimethylamine to give dimethylaminomethylferrocene[194,195]. The quaternary iodide of the latter amine may be converted into ferrocenecarbinol by alkaline hydrolysis[194] and into methylferrocene by reduction with sodium amalgam[194].

Analogues of ferrocene

Indene forms a similar but less stable sandwich compound[196].

The ability to form compounds analogous in structure to ferrocene is common to the transition elements which have two vacant or singly occupied d-orbitals available for bonding. In the first transition series the analogues of all the elements from titanium to nickel (inclusive) are known, but only the compounds derived from Fe^{++} and Co^{+++} are notably stable. Most of them are isomorphous and even have the same melting point[152].

Other metals may form salts, *e.g.* the alkali metal cyclopentadienides, others form covalent compounds, *e.g.* (XLIII)[197]

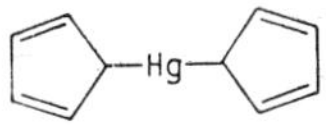

(XLIII)

REFERENCES

1 J. THIELE, *Ber.*, 34 (1901) 68.
2 W. VON E. DOERING AND C. H. DEPUY, *J. Am. Chem. Soc.*, 75 (1953) 5955.
3 D. LLOYD AND J. S. SNEEZUM, *Chem. and Ind.*, (1955) 1221.
4 D. LLOYD AND J. S. SNEEZUM, *Tetrahedron*, 3 (1958) 334.
5 F. RAMIREZ AND S. LEVY, *J. Org. Chem.*, 21 (1956) 488.
6 D. N. KURSANOV, N. K. BARANETSKAIA AND V. N. SETKINA, *Dokl. Akad. Nauk S.S.S.R.*, 113 (1957) 116.
7 J. A. BERSON, E. M. EVLETH AND Z. HAMLET, *J. Am. Chem. Soc.*, 82 (1960) 3793.
8 D. LLOYD AND F. I. WASSON, *Chem. and Ind.*, (1963) 1559.
9 J. THIELE, *Ber.*, 33 (1900) 666.
10 C. H. SCHMIDT, *Angew. Chem.*, 68 (1956) 491; *Chem. Ber.*, 90 (1957) 1352.
11 K. ZIEGLER, H. FROITZHEIM-KÜHLHORN AND K. HAFNER, *Chem. Ber.*, 89 (1956) 434.
12 F. R. GOSS AND C. K. INGOLD, *J. Chem. Soc.*, (1928) 1268.
13 W. VON E. DOERING AND L. H. KNOX, *J. Am. Chem. Soc.*, 76 (1954) 3203.
14 D. N. KURSANOV AND Z. N. PARNES, *Dokl. Akad. Nauk S.S.S.R.*, 109 (1956) 315.
15 G. WILKINSON, F. A. COTTON AND J. M. BIRMINGHAM, *J. Inorg. Nuclear Chem.*, 2 (1956) 95.
16 J. M. BIRMINGHAM AND G. WILKINSON, *J. Am. Chem. Soc.*, 78 (1956) 42.
17 J. R. LETO, F. A. COTTON AND J. S. WAUGH, *Nature*, 180 (1957) 978.
18 G. FRAENKEL, R. E. CARTER, A. MCLACHLAN AND J. H. RICHARDS, *J. Am. Chem. Soc.*, 82 (1960) 5846.
19 R. TKACHUK AND C. C. LEE, *Canad. J. Chem.*, 37 (1959) 1644.
20 J. D. ROBERTS, A. STREITWIESER AND C. M. REGAN, *J. Am. Chem. Soc.*, 74 (1952) 4579.
21 K. ALDER AND H. HOLZRICHTER, *Ann.*, 524 (1936) 145.
22 W. J. LINN AND W. H. SHARKEY, *J. Am. Chem. Soc.*, 79 (1957) 4970.
23 W. F. LITTLE AND R. C. KOESTLER, *J. Org. Chem.*, 26 (1961) 3245.
24 H. A. BRUSON, *J. Am. Chem. Soc.*, 64 (1942) 2457.
25 W. J. HALE, *Ber.*, 45 (1912) 1596; *J. Am. Chem. Soc.*, 34 (1912) 1585.
26 K. HAFNER AND K. H. VÖPEL, *Angew. Chem.*, 71 (1959) 672.
27 D. PETERS, *J. Chem. Soc.*, (1959) 1757.
28 J. M. OSGERBY AND P. L. PAUSON, *J. Chem. Soc.*, (1961) 4606.
29 R. C. COOKSON, J. HUDEC AND B. R. D. WHITEAR, *Proc. Chem. Soc.*, (1961) 117.
30 P. BAMFIELD, R. C. COOKSON, A. CRABTREE, J. HENSTOCK, J. HUDEC, A. W. JOHNSON AND B. R. D. WHITEAR, *Chem. and Ind.*, (1964) 1313.
31 R. WEISSGERBER, *Ber.*, 41 (1908) 2912; 42 (1909) 569.
32 G. WITTIG, M. HEINTZELER AND M. H. WETTERLING, *Ann.*, 557 (1947) 201.
33 G. R. KNOX AND P. L. PAUSON, *Proc. Chem. Soc.*, (1958) 289.
34 R. C. KOESTLER AND W. K. LITTLE, *Chem. and Ind.*, (1958) 1589.
35 R. E. BENSON AND R. V. LINDSEY, *J. Am. Chem. Soc.*, 79 (1957) 5471.
36 T. J. KATZ AND M. ROSENBERGER, *J. Am. Chem. Soc.*, 84 (1962) 865.
37 R. KUHN AND H. FISCHER, *Angew. Chem.*, 73 (1961) 435.
38 C. JUTZ AND H. AMSCHLER, *Angew. Chem.*, 73 (1961) 806.
39 H. STAUDINGER AND O. KUPFER, *Ber.*, 44 (1911), 2197.
40 M. J. HARGER, D. LLOYD AND F. I. WASSON, unpublished work.
41 P. L. PAUSON AND B. J. WILLIAMS, *J. Chem. Soc.*, (1961) 4153.
42 P. L. PAUSON AND B. J. WILLIAMS, *J. Chem. Soc.*, (1961) 4158.
43 P. L. PAUSON AND B. J. WILLIAMS, *J. Chem. Soc.*, (1961) 4162.
44 D. LLOYD AND F. I. WASSON, *J. Chem. Soc.*, (1966) in press.
45 T. WEIL AND M. CAIS, *J. Org. Chem.*, 28 (1963) 2472.

46 H. DISSELNKÖTTER, *Angew. Chem.*, 76 (1964) 431.
47 A. H. REES, *J. Chem. Soc.*, (1963) 2090.
48 D. J. CRAM AND R. D. PARTOS, *J. Am. Chem. Soc.*, 85 (1963) 1273.
49 A. LEDWITH AND E. C. FRIEDRICH, *J Chem. Soc.*, (1964) 504.
50 K. HAFNER, G. SCHULZ AND K. WAGNER, *Ann.*, 678 (1964) 39.
51 G. R. KNOX, *Proc. Chem. Soc.*, (1959) 56.
52 F. RAMIREZ AND S. LEVY, *J. Org. Chem.*, 23 (1958) 2036.
53 W. KIRMSE, L. HORNER AND H. HOFFMANN, *Ann.* 614 (1958) 19.
54 W. B. DE MORE, H. O. PRITCHARD AND N. DAVIDSON, *J. Am. Chem. Soc.*, 81 (1959) 5874.
55 A. SCHÖNBERG AND K. JUNGHAUS, *Chem. Ber.*, 98 (1965) 820.
56 I. B. M. BAND, D. LLOYD AND F. I. WASSON, unpublished work.
57 D. N. KURSANOV AND N. K. BARANETSKAIA, *Izv. Akad. Nauk S.S.S.R., Otdel. khim. Nauk*, (1958) 362.
58 E. M. KOSOWER AND B. G. RAMSEY, *J. Am. Chem. Soc.*, 81 (1959) 856.
59 W. W. SPOONCER, Dissert. Univ. of Washington, 1955; *Chem. Abs.*, 50 (1956) 10664; personal communication.
60 F. RAMIREZ AND S. LEVY, *J. Am. Chem. Soc.*, 79 (1957) 67.
61 D. N. KURSANOV, M. E. VOL'PIN AND Z. N. PARNES, *Khim. Nauka i Prom.*, 3 (1958) 159.
62 D. N. KURSANOV, *Uchenye Zapiski Khar'kovsk. Gosudarst. Univ.*, 110, *Trudy Khim. Fak.: Nauchn.-Issled. Inst. Khim.*, No. 17 (1961) 7; *Chem. Abs.*, 58 (1963) 4398.
63 S. I. ZHDANOV AND L. S. MIRKIN, *Collection Czechoslov. Chem. Commun.*, 26 (1961) 370
64 E. M. KOSOWER AND P. E. KLINEDINST, *J. Am. Chem. Soc.*, 78 (1956) 3493.
65 A. A. BALANDIN AND M. L. KHIDEKEL, *Dokl. Akad. Nauk S.S.S.R.*, 123 (1958) 83.
66 D. N. KURSANOV, N. K. BARANETSKAIA AND Z. N. PARNES, *Izv. Akad. Nauk S.S.S.R., Otdel. khim. Nauk*, (1961) 140.
67 D. LLOYD AND M. J. TODD, unpublished work.
68 D. LLOYD, M. SINGER AND F. I. WASSON, unpublished work.
69 F. RAMIREZ AND S. LEVY, *J. Org. Chem.*, 21 (1956) 1333; 23 (1958) 2035.
70 E. W. ABEL, A. SINGH AND G. WILKINSON, *Chem. and Ind.*, (1959) 1067.
71 F. KRÖHNKE, *Chem. Ber.*, 83 (1950) 253.
72 A. SCHÖNBERG, I. W. AWAD AND N. LATIF, *J. Chem. Soc.*, (1951) 1368.
73 K. D. WARREN, *J. Chem. Soc.*, (1961) 1412.
74 K. D. WARREN, *J. Chem. Soc.*, (1963) 598.
75 N. LATIF AND N. MISHRIKY, *Canad. J. Chem.*, 42 (1964) 2873.
76 H. STAUDINGER AND A. GAULE, *Ber.*, 49 (1916) 1951.
77 H. STAUDINGER, A. GAULE AND J. SIEGWART, *Helv. Chim. Acta.*, 4 (1921) 212.
78 T. KAUFFMANN AND S. M. HAGE, *Angew. Chem.*, 75 (1963) 248.
79 H. STAUDINGER AND J. MEYER, *Helv. Chim. Acta*, 2 (1919) 619.
80 H. WIELAND AND C. REISENEGGER, *Ann.* 401 (1913) 244.
81 E. FUNAKOBO, I. MORITANI, T. NAGAI, S. NISHIDA AND S. MURAHASHI, *Tetrahedron Letters*, (1963) 1069.
82 H. STAUDINGER AND J. SIEGWART, *Helv. Chim. Acta.*, 3 (1920) 840; A. SCHÖNBERG, O. SCHÜTZ AND J. PETER, *Ber.*, 62 (1929) 1663.
83 A. SCHÖNBERG, M. ELKASCHEF, M. NOSSEIR AND M. M. SIDKY, *J. Am. Chem. Soc.*, 80 (1958) 6312.
84 A. SCHÖNBERG AND M. M. SIDKY, *J. Amer. Chem. Soc.*, 81 (1959) 2259.
85 A. SCHÖNBERG, K. H. BROSOWSKI AND E. SINGER, *Chem. Ber.*, 95 (1962) 1910.
86 H. STAUDINGER AND K. MIESCHER, *Helv. Chim. Acta*, 2 (1919) 554.
87 A. WILLIAM JOHNSON, *J. Org. Chem.*, 28 (1963) 252.
88 E. E. SCHWEIZER, G. J. O'NEILL AND J. N. WEMPLE, *J. Org. Chem.*, 29 (1964) 1744.

89 G. WITTIG AND G. PIEPER, *Ber.*, 73 (1940) 295
90 C. K. INGOLD AND J. A. JESSOP, *J. Chem. Soc.*, (1930) 713.
91 C. K. INGOLD AND J. A. JESSOP, *J. Chem. Soc.*, (1929) 2357.
92 G. WITTIG AND G. FELLETSCHIN, *Ann.*, 555 (1944) 133.
93 F. KROLLPFEIFFER AND K. SCHNEIDER, *Ann.*, 530 (1937) 34.
94 L. A. PINCK AND G. E. HILBERT, *J. Am. Chem. Soc.*, 68 (1946) 2011.
95 H. HARTMANN AND H. GROSSEL, *Z. Elektrochem.*, 61 (1957) 337.
96 A. NOVELLI AND A. P. G. DE VARELA, *Ciencia e Invest.*, (*Buenos Aires*), (1948) 82.
97 G. M. PHILLIPS, J. S. HUNTER AND L. E. SUTTON, *J. Chem. Soc.*, (1945) 146.
98 A. WILLIAM JOHNSON AND R. B. LACOUNT, *J. Am. Chem. Soc.*, 83 (1961) 417.
99 E. D. HUGHES AND K. I. KURIYAN, *J. Chem. Soc.*, (1935) 1609.
100 A. WILLIAM JOHNSON AND R. B. LACOUNT, *Chem. and Ind.*, (1958) 1440.
101 L. A. PINCK AND G. E. HILBERT, *J. Am. Chem. Soc.*, 60 (1938) 494; 68 (1946) 751.
102 L. A. PINCK AND G. E. HILBERT, *J. Am. Chem. Soc.*, 69 (1947) 723.
103 G. WITTIG AND H. LAIB, *Ann.*, 580 (1953) 57.
104 A. WILLIAM JONHSON AND R. B. LACOUNT, *Chem. and Ind.*, (1959) 52.
105 A. WILLIAM JOHNSON AND R. B. LACOUNT, *Tetrahedron*, 9 (1960) 130.
106 A. WILLIAM JOHNSON, *J. Org. Chem.*, 24 (1959) 282.
107 A. WILLIAM JOHNSON, *J. Org. Chem.*, 25 (1960) 183.
108 J. W. ARMIT AND R. ROBINSON, *J. Chem. Soc.*, 127 (1925) 1604.
109 F. KRÖHNKE, K. ELLEGAST AND E. BERTRAM, *Ann.*, 600 (1956) 176.
110 J. A. BERSON AND E. M. EVLETH, *Chem. and Ind.*, (1961) 1362.
111 D. N. KURSANOV AND N. K. BARANETSKAIA, *Izv. Akad. Nauk S.S.S.R.*, *Otdel. khim. Nauk*, (1961) 1703.
112 G. V. BOYD, *Proc. Chem. Soc.*, (1960) 253.
113 G. V. BOYD AND L. M. JACKMAN, *J. Chem. Soc.*, (1963) 548.
114 H. MEERWEIN, W. FLORIAN, N. SCHÖN AND G. STOPP, *Ann.*, 641 (1961) 1.
115 D. LLOYD AND F. I. WASSON, *J. Chem. Soc.*, (1966) in press.
116 K. ZIEGLER AND B. SCHNELL, *Ann.*, 445 (1925) 266.
117 W. DILTHEY AND P. HUCHTEMANN, *J. Prakt. Chem.*, 154 (1940) 238.
118 C. H. SCHMIDT, *Angew. Chem.*, 75 (1963) 169.
119 K. HAFNER AND K. WAGNER, *Angew. Chem.*, 75 (1963) 1104.
120 J. A. BERSON AND E. M. EVLETH, *Chem. and Ind.*, (1959) 901.
121 G. V. BOYD, *Proc. Chem. Soc.*, (1959) 93.
122 J. THIELE AND H. BALHORN, *Ann.*, 348 (1906) 1.
123 D. MEUCHE, M. NEUENSCHWANDER, H. SCHALTEGGER AND H. U. SCHLUNEGGER, *Helv. Chim. Acta*, 47 (1964) 1211.
124 E. STURM AND K. HAFNER, *Angew. Chem.*, 76 (1964) 862.
125 G. CRANE, C. E. BOORD AND A. L. HENNE, *J. Am. Chem. Soc.*, 67 (1945) 1237.
126 W. FREIESLEBEN, *Angew. Chem.*, 75 (1963) 576.
127 V. GRIGNARD AND C. COURTOT, *Compt. rend.*, 158 (1914) 1763; C. COURTOT, *Ann. Chim.* (*France*), 4 (1915) 58, 168, 188.
128 J. THIEC AND J. WIEMANN, *Bull. Soc. chim. France*, (1956) 177.
129 G. W. WHELAND AND D. E. MANN, *J. Chem. Phys.*, 17 (1949) 264.
130 E. BERGMANN AND E. FISCHER, *Bull. Soc. chim. France*, (1950) 1084.
131 G. KRESZE AND H. GOETZ, *Chem. Ber.*, 90 (1957) 2161.
132 K. HAFNER, *Angew. Chem.*, 74 (1962) 499.
133 J. H. DAY AND C. OESTREICH, *J. Org. Chem.*, 22 (1957) 214.
134 G. E. CROSS AND D. LLOYD, unpublished work.
135 J. H. DAY AND R. JENKINS, *J. Org. Chem.*, 23 (1958) 2039.
136 G. KRESZE, S. RAU, G. SABELUS AND H. GOETZ, *Ann.*, 648 (1961) 51.

137 W. B. SMITH AND B. A. SHOULDERS, *J. Am. Chem. Soc.*, 86 (1964) 3118.
138 O. DIELS AND K. ALDER, *Ber.*, 62 (1929) 20; E. P. KOHLER AND J. KABLE, *J. Am. Chem. Soc.*, 57 (1935) 917; K. ALDER AND R. RUHMANN, *Ann.*, 566 (1950) 1; K. ALDER, F. W. CHAMBERS AND W. TRIMBORN, *Ann.*, 566 (1950) 27.
139 J. H. DAY AND C. PIDWERBESKY, *J. Org. Chem.*, 20 (1955) 89.
140 E. D. BERGMANN AND E. F. CHRISTIANI, *Ber.*, 63 (1930) 2563.
141 D. LAVIE, *Bull. Res. Council Israel*, 1 (1/2) (1951) 135.
142 D. LAVIE AND E. D. BERGMANN, *Bull. Soc. chim. France* (1951) 250.
143 K. ZIEGLER, H. G. GELLERT, H. MARTIN, K. NAGEL AND J. SCHNEIDER, *Ann.*, 589 (1954) 91.
144 K. ZIEGLER, F. CROESSMANN, H. KLEINER AND O. SCHÄFER, *Ann.*, 473 (1929) 1; K. ZIEGLER AND W. SCHÄFER, *Ann.*, 511 (1934) 101.
145 R. C. FUSON AND F. E. MUMFORD, *J. Org. Chem.*, 17 (1952) 255; R. C. FUSON AND O. YORK, *J. Org. Chem.*, 18 (1953) 570.
146 G. R. KNOX AND P. L. PAUSON, *J. Chem. Soc.*, (1961) 4610.
147 K. HAFNER, K. H. VÖPEL, G. PLOSS AND C. KÖNIG, *Ann.*, 661 (1963) 52.
148 K. HAFNER, K. H. HÄFNER, C. KÖNIG, M. KREUDER, G. PLOSS, G. SCHULZ, E. STURM AND K. H. VÖPEL, *Angew. Chem.*, 75 (1963) 35.
149 K. HAFNER, *Angew. Chem.*, 72 (1960) 574.
150 Z. ARNOLD, *Coll. Czech. Chem. Comm.*, 25 (1960) 1313.
151 K. HAFNER, H. E. A. KRAMER, H. MUSSO, G. PLOSS AND G. SCHULZ, *Chem. Ber.*, 97 (1964) 2066.
152 *inter alia*, P. L. PAUSON, *Quart Rev.*, 9 (1955) 391; P. L. PAUSON in D. GINSBURG (Editor) *Non-benzenoid aromatic compounds*, Interscience, New York, 1959, p. 114; K. PLESSKE, *Angew. Chem.*, 74 (1962) 301, 347.
153 P. L. PAUSON AND T. J. KEALY, *Nature*, 168 (1951) 1039.
154 S. A. MILLER, J. A. TEBBOTH AND J. F. TREMAINE, *J. Chem. Soc.*, (1952) 632.
155 G. WILKINSON, *Org. Synth.*, 36 (1956) 31.
156 J. M. BIRMINGHAM, D. SEYFERTH AND G. WILKINSON, *J. Am. Chem. Soc.*, 76 (1954) 4179.
157 G. WILKINSON, *Org. Synth.*, 36 (1956) 34.
158 O. S. MILLS, *Acta Cryst.*, 11 (1958) 620.
159 B. F. HALLAM AND P. L. PAUSON, *J. Chem. Soc.*, (1956) 3030.
160 G. WILKINSON, M. ROSENBLUM, M. C. WHITING AND R. B. WOODWARD, *J. Am. Chem. Soc.*, 74 (1952) 2125.
161 P. F. EILAND AND R. PEPINSKY, *J. Am. Chem. Soc.*, 74 (1952) 4971; J. D. DUNITZ AND L. E. ORGEL, *Nature*, 171 (1953) 121; J. D. DUNITZ, L. E. ORGEL AND A. RICH, *Acta Cryst.*, 9 (1956) 373.
162 E. R. LIPPINCOTT AND R. D. NELSON, *J. Chem. Phys.*, 21 (1953) 1307; *J. Am. Chem. Soc.*, 77 (1955) 4990; G. WILKINSON, P. L. PAUSON AND F. A. COTTON, *J. Am. Chem. Soc.*, 76 (1954) 1970.
163 C. N. MULAY, E. G. ROCHOW, E. O. STEJSKAL AND N. E. WELIKY, *J. Inorg. Nuclear Chem.*, 16 (1960) 23.
164 J. B. THOMSON, *Tetrahedron Letters*, No. 6 (1959) 26.
165 L. KAPLAN, W. J. KESTER AND J. J. KATZ, *J. Am. Chem. Soc.*, 74 (1952) 5531.
166 R. B. WOODWARD, M. ROSENBLUM AND M. C. WHITING, *J. Am. Chem. Soc.*, 74 (1952) 3458.
167 A. N. NESMEYANOV, E. G. PEREVALOVA, R. V. GOLOVNYA, T. V. NIKITINA AND N. A. SIMUKOVA, *Izv. Akad. Nauk S.S.S.R., Otdel khim. Nauk*, (1956) 739.
168 E. O. FISCHER AND D. SEUS, *Z. Naturforsch.*, 9b (1954) 386.
169 J. M. OSGERBY AND P. L. PAUSON, *J. Chem. Soc.*, (1958) 656.

170 D. S. TRIFAN AND L. NICHOLAS, *J. Am. Chem. Soc.*, 79 (1957) 2746.
171 E. O. FISCHER AND W. PFAB, *Z. Naturforsch.*, 7b (1952) 377.
172 J. THIEC AND J. WIEMANN, *Bull. Soc. chim. France*, (1957) 366; (1960) 1066.
173 T. J. CURPHEY, J. O. SANTER, M. ROSENBLUM AND J. H. RICHARDS, *J. Am. Chem. Soc.*, 82 (1960) 5249; A. BERGER, W. E. MCEWEN AND J. KLEINBERG, *J. Am. Chem. Soc.*, 83 (1961) 2274; E. A. HILL AND J. H. RICHARDS, *J. Am. Chem. Soc.*, 83 (1961) 3840, 4216.
174 M. ROSENBLUM, J. O. SANTER AND W. G. HOWELLS, *J. Am. Chem. Soc.*, 85 (1963) 1450.
175 A. N. NESMEYANOV, E. G. PEREVALOVA AND O. A. NESMEYANOVA, *Dokl. Akad. Nauk S.S.S.R.*, 100 (1955) 1099.
176 V. WEINMAYR, *J. Am. Chem. Soc.*, 77 (1955) 3009.
177 H. GRUBERT AND K. L. RINEHART, *Tetrahedron Letters*, No. 12 (1959) 16.
178 J. F. HELLING AND H. SHECHTER, *Chem. and Ind.*, (1959) 1157.
179 R. A. BENKESER, D. GOGGIN AND G. SCHROLL, *J. Am. Chem. Soc.*, 76 (1954) 4025.
180 A. N. NESMEYANOV, E. G. PEREVALOVA, R. V. GOLOVNYA AND O. A. NESMEYANOVA, *Dokl. Akad. Nauk S.S.S.R.*, 97 (1954) 459.
181 D. W. MAYO, P. SHAW AND M. RAUSCH, *Chem. and Ind.*, (1957) 1388.
182 P. J. GRAHAM, R. V. LINDSEY, G. W. PARSHALL, K. L. PETERSON AND G. M. WHITMAN, *J. Am. Chem. Soc.*, 79 (1957) 3416.
183 G. D. BROADHEAD, J. M. OSGERBY AND P. L. PAUSON, *J. Chem. Soc.*, (1958) 650.
184 J. H. RICHARDS AND T. J. CURPHEY, *Chem. and Ind.*, (1956) 1456.
185 F. S. ARIMOTO AND A. C. HAVEN, *J. Am. Chem. Soc.*, 77 (1955) 6295; A. N. NESMEYANOV AND N. A. VOL'KENAU, *Dokl. Akad. Nauk S.S.S.R.*, 107 (1956) 262; N. WELIKY AND E. S. GOULD, *J. Am. Chem. Soc.*, 79 (1957) 2742.
186 G. D. BROADHEAD AND P. L. PAUSON, *J. Chem. Soc.*, (1955) 367.
187 A. N. NESMEYANOV, E. G. PEREVALOVA AND R. V. GOLOVNYA, *Dokl. Akad. Nauk S.S.S.R.*, 99 (1954) 539.
188 V. WEINMAYR, *J. Am. Chem. Soc.*, 77 (1955) 3012.
189 M. ROSENBLUM, *Chem. and Ind.*, (1957) 72.
190 G. D. BROADHEAD, J. M. OSGERBY AND P. L. PAUSON, *Chem. and Ind.*, (1957) 209; *J. Chem. Soc.*, (1958) 650.
191 A. N. NESMEYANOV AND E. G. PEREVALOVA, *Dokl. Akad. Nauk S.S.S.R.*, 112 (1957) 439.
192 K. SCHLÖGL, *Monatsh.*, 88 (1957) 601.
193 C. R. HAUSER, J. K. LINDSAY, D. LEDNICER AND C. E. CAIN, *J. Org. Chem.*, 22 (1957) 717.
194 C. R. HAUSER AND J. K. LINDSAY, *J. Org. Chem.*, 22 (1957) 355, 906.
195 C. R. HAUSER AND J. K. LINDSAY, *J. Org. Chem.*, 21 (1956) 382.
196 E. O. FISCHER AND R. JIRA, *Z. naturforsch.*, 8b (1953) 692; E. O. FISCHER AND D. SEUS, *Z. Naturforsch.*, 8b (1953) 694; P. L. PAUSON AND G. WILKINSON, *J. Am. Chem. Soc.*, 76 (1954) 2024.
197 G. WILKINSON AND T. S. PIPER, *J. Inorg. Nuclear Chem.*, 2 (1956) 32.
198 V. FRANZEN, *Chem. Ber.*, 93 (1960) 557.
199 J. M. BLAIR AND D. BRYCE-SMITH, *Chem. and Ind.*, (1957) 287; H. J. F. ANGUS, J. M. BLAIR AND D. BRYCE-SMITH, *J. Chem. Soc.*, (1960) 2003.
200 O. W. WEBSTER, *J. Am. Chem. Soc.*, 87 (1965) 1820.
201 H. RAPOPORT AND G. SMOLINSKY, *J. Am. Chem. Soc.*, 82 (1960) 934.
202 F. KLAGES AND K. BOTT, *Chem. Ber.*, 97 (1964) 735.
203 S. MCLEAN AND P. HAYNES, *Tetrahedron*, 21 (1965) 2313.
204 H. BEHRINGER AND F. SCHEIDL, *Tetrahedron Letters*, (1965) 1757.
205 K. D. WARREN AND J. R. YANDLE, *J. Chem. Soc.*, (1965) 4221.

206 M. JONES AND K. R. RETTIG, *J. Am Chem. Soc.*, 87 (1965) 4013, 4015.
207 J. A. BERSON, E. M. EVLETH AND Z. HAMLET, *J. Am. Chem. Soc.*, 87 (1965) 2887
208 J. A. BERSON, E. M. EVLETH AND S. L. MANATT, *J. Am. Chem. Soc.*, 87 (1965) 2901, 2908.
209 A. LÜTTRINGHAUS, E. FUTTERER AND H. PRINZBACH, *Tetrahedron Letters*, (1963) 1209; A. LÜTTRINGHAUS, H. BERGER AND H. PRINZBACH, *Tetrahedron Letters*, (1965) 2121; R. GOMPPER AND E. KUTTER, *Chem. Ber.*, 98 (1965) 2825.
210 H. SCHALTEGGER, M. NEUENSCHWANDER AND D. MEUCHE, *Helv. Chim. Acta*, 48 (1965) 955.
211 Y. KITAHARA, I. MURATA, K. SHIRAHATA, S. KATAGIRI AND H. AZUMI, *Bull. Chem. Soc. Japan*, 38 (1965) 780.

CHAPTER V

Tropylium salts

INTRODUCTION

The positively charged ion $C_7H_7^+$ derived from cycloheptatriene by loss of a hydride ion has a closed shell of six π-electrons. Its salts are known as *tropylium* or *tropenium* salts. In formulating his rule in 1931, Hückel predicted that these salts would have aromatic character. Tropylium bromide was in fact first prepared in 1891[1] but its identity was not recognised. Cycloheptatriene had been converted into its dibromo-derivative; on attempted distillation of the latter a yellow solid was formed. This work was repeated in 1954[2] and the yellow solid shown to be tropylium bromide.

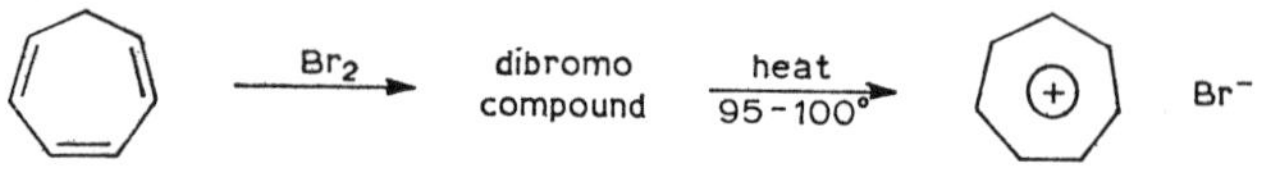

$$\text{cycloheptatriene} \xrightarrow{Br_2} \text{dibromo compound} \xrightarrow[95-100°]{\text{heat}} \text{tropylium}^+ \; Br^-$$

PREPARATION OF TROPYLIUM SALTS

(a) From cycloheptatriene

Cycloheptatriene is an obvious starting material for the preparation of tropylium salts. Recent new routes for preparing it either from benzoic acid[3] or from cyclohexene[85], or by addition of diazomethane to benzene in the presence of cuprous bromide[4] have greatly improved its accessibility.

As already described, cycloheptatriene may be converted into tropylium bromide by the action of heat on its dibromo derivative[1,2,5]. The same method has been used to prepare carboxytropylium[6-8], bromotropylium[9], phenyltropylium[9] and tertiary butyltropylium[9] salts. Dibromocycloheptatriene also decomposes to tropylium bromide on standing in liquid sulphur dioxide[10,11]. Heptaphenyltropylium bromide has been prepared by the action of bromine on heptaphenylcycloheptatriene[37].

A wide variety of electrophilic reagents extract hydride ions directly from cycloheptatriene, despite the fact that many of these reagents react vigorously with the double bonds. For example the action of concentrated sulphuric acid produces an 18% yield of tropylium hydrogen sulphate[12]. The most valuable reagents of this sort are triphenylmethyl salts. By reaction of the appropriate salt with cycloheptatriene in acetonitrile or liquid sulphur dioxide, tropylium chloride, bromide, iodide, tetrafluoroborate, perchlorate[13] and tetrabromoborate[14] have been prepared in high yields.

$$\bigcirc + (C_6H_5)_3C^+X^- \longrightarrow \bigoplus \quad X^- + (C_6H_5)_3CH$$

Various substituted tropylium salts have been prepared similarly[13,15]. A p-phenylene-bistropylium cation has also been obtained in this way from 1,4-ditropylbenzene[86].

Among other reagents which convert cycloheptatriene to tropylium salts are Lewis acids such as stannic chloride[16,17], boron tribromide[14], boron trichloride[16,83], boron trifluoride[12] and aluminium chloride[12]. Strong Lewis acids also react vigorously with the double bonds and the yields are not good.

Halides of elements of variable valency such as phosphorus pentachloride[12,18], sulphuryl chloride, ferric chloride and titanium tetrachloride[19] may be used effectively. Thus phosphorus pentachloride produces an almost quantitative yield at room temperature of what was originally described as tropylium chloride[18,87]; it is probably the double salt $[C_7H_7]^+Cl^-$; $[C_7H_7]^+$ $[PCl_6]^-$ (Ref. 20). Substituted tropylium salts have also been obtained by this method[12,18].

Various oxidising agents e.g. concentrated nitric acid, chromium trioxide and selenium dioxide convert cycloheptatriene to the tropylium ion to some extent[12]; electrolytic oxidation has also been used[21]. The most successful oxidative methods involve the use of quinones, especially those of higher redox potential such as tetrachloro-1,2-benzoquinone and 2,3-dichloro-5,6-dicyano-1,4-benzoquinone[22], and of autoxidation in acetic acid in the presence of a strong acid such as perchloric acid[23]. Both of these methods give excellent yields of tropylium salts.

(b) From cycloheptatrienecarboxylic acid

Cycloheptatrienecarboxylic acid and its related nitrile have been utilised as intermediates in the preparation of tropylium salts since they can be readily obtained by the reaction of diazoacetic ester or diazoacetonitrile with benzene.

The first synthesis[24,25] from these compounds involved conversion of the acid to its acid azide which was then made to undergo a Curtius reaction by heating in benzene. Two products were obtained which were described respectively as a covalent isocyanato derivative and tropylium cyanate, but the latter has subsequently been shown to be ditropylurea[26,27] (see reaction chart below). Treatment of the isocyanatocycloheptatriene with hydrogen bromide produced tropylium bromide[24,25].

$$C_7H_7CO_2H \to C_7H_7CON_3 \xrightarrow[C_6H_6]{\text{heat in}} \begin{cases} C_7H_7NCO \xrightarrow{HBr} [C_7H_7]^+ Br^- \\ + \\ (C_7H_7NH)_2CO \end{cases}$$

Methyltropylium bromide was obtained similarly[24,25]. Cyanocycloheptatriene was converted into a tropylium salt by the action of boron trifluoride and aluminium chloride[25].

Acidic permanganate converts cycloheptatrienecarboxylic acid into a mixture of benzaldehyde and tropylium salts, the latter being obtained in up to 40% yield[28]. Periodic acid, potassium persulphate, lead tetracetate and ceric ammonium nitrate also effect the same oxidation[28]. The benzaldehyde which is also produced probably arises from further oxidation of the tropylium cation (see below, p. 106).

Cycloheptatrienecarboxylic acid has also been converted into tropylium salts by bromination and subsequent dehydrobromination[8], and by treatment of the free acid or its acid chloride with acetyl fluoroborate[8]. Reaction of the acid chloride or of cyanocycloheptatriene with silver perchlorate in nitromethane gives a solution of tropylium perchlorate, which can be precipitated by addition of ether[8]. A Hunsdiecker reaction on the silver salt of the acid (*i.e.* treatment with iodine) precipitates a mixture of tropylium iodide and silver iodide which proved difficult to separate[8].

(c) From benzene and halocarbenes

The reaction of benzene with halocarbenes derived from methylene halides and potassium tertiary butoxide[29] or methyl lithium[30] have been shown to produce tropylium halides but only in small yield. Reaction presumably proceeds *via* the halobicyclo[4,1,0]heptadiene.

(d) From cyclooctatetraene

Cyclooctatetraene is oxidised by acidic permanganate to benzaldehyde and benzoic acid together with about 5% tropylium salts[31]. The proposed mechanism of this reaction involves pinacol rearrangement of the glycol which is formed first, giving the protonated form of cycloheptatrienealdehyde; this is oxidised at once to cycloheptatrienecarboxylic acid. The latter compound is known to be converted into benzaldehyde and tropylium salts on oxidation (see above).

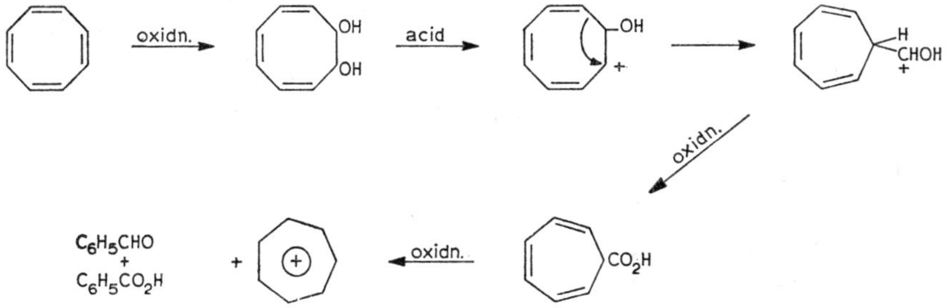

(e) From tropyl (cycloheptatrienyl) compounds

Tropyl ethers, such as methyltropyl ether or ditropyl ether, and tropyl esters are unusually easily cleaved by mineral acid to give tropylium salts[32,33]. Methyl tropyl ether is also cleaved by dry hydrogen chloride or bromide in ether[28,34] to give double salts, e.g.

$$C_7H_7OCH_3 \xrightarrow[Et_2O]{dry\ HCl} [C_7H_7]^+[H]^+2[Cl]^-$$

These salts are obtained in good yield (ca. 80%) and are useful in that they are less hygroscopic and less sensitive to light than the simple tropylium halides. They may be converted into the latter by the action of water.

Tropone (cycloheptatrienone; see Chapter VI) reacts with strong acids to form hydroxy tropylium salts; with triethyloxonium fluoroborate it gives an ethoxytropylium salt[35].

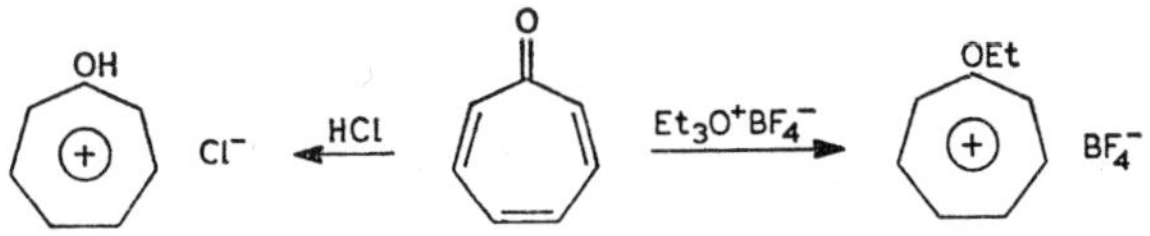

Tropylium salts may also be prepared from tropyl derivatives by fragmentation reactions involving rupture of C–C bonds. Thus treatment of the alcohol (I) with perchloric acid produces tropylium perchlorate and 2-methylpropene[36].

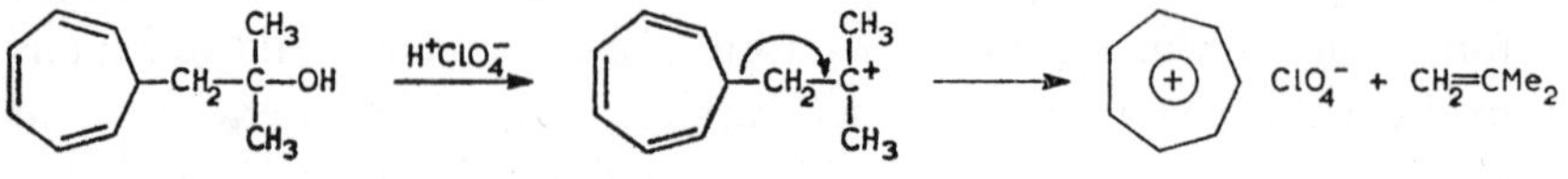

In a similar fashion the alcohol (II) also gives rise to tropylium salts and 2-methylpropene on treatment with a variety of acids or with thionyl chloride[37].

The chloro-compound (III), but not the corresponding bromide, reacts with silver perchlorate to give tropylium perchlorate and 2-methylpropene[36];

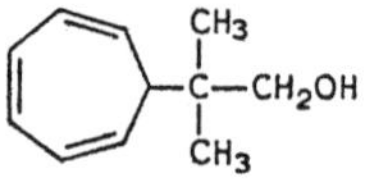

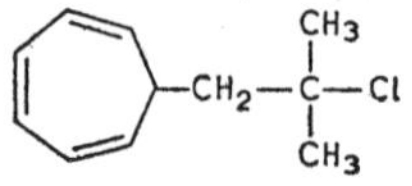

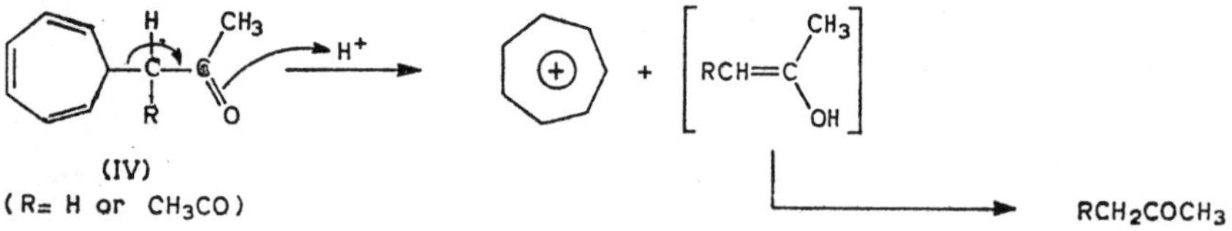

the aldehyde (IV) also reacts with acids with rupture of a carbon–carbon bond giving a tropylium salt[30].

A carbon–carbon bond in ditropyl is cleaved by electrolytic oxidation with formation of tropylium ions[21].

(f) Other methods of formation

It has been shown that various deuterium labelled alkylbenzenes on collision with electrons in the gas phase in a mass spectrometer give rise not to the benzyl cation but to the isomeric tropylium cation[38]:

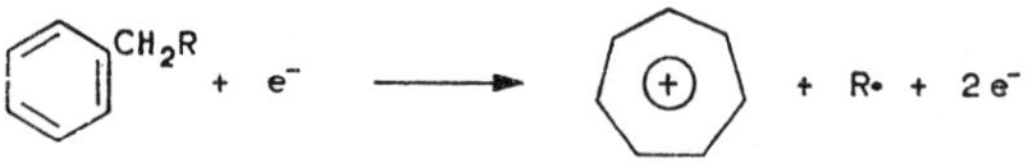

STRUCTURE OF THE TROPYLIUM ION

Evaluation of the energy changes in this process shows that the transition of a benzyl cation into a tropylium cation is thermodynamically probable. Such a rearrangement has not yet been observed in solution, apparently because the benzyl cation is stabilised by solvation to a considerably greater extent than the tropylium ion.

The ion of mass 91 obtained from benzyl alcohol or benzyl chloride in a mass spectrometer has also been shown to be the tropylium cation[39].

STRUCTURE OF THE TROPYLIUM ION

Physical data confirm the ionic structure of tropylium salts, *e.g.* the infra-red, ultra-violet, Raman and n.m.r. spectra (see below) are all consistent with a molecular structure involving a symmetrical planar seven-membered ring. X-ray crystallographic examination of tropylium iodide and perchlorate showed that these compounds were ionic. It was not possible to obtain a detailed structure of the tropylium ring since the rings apparently occupy statistically disordered positions in the crystal[84].

An elegant demonstration of the symmetry of the tropylium cation was carried out as follows[40]. A sample of tropylium bromide was prepared having one of its seven carbon atoms labelled by a ^{14}C atom:

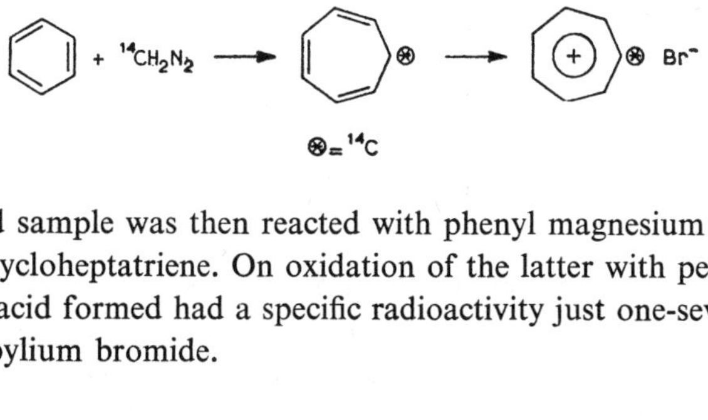

[benzene] + $^{14}CH_2N_2$ ⟶ [cycloheptatriene with label] ⟶ [tropylium cation with label] Br^-

⊛ = ^{14}C

This labelled sample was then reacted with phenyl magnesium bromide to give phenylcycloheptatriene. On oxidation of the latter with permanganate the benzoic acid formed had a specific radioactivity just one-seventh of the original tropylium bromide.

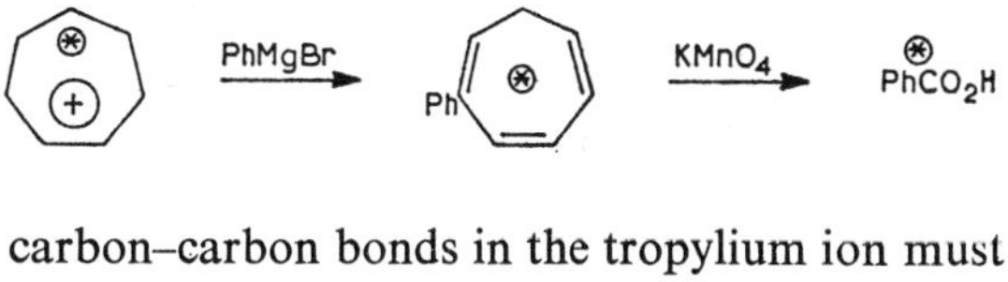

[tropylium cation with label] —PhMgBr→ [phenylcycloheptatriene with label] —$KMnO_4$→ $PhCO_2H$ (labelled)

Hence all the carbon–carbon bonds in the tropylium ion must be equivalent and the positive charge uniformly distributed around the ring.

SPECTRA

The ultra-violet spectrum of the tropylium ion has maxima at $\lambda = 275, 247$ mμ (log ε_{max} = 3.64, 3.06)[2], and is in accord with theory[43]. Many tropylium salts having colourless anions are themselves colourless but the bromide and iodide are yellow and red respectively. The colour is due to long tailing of the u.v. spectra into the visible region and not to a discrete maximum as in the case of the similar azulenium salts. (See Chapter VIII.) The colour of these halides has been attributed to charge-transfer complex formation[5,9,41]. On addition of benzenoid hydrocarbons to solutions of tropylium tetrafluoroborate or perchlorate new absorption bands appear; this also has been attributed to the formation of charge transfer complexes with the hydrocarbons[42].

The infra-red spectrum[2,44] of the tropylium ion is remarkably simple, showing only four bands of reasonable intensity. The Raman spectrum is also very simple and both are consistent with a structure of high symmetry. Furthermore the infra-red and Raman spectra of tropylium bromide in hydrobromic acid have no common bands, which again indicates that the ion exists in a highly symmetrical form in this solution. These spectra are thus readily interpreted in terms of an aromatic system.

The n.m.r. spectrum of the tropylium ion consists of a single line at about 0.8 τ [45]. This again shows the equivalence of all the hydrogen atoms and hence the symmetry of the ring. The shift in the position of the peak from that of benzene is reasonable allowing for the charge on the ions and compares favourably with results obtained from molecular orbital calculations.

PROPERTIES OF TROPYLIUM SALTS

Tropylium salts having a variety of anions are known. Whereas salts such as the perchlorate are completely dissociated, the compounds derived from weak acids such as acetic, benzoic or hydrocyanic exist in a covalent form as cycloheptatriene derivatives, rather than as salts, e.g. cyanocycloheptatriene rather than tropylium cyanide. The degree of dissociation is dependent on the nature of the anion[33].

The salts are solids of high melting point. The majority are readily soluble in water and insoluble in organic solvents of low polarity. They may however be recrystallised from polar solvents such as acetonitrile or nitromethane. Solutions of tropylium salts having anions which form insoluble silver salts give instantaneous precipitates with silver nitrate, again illustrating the ionic nature of these salts. In many ways they resemble alkali metal salts, for example, they give a difficultly soluble perchlorate and very insoluble hexachloroplatinate; the latter salt has been used for the analytical determination of tropylium salts in solution[46]. The perchlorate may explode on heating[23,88].

In general tropylium salts fall into two distinct groups[41]: (1) those like the halides which show a range of colour, sublime easily, are not stable in the atmosphere and are sensitive to heat and light, and (2) those like the perchlorate or tetrafluoroborate which are colourless, of high melting point, stable in air, and sublime only with difficulty if at all. These salts have anions of low nucleophilicity.

The chloride and bromide deliquesce very readily with decomposition; the iodide is slowly converted to the triiodide. These salts are also soluble in methylene chloride giving solutions which are unstable unless rigorously protected from the atmosphere. In air such solutions are decomposed by hydrolytic attack giving rise to the formation of cycloheptatriene, tropone, and benzaldehyde[41].

CHEMICAL REACTIONS OF TROPYLIUM SALTS

(a) Reduction

Polarographic studies[46,47] show that reduction of the tropylium ion begins at a very low potential (0.2 V), the character of the reduction curves being independent of the identity of the anion. A remarkable fact is the adsorption of the tropylium ions on the positively charged mercury surface, despite electrostatic repulsion. This points to the polarisability of the π-electron system in the tropylium ion and indicates a similarity to benzene and other aromatic compounds, which are also adsorbed on a positively charged mercury surface[48].

Catalytic hydrogenation of tropylium salts produces cycloheptane[2,5]. Bimolecular reduction is brought about by the action of zinc dust on an aqueous solution of tropylium ions[2,5]. Since this reduction probably proceeds by a one electron transfer to give the cycloheptatriene radical as an intermediate, it may be concluded that this radical is not stable *vis-a-vis* its dimer at room temperature[5] (*cf*. p. 7).

Tropylium salts are reduced to cycloheptatriene by means of lithium aluminium hydride or sodium borohydride[49,37]. Alcohols also may act as reducing agents[49,50], *e.g.*

$$\text{[tropylium]}^+ X^- + PhCH_2OH \longrightarrow \text{[cycloheptatriene]} + PhCHO + HX$$

Studies have been made[49,51] of the equilibrium tropylium ion $+ H^- \rightleftharpoons$ cycloheptatriene. By using deuterium labelled cycloheptatriene it has been shown that there is a reversible hydride ion transfer between the two species[49]. From a study of equilibrium constants it has been inferred that the methyltropylium ion is more stable than the unsubstituted ion by *ca*. 4 kcal/mole[51]. On the other hand the heptaphenyltropylium ion is less stable than the parent ion, presumably owing to the inductive effects of the phenyl groups swamping the effect due to conjugation of the π-electrons, the latter effect being reduced by the inevitable lack of coplanarity (due to crowding) of the phenyl groups and the seven-membered ring[37].

(b) Oxidation

Most oxidising agents degrade the tropylium ion with simultaneous molecular rearrangement to give benzene derivatives. With many reagents, *e.g.* chromic oxide in acetic acid[5], silver oxide[5] or dilute neutral permanganate[25,52], the first product obtained is benzaldehyde. The type of mechanism proposed for this reaction is as follows:

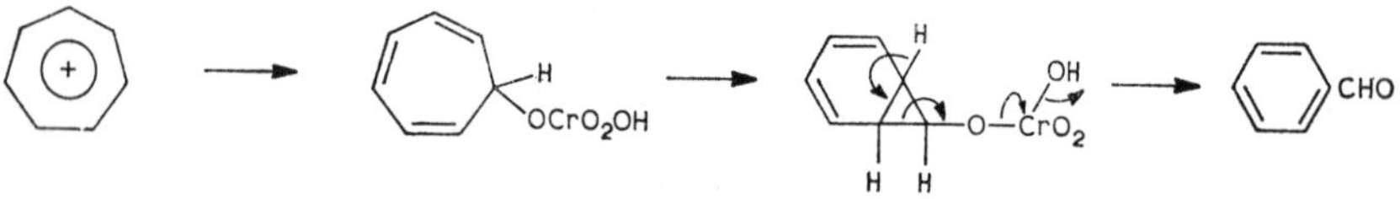

$$\text{[tropylium]}^+ \longrightarrow \text{[cycloheptatriene-OCrO_2OH]} \longrightarrow \text{[intermediate]} \longrightarrow \text{[PhCHO]}$$

The action of bromine on tropylium bromide gives a tribromocycloheptadiene, which on shaking with water at once produces benzaldehyde[25,52].

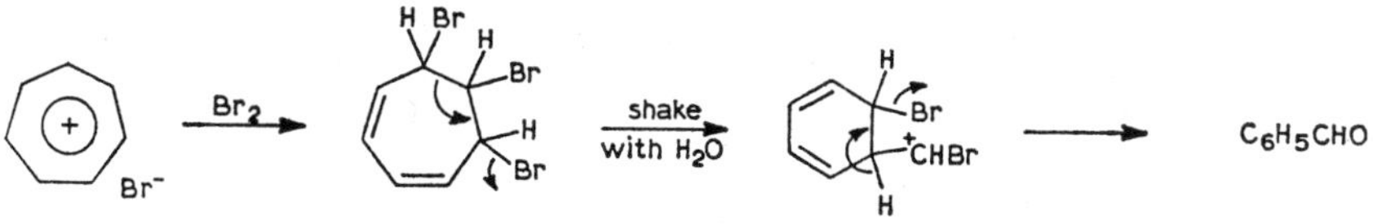

Aqueous hydrogen peroxide oxidises tropylium bromide to a mixture of benzene, formic acid and carbon monoxide in good yield, plus a little phenol[53].

$$[C_7H_7]^+ \ Br^- \xrightarrow{aq. \ H_2O_2} C_6H_6 + CO + H_2O + HBr$$
$$C_6H_6 + HCO_2H + HBr$$

The reaction does not appear to proceed *via* tropone or to involve free radicals[54]. A mechanism which has been suggested is as follows[54].

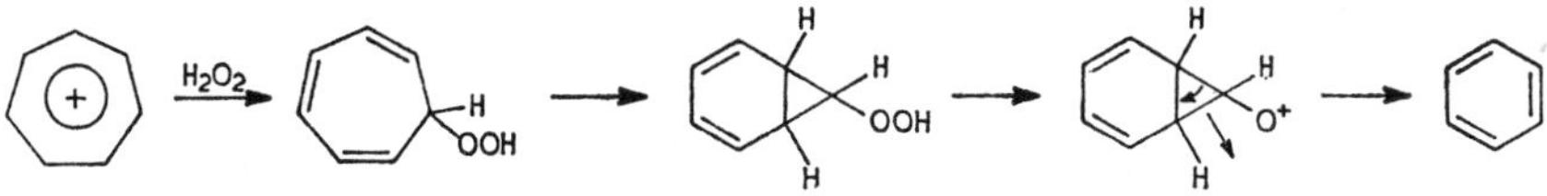

It has been reported that oxidation by chromic oxide in pyridine does not give rise to rearrangement leading to a benzene derivative, but that instead tropone (cycloheptatrienone) is formed[55]. Aqueous hydrogen peroxide converts *p*-phenylene-bistropylium fluoroborate into terphenyl in quantitative yield[86].

(c) Reaction with nucleophilic reagents

Although the tropylium ion is a very stable carbonium ion, as such it naturally reacts very readily with nucleophilic reagents. The general pattern of these reactions may be formulated by the reaction schemes:

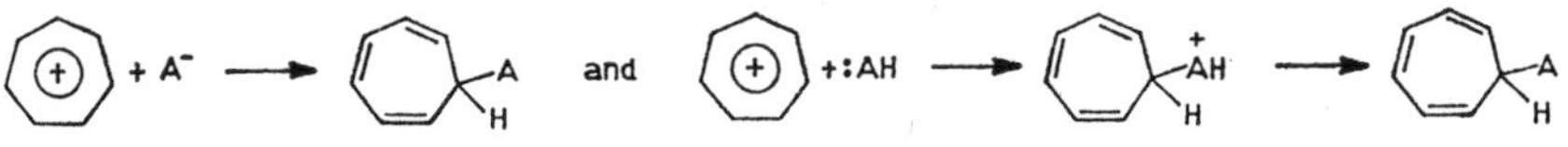

In many instances these reactions are governed by an equilibrium between the tropylium ion and the covalent compound. Thus in water the ion behaves as a Lewis acid and is in equilibrium with cycloheptatrienol.

References p. 114

$$\text{tropylium}^+ + 2\,H_2O \rightleftharpoons \text{cycloheptatrienol-OH} + H_3O^+$$

The pK value of tropylium bromide determined at 25° by potentiometric titration[2] is 4.75; in other words its acidity is comparable to that of acetic acid when water is the reference base. This result is in excellent agreement with theoretical predictions based on molecular orbital calculations[56].

It is not possible to isolate the cycloheptatrienol as such for it reacts with further tropylium ion to form ditropyl ether, which may be isolated[2,5].

$$\text{C}_7\text{H}_7\text{OH} + \text{C}_7\text{H}_7^+ \longrightarrow [\text{C}_7\text{H}_7\text{-O(H)-C}_7\text{H}_7]^+ \longrightarrow \text{C}_7\text{H}_7\text{-O-C}_7\text{H}_7$$

This ether is also obtained by treating tropylium salts with aqueous alkali (sodium carbonate or bicarbonate)[2,5,57,58]; by using concentrated sodium hydroxide solution a 96% yield is obtained[59]. On treating this ether with hydrobromic acid, the tropylium ion is reformed[2], but on distilling it from acid-treated silica gel[57] or by allowing it to stand with a very small quantity of either concentrated hydrochloric acid or of tropylium ions[58] a disproportionation reaction takes place and tropone and cycloheptatriene are formed.

$$\text{C}_7\text{H}_7\text{-O-C}_7\text{H}_7 \underset{}{\overset{H^+}{\rightleftharpoons}} [\text{C}_7\text{H}_7\text{-O(H)-C}_7\text{H}_7]^+ \rightleftharpoons \text{C}_7\text{H}_7\text{-OH} + \text{C}_7\text{H}_7^+$$

$$\downarrow$$

$$\text{tropone}\,(=O) + \text{cycloheptatriene} + H^+$$

This provides a very convenient way of obtaining tropone from tropylium salts and hence from cycloheptatriene.

The tropylium ion reacts with sodium methoxide to form methoxycycloheptatriene; this product is also reconverted to the tropylium ion by the action of hydrobromic acid[2].

Nucleophilic attack both by tertiary alcohols and by ethylene oxide has been shown to lead ultimately to a disproportionation reaction and the formation of a mixture of tropone and cycloheptatriene[60]:

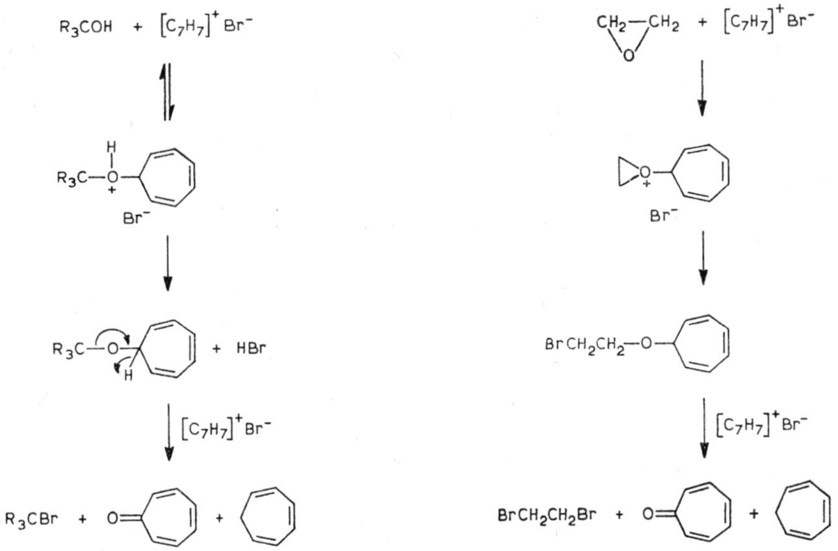

CHEMICAL REACTIONS 109

[Reaction schemes showing R₃COH + [C₇H₇]⁺Br⁻ and CH₂—CH₂ (epoxide) + [C₇H₇]⁺Br⁻ pathways leading to R₃CBr and BrCH₂CH₂Br respectively, via tropyl ether intermediates]

Tropylium salts react with hydrogen sulphide to give ditropylsulphide[5], with aqueous ammonia to give ditropylamine, with ethanolic ammonia to give tritropylamine, and with dimethylamine to give dimethyltropylamines[5]. They also react with amides, thus with acetamide and succinimide, N-tropylacetamide and N-tropylsuccinimide, respectively, are formed[5].

Potassium cyanide reacts to form cyanocycloheptatriene[5,18,24,25]; the product is identical with that obtained from the reaction between benzene and diazoacetonitrile[25]. This normally covalent nitrile dissociates in the presence of aluminium chloride or boron trifluoride[25].

Tropylium salts react with Grignard reagents or with organo-lithium derivatives to form alkyl- or aryl-cycloheptatrienes[2,5,9,30]. For example phenylcycloheptatriene is obtained by the action of either phenyllithium or phenylmagnesium bromide[2,5]. By using vinyl magnesium chloride, vinylcycloheptatriene and hence vinyltropylium salts have been prepared[15]. With lithium cyclopentadienide cyclopentadienyl cycloheptatriene is formed, and not a tropylium cyclopentadienide[9,32,33]:

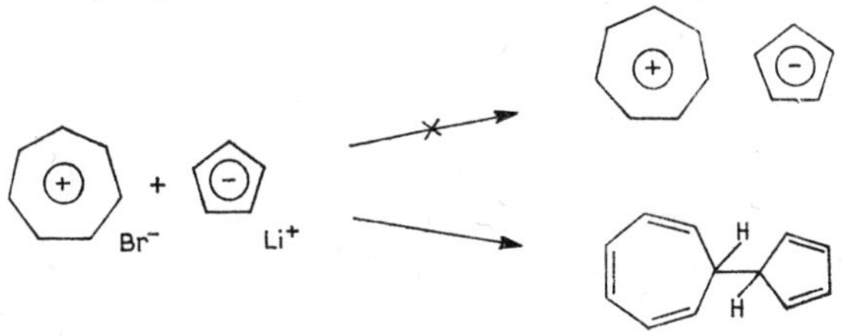

[Reaction scheme: tropylium Br⁻ + cyclopentadienide Li⁺ does not give the ion pair but gives the covalent cyclopentadienyl cycloheptatriene]

References p. 114

The formation of the covalent compound is explicable if it is remembered that cyclopentadiene is a very weak acid and, as discussed above (p. 104), the tropylium ion tends to form covalent derivatives rather than salts with anions of weak acids. Tropylium salts react readily with some organomercury compounds[61], *e.g.*

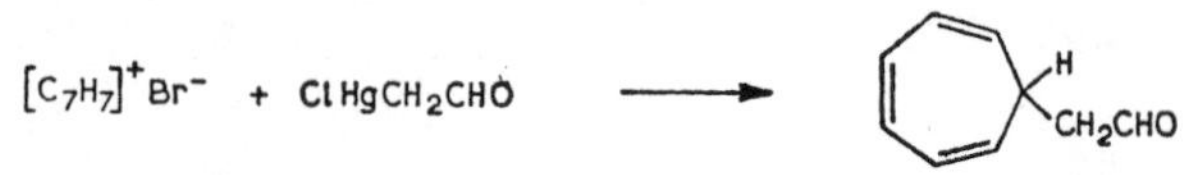

$[C_7H_7]^+ Br^-$ + $ClHgCH_2CHO$ ⟶ (cycloheptatrienyl-CH₂CHO)

Tropylium ions also react readily with anions derived from compounds having reactive methylene groups such as β-ketoacids and esters, β-cyano esters, β-dicarboxylic acids and their derivatives[36,62–64]. Reaction frequently proceeds with ease in the cold, with evolution of heat.

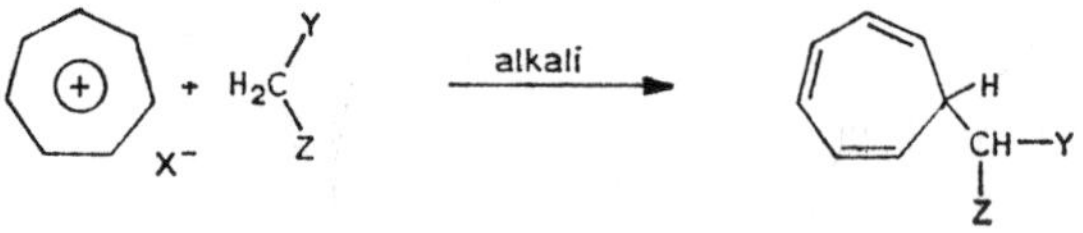

(Y and Z = $-CO_2H$, $-CO_2Et$, $-COMe$, $-NO_2$, $-CN$, etc.)

Ditropyl derivatives may also be formed. Cyclopentadiene in the presence of bases reacts with the tropylium ion as follows[19]:

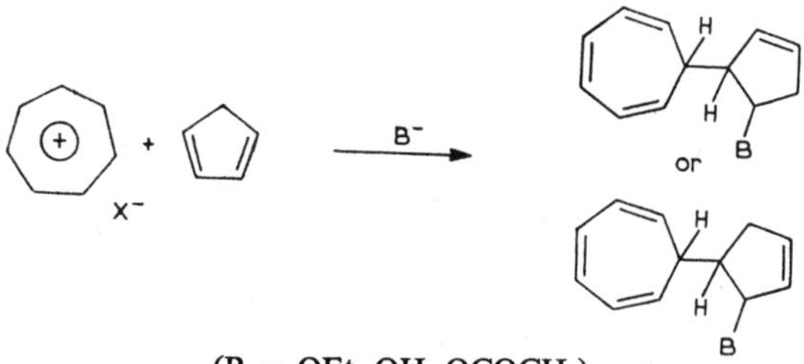

(B = OEt, OH, OCOCH₃)

Other dienes do not react.

Aliphatic aldehydes react similarly, *e.g.*

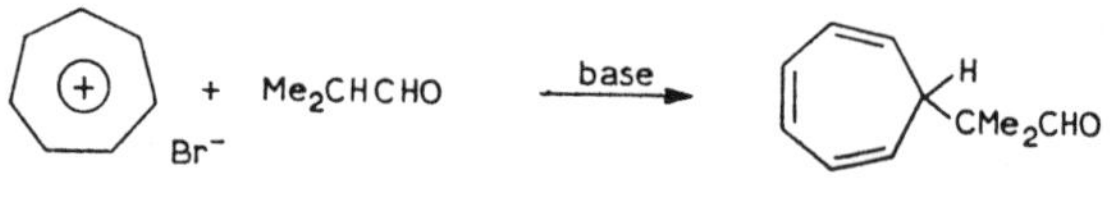

(tropylium)Br⁻ + Me_2CHCHO —base→ (cycloheptatrienyl-CMe₂CHO)

Aliphatic and alkyl aryl ketones also react, but only on heating[62].

The tropylium ion attacks phenols in the presence of alkali[62,65–69]. Thus with phenol itself a tropylphenol is formed which has been shown to consist mainly of the *para*-isomer plus a little of the *ortho*-isomer. This indicates that reaction proceeds by direct electrophilic attack by the tropylium ion on the sodium phenoxide. In the case of β-naphthol a more complicated reaction ensues in which the naphtholate ion is oxidised as well as substituted[68]. Its course has been formulated as follows:

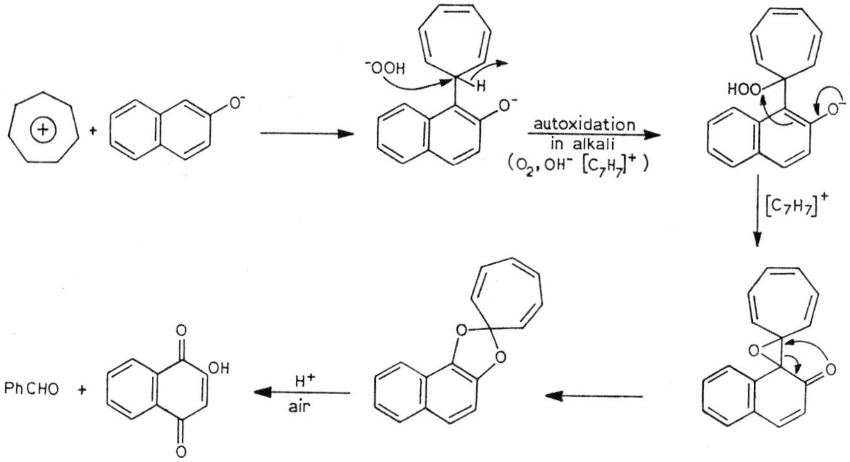

The tropylium ion also attacks the tropolonate ion (see Chapter VI), giving a tropyltropolone[65-67]. Reaction of 2,6-dimethylphenol with two molar equivalents of the tropylium cation produces the substituted tropylium salt (V), presumably by the mechanism shown. This stable orange-red salt is converted by base into a somewhat unstable purple quinocycloheptatriene (VI)[82].

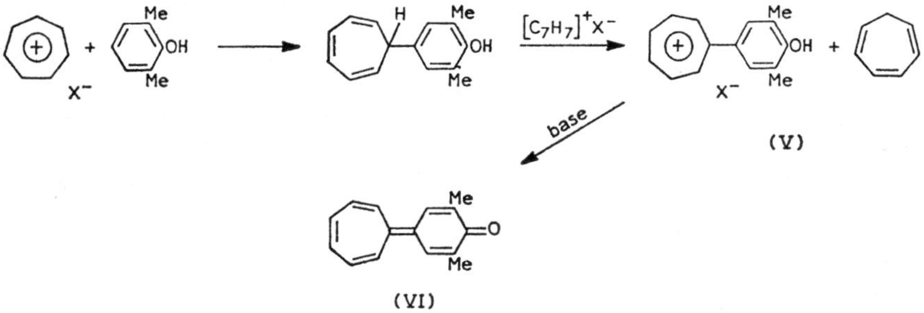

The highly electrophilic character of the tropylium ion is again demonstrated by the fact that it readily attacks a number of olefinic compounds which are suitably activated by neighbouring electron-donating groups[61,62]. Thus it reacts with vinyl ethers in the cold[61].

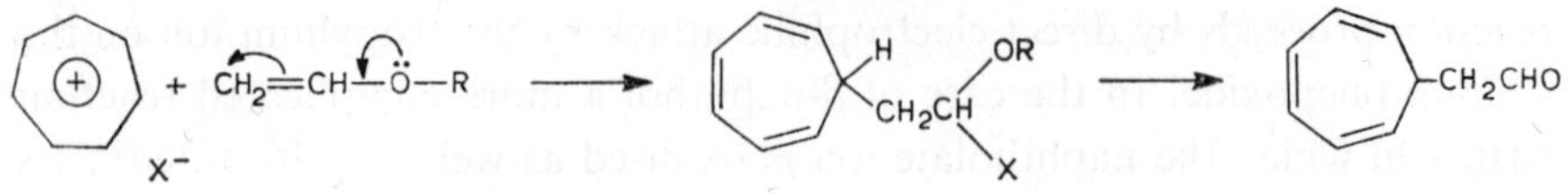

Compounds having double-bonds with adjacent electron-withdrawing groups, e.g. acrylic acid, do not react with tropylium salts even on heating. These alkylation reactions are the reverse of the fragmentation reactions mentioned above (p. 102) which give rise to tropylium ions.

Substituent groups attached to the tropylium ring may also be replaced by nucleophilic substitution. Thus chloro- and bromo-tropylium salts react with water to form hydroxytropylium salts which in turn generate tropone[9,18].

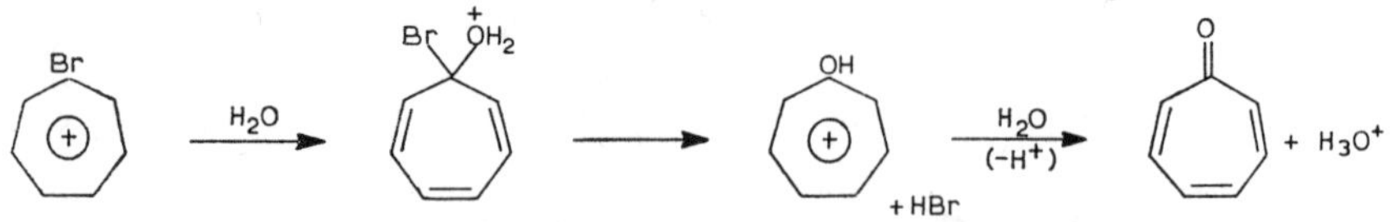

(d) Attempted electrophilic substitution of the tropylium ion

Not surprisingly tropylium salts are extremely inert towards electrophilic reagents and it has not yet proved possible to carry out any of the electrophilic substitution reactions such as nitration, sulphonation and Friedel–Crafts acylation or alkylation. There is no deuterium exchange either on prolonged treatment with concentrated deuterosulphuric acid or even with a solution of aluminium bromide in deuterobromic acid, which is one of the strongest of all deuteroacids. Under these conditions benzene undergoes deuterium exchange almost instantaneously and even saturated hydrocarbons undergo rapid exchange[71]. Hence the tropylium ion is one of the most inert of aromatic systems towards typical aromatic substitution reactions.

(e) Reactions of the alkyl groups in alkyltropylium salts

The alkyl groups in alkyltropylium salts will be somewhat activated towards electrophilic attack. Orthoformic ester, in the presence of acetic anhydride reacts with methyltropylium perchlorate to form the deep blue salt (VII), while with 1,2-dimethyltropylium tetrafluoroborate the azulenium salt (VIII) is obtained (*cf.* Chapter VIII)[35].

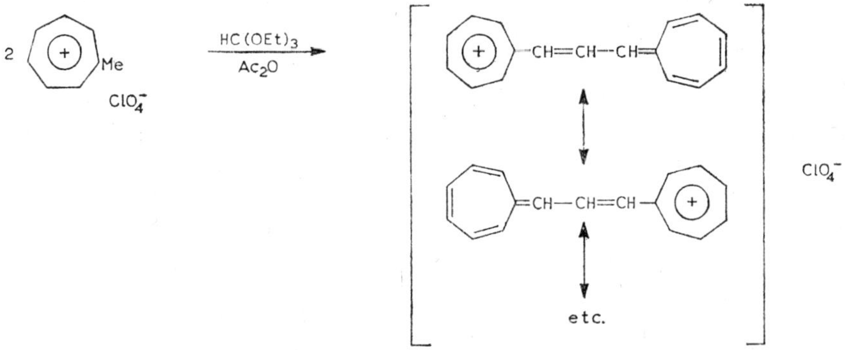

METAL COMPLEXES

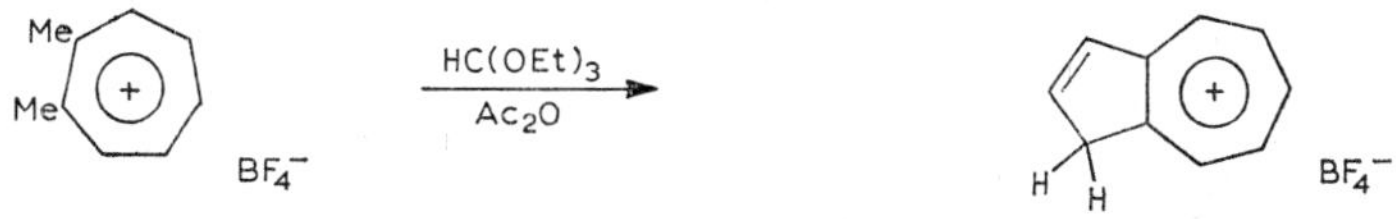

Tropylium metal complexes such as (IX) have been prepared from the corresponding cycloheptatrienyl-metal complexes[72,73].

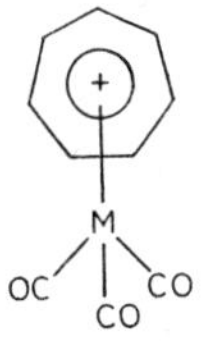

(IX)
M = Mo, Cr

The chromium complex reacts with nucleophilic reagents to form in most instances substituted cycloheptatrienyl-metal complexes[73].

BENZOTROPYLIUM SALTS

A benzotropylium salt was first isolated in 1955[74], but the formation of a tetramethoxybenzotropylium salt was postulated two years earlier (*i.e.* before

References p. 114

the recognition of the simple tropylium ion) in the following reaction sequence[75], although the salt was not isolated from the red solution:

The unsubstituted benzotropylium salt was prepared by an identical route from 2,3-benzotropone and isolated as a yellow solid[74]. Since then a number of other benzotropylium[56,76,77], dibenzotropylium[56,78,79], tribenzotropylium[80,81], and naphthotropylium[56] salts have been prepared.

Increasing annellation leads to greatly diminished acid strength, e.g. the pK values for the benzo-, 1,2,4,5-dibenzo-, and tribenzotropylium salts are respectively 1.7[56], −3.7[78], and ∼−15[80]. This suggests that there is little if any aromatic stabilisation in the tribenzotropylium cation[80].

The position of the longest wavelength maximum in the ultra-violet/visible spectrum shifts to longer wavelengths with increased annellation, e.g. λ_{max} benzo- and dibenzo-tropylium cations, respectively, = 426 and 540 mμ[76,78]. In consequence the benzotropylium salts are coloured.

The pK values and the positions of the long wavelength absorption correlate closely with the calculations from simple molecular orbital models[56].

REFERENCES

1 G. MERLING, Ber., 24 (1891) 3108.
2 W. VON E. DOERING AND L. H. KNOX, J. Am. Chem. Soc., 76 (1954) 3203.
3 N. A. NELSON, J. H. FASSNACHT AND J. U. PIPER, J. Am. Chem. Soc., 81 (1959) 5009; 83 (1961) 206.
4 E. MÜLLER AND H. FRICKE, Ann., 661 (1963) 38.
5 W. VON E. DOERING AND L. H. KNOX, J. Am. Chem. Soc., 79 (1957) 352.
6 A. W. JOHNSON AND M. TISLER, Chem. and Ind., (1954) 1427.
7 A. W. JOHNSON, A. LANGEMANN AND M. TISLER, J. Chem. Soc., (1955) 1622.
8 M. J. S. DEWAR AND C. R. GANELLIN, J. Chem. Soc., (1959) 2438.
9 W. VON E. DOERING AND H. KRAUCH, Angew. Chem., 68 (1956) 661.
10 K. KITAHARA AND M. FUNAMIZU, Annual Meeting Chem. Soc. Japan, (1959) Abs. 230; quoted in ref 11.
11 T. NOZOE, in J. W. COOK (Editor), Progress in organic chemistry, Vol. 5, Butterworth, London, 1961, p. 132.
12 D. N. KURSANOV AND M. E. VOL'PIN, Dokl. Akad. Nauk S.S.S.R., 113 (1957) 339.
13 H. J. DAUBEN, F. A. GADECKI, K. M. HARMON AND D. L. PEARSON, J. Am. Chem. Soc., 79 (1959) 4557.
14 K. M. HARMON AND A. B. HARMON, J. Am. Chem. Soc., 83 (1961) 865.

REFERENCES

15 G. A. GLADKOVSKII, S. S. SKOROKHODOV, S. G. SLYVINA AND A. S. KHACHATUROV, *Izv. Akad. Nauk S.S.S.R., Otdel Khim. Nauk*, (1963) 1273.
16 K. M. HARMON, A. B. HARMON AND F. E. CUMMINGS, *J. Am. Chem. Soc.*, 83 (1961) 3912
17 D. BRYCE-SMITH AND N. A. PERKINS, *Chem. and Ind.*, (1959) 1022.
18 M. E. VOL'PIN, I. S. AKHREM AND D. N. KURSANOV, *Izv. Akad. Nauk S.S.S.R., Otdel. Khim. Nauk*, (1957) 760.
19 D. N. KURSANOV, M. E. VOL'PIN AND Z. N. PARNES, *Khim. Nauka i Prom.*, 3 (1958) 159.
20 D. BRYCE-SMITH AND N. A. PERKINS, *J. Chem. Soc.*, (1962) 1339.
21 D. H. GESKE, *J. Am. Chem. Soc.*, 81 (1959) 4145.
22 D. H. REID, M. FRASER, B. B. MOLLOY, H. A. S. PAYNE AND R. G. SUTHERLAND *Tetrahedron Letters*, (1961) 530.
23 A. P. TER BORG, R. VAN HELDEN AND A. F. BICKEL, *Rec. Trav. chim.*, 81 (1962) 164.
24 M. J. S. DEWAR AND R. PETTIT, *Chem. and Ind.*, (1955) 199.
25 M. J. S. DEWAR AND R. PETTIT, *J. Chem. Soc.*, (1956) 2021.
26 C. E. WULFMAN, C. F. YARNELL, AND D. S. WULFMAN, *Chem. and Ind.*, (1960) 1440.
27 W. VON E. DOERING AND L. E. HELGEN, *J. Chem. Soc.*, (1961) 482.
28 M. J. S. DEWAR, C. R. GANELLIN AND R. PETTIT, *J. Chem. Soc.*, (1958) 55.
29 M. E. VOL'PIN, V. G. DULOVA AND D. N. KURSANOV, *Dokl. Akad. Nauk S.S.S.R.*, 128 (1959) 951; M. E. VOL'PIN, D. N. KURSANOV AND V. G. DULOVA, *Tetrahedron Letters*, No. 8 (1960) 33.
30 G. L. CLOSS AND L. E. CLOSS, *Tetrahedron Letters*, No. 10 (1960) 38.
31 C. R. GANELLIN AND R. PETTIT, *J. Am. Chem. Soc.*, 79 (1957) 1767; *Chem. Ber.*, 90 (1957) 2951.
32 M. E. VOL'PIN, I. S. AKHREM AND D. N. KURSANOV, *Khim. Nauka i Prom.*, 2 (1957) 656.
33 M. E. VOL'PIN, I. S. AKHREM AND D. N. KURSANOV, *Zh. obshch. Khim.*, 28 (1958) 330.
34 K. M. HARMON AND D. A. DAVIS, *J. Am. Chem. Soc.*, 84 (1962) 4359.
35 K. HAFNER, H. W. RIEDEL AND M. DANIELISZ, *Angew. Chem.*, 75 (1963) 344.
36 K. CONROW, *J. Am. Chem. Soc.*, 81 (1959) 5461.
37 M. A. BATTISTE, *Chem. and Ind.*, (1961) 550; *J. Am. Chem. Soc.*, 83 (1961) 4101.
38 P. N. RYLANDER, S. MEYERSON AND H. M. GRUBB, *J. Am. Chem. Soc.*, 79 (1957) 842; S. MEYERSON AND P. N. RYLANDER, *J. Chem. Phys.*, 27 (1957) 901; V. HANUS, *Nature*, 164 (1959) 1796.
39 S. MEYERSON, P. N. RYLANDER, E. L. ELIEL AND J. D. MCCOLLUM, *J. Am. Chem. Soc.*, 81 (1959) 2606.
40 M. E. VOL'PIN, D. N. KURSANOV, M. M. SHEMYAKIN, V. I. MAIMIND AND L. A. NEIMAN, *Chem. and Ind.*, (1958) 1261; *Zh. obshch. Khim.*, 29 (1959) 3711.
41 K. M. HARMON, F. E. CUMMINGS, D. A. DAVIS AND D. J. DIESTLER, *J. Am. Chem. Soc.*, 84 (1962) 120, 3349.
42 M. FELDMAN AND S. WINSTEIN, *J. Am. Chem. Soc.*, 83 (1961) 3338.
43 J. N. MURRELL AND H. C. LONGUET-HIGGINS, *J. Chem. Phys.*, 23 (1955) 2347.
44 W. G. FATELEY AND E. R. LIPPINCOTT, *J. Am. Chem. Soc.*, 77 (1955) 249.
45 G. FRAENKEL, R. E. CARTER, A. MCLACHLAN AND J. H. RICHARDS, *J. Am. Chem. Soc.*, 82 (1960) 5846; T. SCHAEFER AND W. G. SCHNEIDER, *Canad. J. Chem.*, 41 (1963) 966; J. R. LETO, F. A. COTTON AND J. S. WAUGH, *Nature*, 180 (1957) 978.
46 M. E. VOL'PIN, S. I. ZHDANOV AND D. N. KURSANOV, *Dokl. Akad. Nauk S.S.S.R.*, 112 (1957) 264.
47 A. N. FRUMKIN AND S. I. ZHDANOV, *Dokl. Akad. Nauk S.S.S.R.*, 122 (1958) 412.
48 M. A. GEROVICH, *Dokl. Akad. Nauk S.S.S.R.*, 96 (1954) 543; 105 (1955) 1278.
49 Z. N. PARNES, M. E. VOL'PIN AND D. N. KURSANOV, *Tetrahedron Letters*, No. 21 (1960) 20.

50 Z. N. PARNES, G. D. MUR, P. B. KUDRAVTSEV AND D. N. KURSANOV, *Dokl. Akad. Nauk S.S.S.R.*, 155 (1964) 1371.
51 K. CONROW, *J. Am. Chem. Soc.*, 83 (1961) 2343.
52 M. J. S. DEWAR AND R. PETTIT, *J. Chem. Soc.*, (1956) 2026.
53 M. E. VOL'PIN AND D. N. KURSANOV, *Dokl. Akad. Nauk S.S.S.R.*, 126 (1959) 780.
54 M. E. VOL'PIN, D. N. KURSANOV AND V. G. DULOVA, *Tetrahedron Letters*, No. 8 (1960) 33.
55 T. NOZOE, T. MUKAI, T. TEZUKA AND K. OSAKA, quoted in ref. 11.
56 D. MEUCHE, H. STRAUSS AND E. HEILBRONNER, *Helv. Chim. Acta*, 41 (1958) 57.
57 A. P. TER BORG, R. VAN HELDEN, A. F. BICKEL, W. RENOLD AND A. S. DREIDING, *Helv. Chim. Acta*, 43 (1960) 457.
58 T. IKEMI, T. NOZOE AND H. SUGIYAMA, *Chem. and Ind.*, (1960) 932.
59 A. P. TER BORG, R. VAN HELDEN AND A. F. BICKEL, *Rec. Trav. chim.*, 81 (1962) 177.
60 Z. N. PARNES, G. D. MUR, P. B. KUDRAVTSEV AND D. N. KURSANOV, *Dokl. Akad. Nauk S.S.S.R.*, 159 (1964) 857.
61 M. E. VOL'PIN, I. S. AKHREM AND D. N. KURSANOV, *Dokl. Akad. Nauk S.S.S.R.*, 120 (1958) 531.
62 M. E. VOL'PIN, I. S. AKHREM AND D. N. KURSANOV, *Izv. Akad. Nauk S.S.S.R., Otdel. khim. Nauk*, (1957) 1501.
63 M. W. JORDAN AND I. W. ELLIOT, *J. Org. Chem.*, 27 (1962) 1445.
64 F. KORTE, K-H. BÜCHEL AND F. F. WEISE, *Ann.*, 664 (1963) 114.
65 M. E. VOL'PIN, I. S. AKHREM AND D. N. KURSANOV, *Zh. obshch. Khim.*, 29 (1959) 2855.
66 M. E. VOL'PIN, I. S. AKHREM AND D. N. KURSANOV, *Izv. Akad. Nauk S.S.S.R., Otdel. khim. Nauk*, (1957) 1905.
67 M. E. VOL'PIN, I. S. AKHREM AND D. N. KURSANOV, *Zh. obshch. Khim.*, 30 (1960) 159, 1187.
68 T. NOZOE, S. ITO AND T. TEZUKA, *Chem. and Ind.*, (1960) 1088.
69 T. NOZOE AND K. KITAHARA, *Chem. and Ind.*, (1962) 119.
70 M. E. VOL'PIN, K. I. ZHDANOVA, D. N. KURSANOV, V. N. SETKINA AND A. I. SHATENSTEIN, *Izv. Akad. Nauk S.S.S.R., Otdel. khim. Nauk* (1959) 754.
71 K. I. ZHDANOVA, L. N. VINOGRADOV AND V. R. KALINACHENKO, *Dokl. Akad. Nauk S.S.S.R.*, 102 (1955) 779.
72 H. J. DAUBEN AND L. R. HONNEN, *J. Am. Chem. Soc.*, 80 (1958) 5570.
73 J. D. MUNRO AND P. L. PAUSON, *Proc. Chem. Soc.*, (1959) 267; *J. Chem. Soc.*, (1961) 3475, 3479, 3484.
74 H. H. RENNHARD, E. HEILBRONNER AND A. ESCHENMOSER, *Chem. and Ind.*, (1955) 415.
75 A. ESCHENMOSER AND H. H. RENNHARD, *Helv. Chim. Acta*, 36 (1953) 290.
76 H. H. RENNHARD, G. DI MODICA, W. SIMON, E. HEILBRONNER AND A. ESCHENMOSER, *Helv. Chim. Acta*, 40 (1957) 957.
77 D. MEUCHE, W. SIMON AND E. HEILBRONNER, *Helv. Chim. Acta*, 42 (1959) 452.
78 G. BERTI, *J. Org. Chem.*, 22 (1957) 230.
79 R. W. MURRAY, *Tetrahedron Letters*, No. 7, (1960) 27.
80 M. STILES AND A. J. LIBBEY, *J. Org. Chem.*, 22 (1957) 1243.
81 G. WITTIG, E. HAHN AND W. TOCHTERMANN, *Chem. Ber.*, 95 (1962) 431.
82 A. F. BICKEL, A. P. TER BORG AND R. VAN HELDEN, *Rec. Trav. chim.*, 81 (1962) 599.
83 K. M. HARMON, A. B. HARMON AND F. E. CUMMINGS, *J. Am. Chem. Soc.*, 86 (1964) 5511.
84 A. I. KITAIGORODSKII, Y. T. STRUCHKOV, T. L. KHOTSYANOVA, M. E. VOL'PIN AND D. N. KURSANOV, *Izv. Akad. Nauk S.S.S.R., Otdel khim. Nauk*, (1960) 39.
85 D. G. LINDSAY AND C. B. REESE, *Tetrahedron*, 21 (1965) 1673.
86 R. W. MURRAY AND M. L. KAPLAN, *Tetrahedron Letters*, (1965) 2903.
87 K. CONROW, *Org. Synth.*, 43 (1963) 101.
88 P. G. FERRINI AND A. MARXER, *Angew. Chem.*, 74 (1962) 488.

CHAPTER VI

Tropones, tropolones and related compounds

INTRODUCTION

A cycloheptatrienolone or *tropolone* type formula (I) was first put forward in 1945 to account for the properties of the mould product stipitatic acid[1].

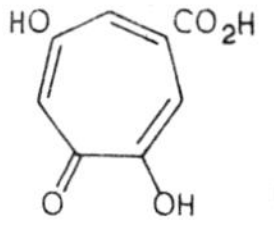

(I)

It was suggested that the cycloheptatrienolone ring might represent a new type of aromatic system which would be stabilised by resonance and tautomerism:

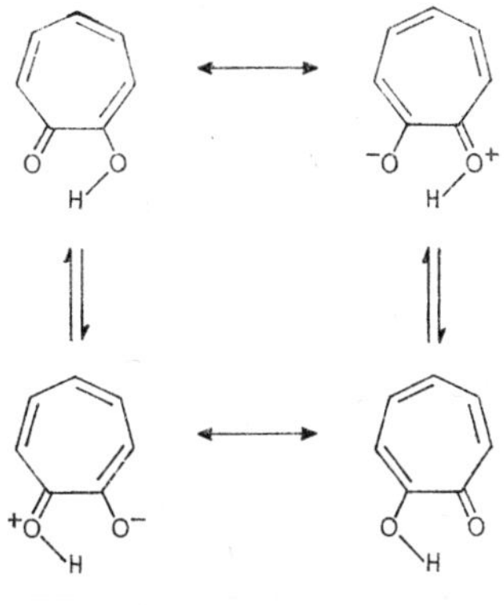

Independent work at about the same time on the compound hinokitiol, an antibiotic obtained from the heartwood of Cupressaceae led to its being described initially as an isopropylcycloheptenedione (or enol form), but this formula was later corrected to an isopropyltropolone[2].

Since then a number of tropolone derivatives have been found to occur naturally, for example the α-, β- and γ-thujaplicins which are respectively α-, β- and γ-isopropyltropolones (II)(III)(IV) (β-thujaplicin ≡ hinokitiol) in the heartwood or essential oil of various trees of the Cupressaceae family,

References p. 155

e.g. western red cedar[3-5]; hydroxytropolones such as puberulic acid (V)[6,7] as well as stipitatic acid (I)[1,6,8] in mould products; and colchicine (VI)[9,10] in various parts of plants of the Liliaceae family, e.g. autumn crocuses. In addition the red colouring matter purpurogallin, formed by oxidation of pyrogallol in neutral or weakly acid medium with a variety of oxidising agents (most satisfactorily sodium iodate[11]) (and also occurring naturally in various galls) was shown to be the benzotropolone derivative (VII)[12,13].

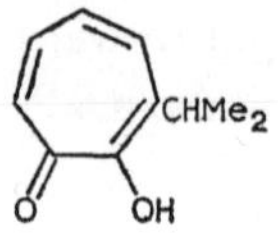

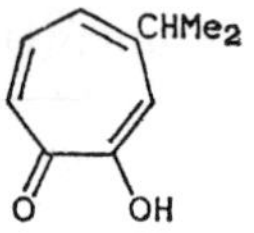

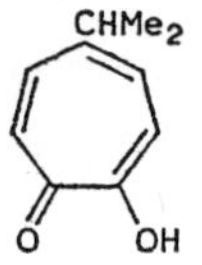

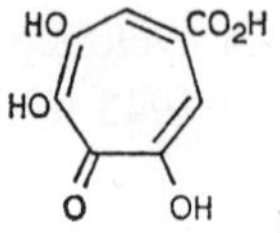

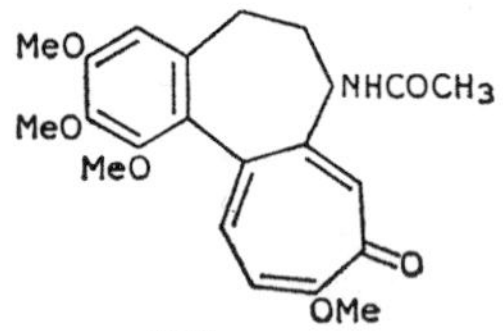

Tropolone derivatives are either named as α-, β- or γ-derivatives (cf. the thujaplicins, above) or numbered as follows:

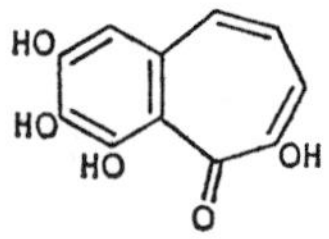

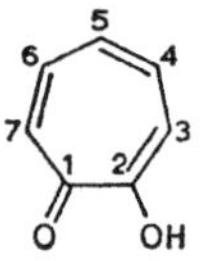

Since tautomeric exchange between the carbonyl and hydroxyl groups takes place it is not possible to specify which carbon atom bears the hydroxyl group and which is part of a carbonyl group. Thus compound (VIII), 3-methyltropolone, may be written in either form. On the other hand methylation of the hydroxyl group in (VIII) leads to distinct isomers (IX) and (X) since tautomeric change is no longer possible. These compounds are called respectively, 2-methoxy-7-methyltropone (IX) and 2-methoxy-3-methyltropone (X), *tropone* being the abbreviated name of cycloheptatrienone.

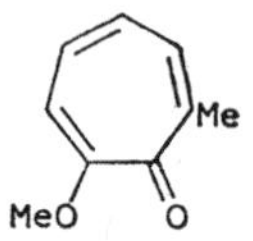

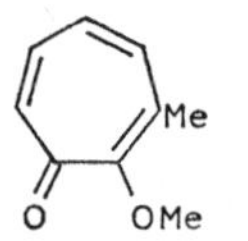

INTRODUCTION 119

The first formulations of tropolone made no mention of possible contributions from dipolar forms such as (XI) in which a delocalised positive charge resides on the seven-membered ring, making the ring analogous to a tropylium ion.

(XI)

It was however, pointed out that tropolone resembles an extended carboxyl group:

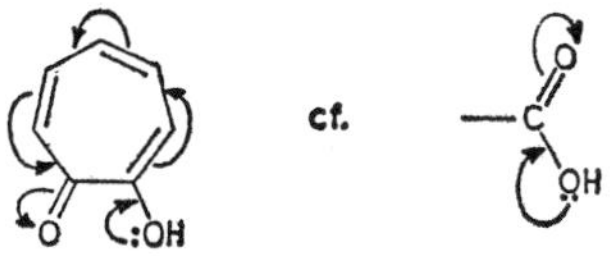

Gradually the significance of Hückel's rule to the structure of tropolone was appreciated and the importance of contributions from structure (XI) was suggested. This led to the further idea that the parent compound of this series was not tropolone but tropone or cycloheptatrienone which could also be represented as a hybrid between covalent and dipolar forms (XII).

(XII)

Evidence which supported this concept included the dipole moment of tropone and the abnormal position of the infra-red absorption peak due to the carbonyl group. (For details see later in the chapter.)

Furthermore the original formulation of tropolone did not involve the 1,2-carbon–carbon bond in the resonating system and this bond might therefore be expected to show single bond properties. It appeared however to be too short for this to be the case; this was completely in accord with a molecule having a hybrid structure involving as one contributing form a tropylium type ion as in (XI) for in such an ion the ring bonds would be substantially equivalent and intermediate in length between normal single and double bonds.

On this basis tropolone might be regarded as a "phenol" derived from

References p. 155

tropone. Furthermore the electron-donating hydroxyl group might be expected to stabilise the positively charged aromatic seven-membered ring, as negatively charged cyclopentadienide ions are stabilised by electron-withdrawing groups. This is typified by the greater difficulty of bringing about molecular rearrangement of the tropone skeleton in the presence of electron-donating groups.

Following the development of these ideas it seemed reasonable to look for chemical evidence of aromatic character in the tropones and tropolones. Such evidence was easy to find in the case of the tropolones, which underwent typical aromatic electrophilic substitution reactions but less evident in the case of tropones, whose properties resembled rather those of an unsaturated ketone.

In fact the whole attempt to seek evidence for the aromaticity of tropones and tropolones in their chemical reactions is now seen to be misleading in the context of present day thinking on aromaticity (cf. Chapter I).

Spectral determinations (see below) suggest aromaticity of the ring in tropone and tropolone arising from the contribution of dipolar forms, but the extent of this contribution cannot be great in view of their relatively low dipole moments.

It has already been suggested that the chemical properties of tropone can be explained without recourse to any particular aromatic character[14]. Indeed it would be odd if the contribution of the dipolar form did result in marked aromatic character as evinced by the molecule undergoing electrophilic substitution, for in this dipolar form the seven-membered ring has the character of a tropylium ion and tropylium ions are notably resistant to electrophilic attack, as would be expected since they themselves are electrophilic reagents. The main contribution of this dipolar form to the chemical properties of tropones is possibly shown in the somewhat reduced reactivity of the carbonyl group towards normal carbonyl reagents; it may also contribute to the slow elimination reactions which the halogen addition products from tropone undergo on standing.

Although it has been generally recognised that the chemical properties of tropone can be adequately explained by the covalent cycloheptatrienone form, tropolones have continued to be considered as compounds whose properties are best explained in terms of aromatic character. It is doubtful in this case also whether it is necessary to consider the dipolar form to explain their chemical reactions.

Interaction between the neighbouring enol and ketone groups due to hydrogen bonding has sometimes been seen as the source of difference between tropolone and tropone. This concept is inadmissible since 3- and 4-hydroxytropones closely resemble tropolone in their chemical properties, points of difference arising rather in their physical properties where the intramolecular hydrogen bonding in tropolone, which is not possible in its isomers, leads to big differences in volatility, solubility, etc.

The properties of the hydroxyl group in tropolone have been explained by comparing the compound to an extended carboxylic acid (see above) — in this way the ease of ether formation and of replacement of the hydroxyl group by halogen atoms is likened to ester and acid chloride formation respectively. Further the ready hydrolysis of tropolone ethers by either acid or alkali, which is in contrast to the behaviour of phenol ethers, can be compared to the hydrolysis of esters.

Not only is tropolone an extended carboxylic acid, it is also a vinylogue of the mono-enol form of a β-diketone:

Electronic interaction between the carbonyl and enol groups is facilitated by the planarity of the molecule. Hence it is not unreasonable that tropolone might show reactions akin to β-diketones. This relationship is sufficient to explain many of the reactions of tropolones which have been seen as manifestations of the aromatic character of the ring; for example β-diketones can be nitrosated[15], nitrated[16], and couple with diazonium salts[17] under mild conditions. Like β-diketones tropolones are unstable to alkali; they are stable to acids since protonation of the carbonyl group leads to the formation of a stable dihydroxytropylium salt. To attribute the attack of electrophiles on tropolones to the contribution of the dipolar form, seems, as in the case of tropones, unlikely since the ring bears a positive charge in this form. That such reactions proceed by substitution illustrates, however, the stability of the tropolone structure.

It seems in general reasonable to expect some contribution of the dipolar forms of tropone and tropolone to their overall structure and this manifests itself in various physical properties and in the diminished reactivity of the carbonyl group. In the case of both tropolones and 3- and 4-hydroxy-tropones the reactivity of the molecule towards electrophilic reagents is greatly enhanced as compared to tropones for in the transition state the positive charge can be delocalised over the hydroxyl group as well as the ring, thereby stabilising the transition state and facilitating reaction.

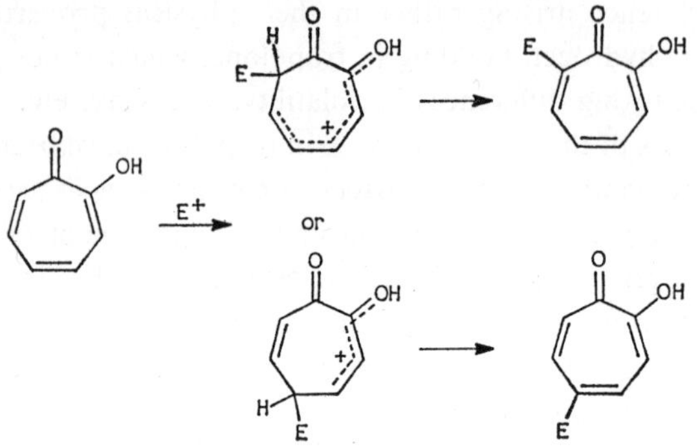

The preparation and properties of tropones and tropolones and also of some of their derivatives, such as benzotropones, benzotropolones, aza- and thia-tropolones and heptafulvenes are now considered in more detail.

PREPARATION OF TROPONES

Three independent syntheses of tropone were announced almost simultaneously in 1951, one starting from anisole[18] and the others from cycloheptenone[19] and cycloheptanone[20] respectively. 4,5-Benzotropone was first prepared as long ago as 1909[21,22], and alkyl derivatives of this benzotropone in 1906[22,23]. A survey of methods which have been used to prepare tropone is now given; the synthesis of benzotropones is considered later in the chapter.

(a) From benzene derivatives

Anisole reacts with diazomethane in the presence of ultra-violet light to give a methoxycycloheptatriene; treatment of the latter compound with in turn, acid, bromine, and aqueous sodium carbonate leads to the formation of tropone[18].

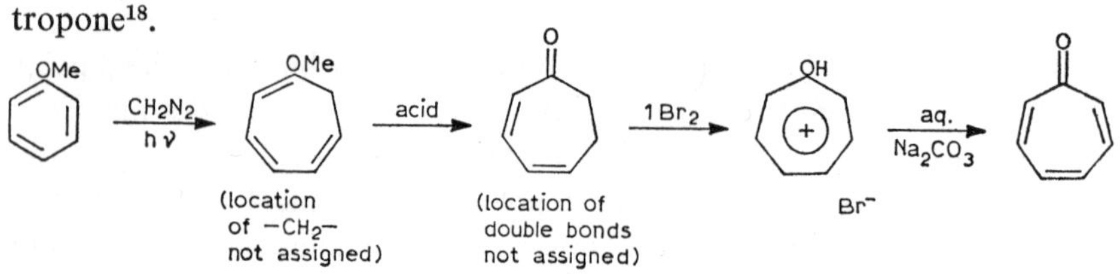

4-Carboxytropone has been prepared by the reaction of ethyl diazoacetate with anisole, followed by hydrolysis[24,25], while tropone itself has also been prepared by hydrolysis of the bromotropylium ion initially formed from bromobenzene and diazomethane[26]. Reaction of o-[27] and p-benzoquinole acetates[28] with diazomethane in the presence of boron trifluoride leads to tropones.

Ring-expansion of the benzene ring by means of a chlorocarbene has also been effected[29]:

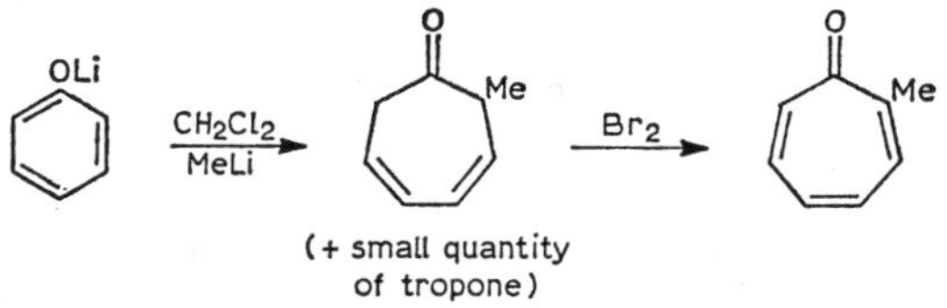

$$\text{PhOLi} \xrightarrow[\text{MeLi}]{CH_2Cl_2} \text{(methylcycloheptadienone)} \xrightarrow{Br_2} \text{(methyltropone)}$$

(+ small quantity of tropone)

Although methylcycloheptadienone is the principal product rather than tropone in this instance, if ditertiarybutylphenol is used ditertiarybutyltropone is the major product[29].

Particularly effective methods involve reduction of anisole[30] or methoxybenzoic acid[31,32] to 1,4-cyclohexadiene derivatives. In the first case[30] the product so obtained is reacted with dibromocarbene, followed by silver nitrate:

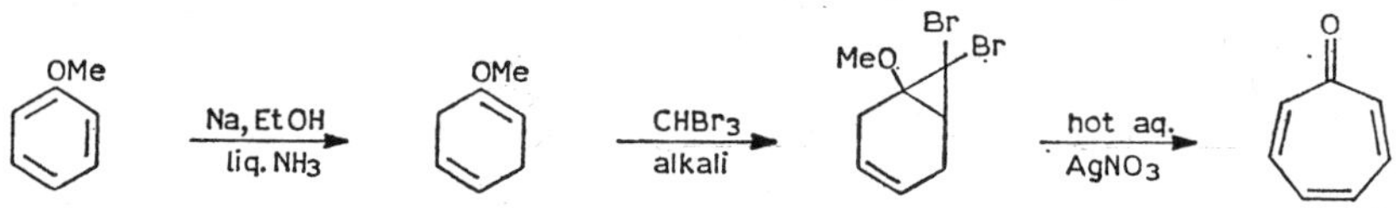

$$\text{OMe} \xrightarrow[\text{liq. NH}_3]{Na, EtOH} \text{OMe} \xrightarrow[\text{alkali}]{CHBr_3} \text{MeO, Br, Br} \xrightarrow[\text{AgNO}_3]{hot\ aq.} \text{tropone}$$

When dichlorocarbene was used instead of dibromocarbene, treatment of the resultant bicyclic intermediate with aqueous silver nitrate resulted for the most part in the regeneration of starting material rather than the formation of tropone. In the other case[31,32] the methoxycyclohexadienecarboxylic acid is reduced to the corresponding carbinol, whose p-toluenesulphonyl derivative undergoes rearrangement to methoxycycloheptatriene on treatment with pyridine:

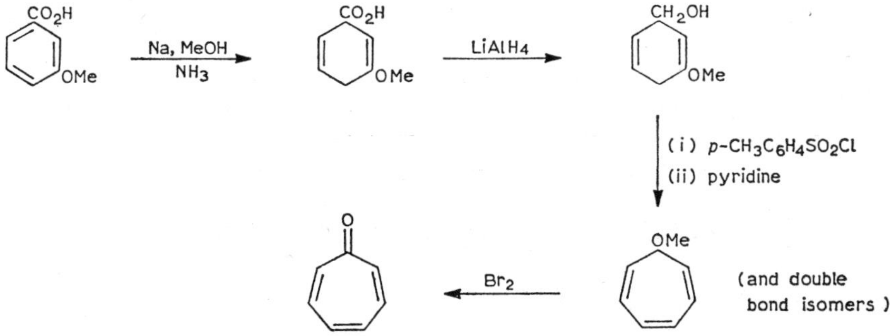

$$\text{CO}_2\text{H, OMe} \xrightarrow[\text{NH}_3]{Na, MeOH} \text{CO}_2\text{H, OMe} \xrightarrow{LiAlH_4} \text{CH}_2\text{OH, OMe}$$

(i) p-CH$_3$C$_6$H$_4$SO$_2$Cl
(ii) pyridine

$$\text{tropone} \xleftarrow{Br_2} \text{OMe (and double bond isomers)}$$

References p. 155

Resorcinol-5-carboxylic acid has served as a starting material for another synthetic route involving ring-expansion of the *p*-toluenesulphonyl-derivative of a carbinol[33,34], *viz.*:

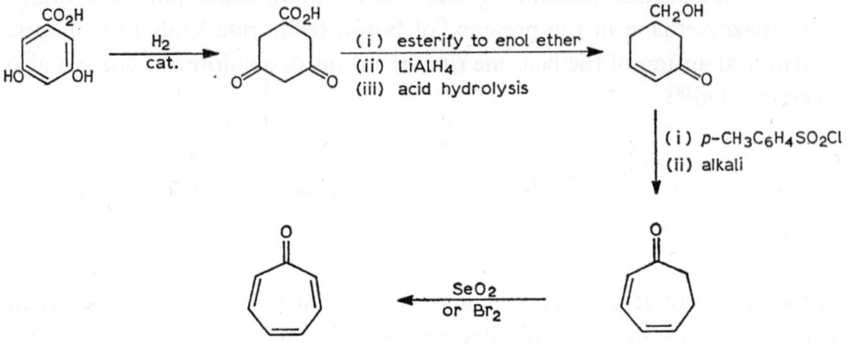

[reaction scheme: resorcinol-5-carboxylic acid → (H₂ cat.) → dihydro diketo acid → (i) esterify to enol ether (ii) LiAlH₄ (iii) acid hydrolysis → hydroxymethyl cyclohexenone → (i) *p*-CH₃C₆H₄SO₂Cl (ii) alkali → cycloheptadienone → (SeO₂ or Br₂) → tropone]

The oxime of the Reimer–Tiemann reaction product obtained from *p*-cresol also undergoes a ring expansion in the presence of alkali to give 4-methyltropone oxime[35].

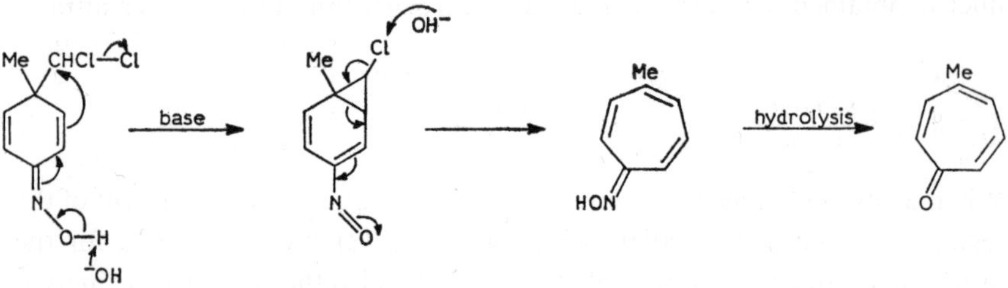

[reaction scheme showing base-induced ring expansion of dichloromethyl cresol oxime → 4-methyltropone oxime → (hydrolysis) → 4-methyltropone]

A further route involving addition of dichlorocarbene to a cyclohexene derivative starts from an enol ether of cyclohexanone. The adduct reacts with pyridine to give a chloroethoxycycloheptadiene which may be dehydrohalogenated, best by potassium tertiary butoxide, to an ethoxycycloheptatriene[246]:

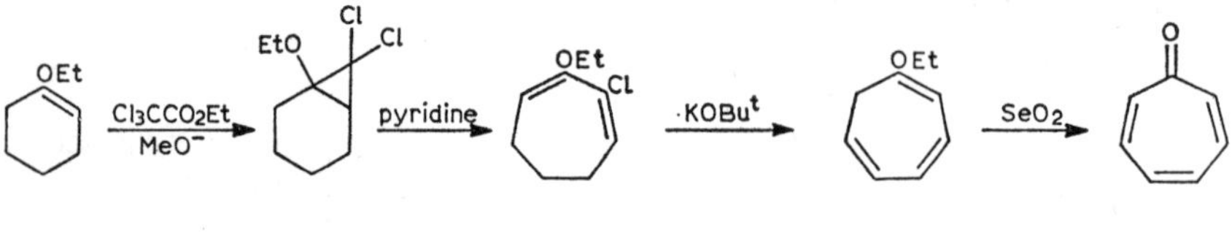

[reaction scheme: 1-ethoxycyclohexene → (Cl₃CCO₂Et, MeO⁻) → dichlorocarbene adduct → (pyridine) → chloroethoxycycloheptadiene → (KOBuᵗ) → ethoxycycloheptatriene → (SeO₂) → tropone]

(b) From cycloheptanone and cycloheptenone

Treatment of cycloheptanone[20,36] or cycloheptenone[19] with bromine in glacial acetic acid produces 2,4,7-tribromotropone, which may be reduced catalytically to tropone[19,20,36-38]. 2-Phenyltropone has been prepared similarly[39]. Alternative routes from cycloheptanone involve its reaction either with benzaldehyde[40] or in a Mannich reaction[41] to give 2,7-dimethylenecycloheptanones which may be dehydrogenated over palladium-charcoal to tropones. The exocyclic double bonds migrate into the ring during the dehydrogenation process. Dibenzyltropones have been prepared by the same route but starting from cycloheptenone[42].

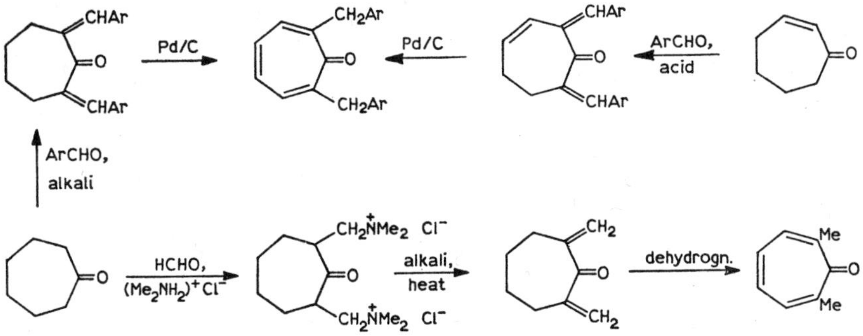

(c) From tropinone derivatives

Tropone was actually prepared[43] in 1887 by reacting tropinone methiodide with alkali and treating the product with bromine; the structure of the product was not recognised until the work was repeated in 1953[44,45].

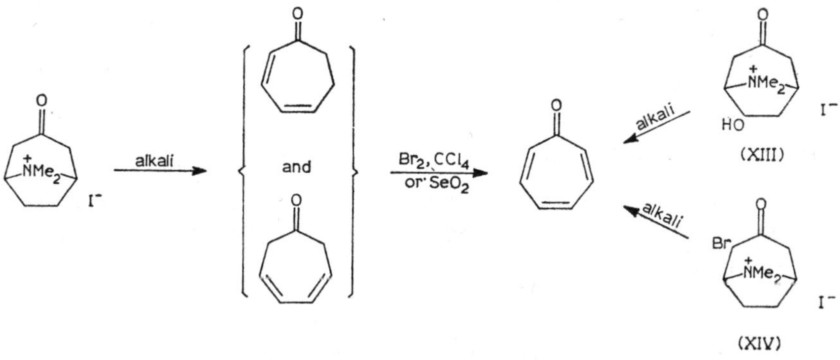

An analogous reaction has been used to prepare 2-phenyltropone[46]. Reaction of the hydroxy- and bromo-derivatives (XIII) and (XIV) of tropinone methiodide with alkali gave tropone in one step[47].

(d) From cycloheptatriene

Oxidation of cycloheptatriene by means of either chromium trioxide in pyridine or selenium dioxide produces tropone in 30–40% yield[48,49].

(e) From tropylium salts

It has been mentioned above that tropone is obtained by hydrolysis of the bromotropylium ion[26,50]. (See above, p. 123 and Chapter V.) Halo- or methoxy-tropylium salts can in fact be converted into tropone merely by pouring solutions of them into water[50,51].

An excellent and convenient method for the preparation of tropone from tropylium salts (and hence from cycloheptatriene) consists in first of all treating the tropylium salt with aqueous alkali to give ditropyl ether; by using concentrated sodium hydroxide solution a 96% yield is obtained[52]. If this ether is either distilled from acid-treated silica gel[53] or allowed to stand with a very small quantity of either concentrated hydrochloric acid or of tropylium ions[54] a disproportionation reaction takes place and tropone and cycloheptatriene are formed.

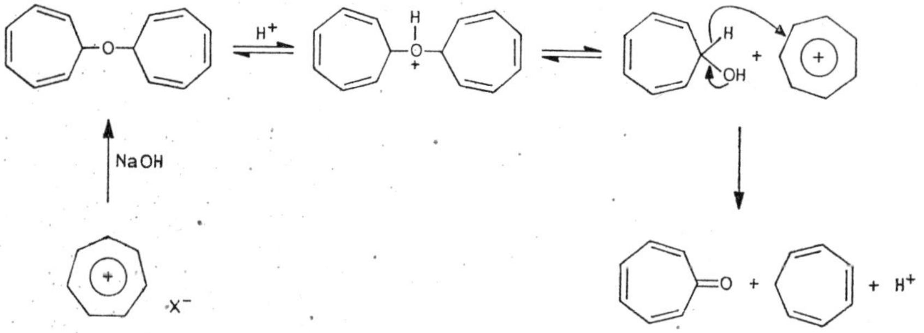

Tropylium bromide may also be converted into tropone and cycloheptatriene by the action of tertiary butanol[55]. Two moles of the bromide react with one mole of the alcohol to produce 0.6 mole of tropone and 0.7–0.8 mole of the triene.

PREPARATION OF TROPOLONES

A benzotropolone was first prepared in 1949[56]; three separate syntheses of tropolone itself[57–59] were published in 1950, one using benzene as starting

PREPARATION OF TROPOLONES 127

point and the others starting from cycloheptanone. Some methods of preparation of tropolones are now described; benzotropolones are discussed later in the chapter.

(a) From cycloheptanone and its derivatives

Cycloheptanone on oxidation with selenium dioxide gives cycloheptane-1,2-dione. This dione has the same carbon-oxygen skeleton as tropolone and its conversion to tropolone involves suitable dehydrogenation procedure. Treatment of the dione with bromine in acetic acid followed by alkali produces 3-bromotropolone; this may be catalytically reduced over palladium-charcoal to give tropolone itself[57,60].

Treatment of the dione with N-bromosuccinimide leads to tropolone directly together with a little 5-bromotropolone[58]. The same method starting from alkylcycloheptanones has been used to prepare a variety of alkyltropolones; (*inter alia*[61–71]) for example the naturally occurring α-[62,63], β-[61], and γ-[65] thujaplicins (see p. 117) have been prepared in this way. A mixture of 4- and 5-phenyltropolones have been prepared similarly from 4-phenylcycloheptanone[72]. 2-Hydroxycycloheptanone, obtained by intramolecular acyloin condensation of diethyl pimelate, has also been used instead of cycloheptane-1,2-dione[73]. Reaction of 3,7-dicarbethoxycycloheptane-1,2-dione with bromine caused replacement of the ester groups and the product was 3,7-dibromotropolone; if iodine was used instead of bromine 3-carboxytropolone was obtained[74].

An alternative mode of approach involving an isomerisation instead of a dehydrogenation consisted of treating 3,7-dibenzylidenecycloheptane-1,2-dione with either hydrogen bromide in acetic acid or palladium/charcoal to give 3,7-dibenzyltropolone[75,76].

(b) From benzene and other benzenoid compounds

A synthesis of tropolone which is elegant in its simplicity involves ring-expansion of benzene with diazomethane to give cycloheptatriene, followed by oxidation of the latter compound with potassium permanganate[59,77].

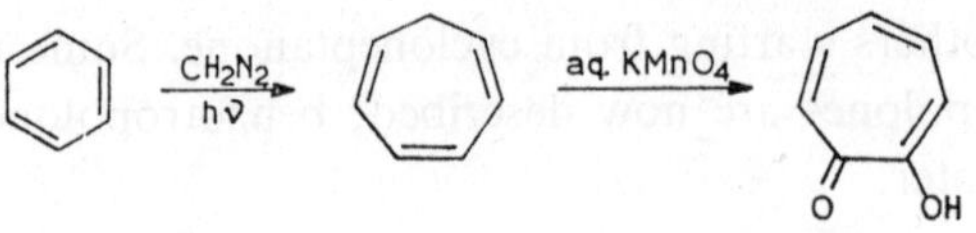

The overall yield was only 1% but the ready availability of the reactants and its simplicity made it none the less a practical method. This method has also been used to prepare substituted tropolones, including thujaplicins[78].

The reactions of veratrole[79,80] and of 1,2,4-trimethoxybenzene[80,81] with diazoacetic ester, followed by treatment with bromine have been used as methods for the preparation of tropolone-4-carboxylic acid and stipitatic acid (see p. 117) respectively.

Two other methods starting from benzenoid compounds involve prior reduction of the benzene ring. In the first resorcinol-5-carboxylic acid is converted into cycloheptadienone as described above (p. 124). Treatment of the enol acetate of this dienone with perbenzoic acid or, better, pertrifluoracetic acid results in the formation of tropolone[34].

In the second method, 2,3-dimethoxybenzoic acid is reduced in two steps to a cyclohexadienylcarbinol. The *p*-toluenesulphonyl derivative undergoes ring-expansion on treatment with pyridine giving 1,2-dimethoxycycloheptatrienes which are converted into tropolone by the action of bromine[32].

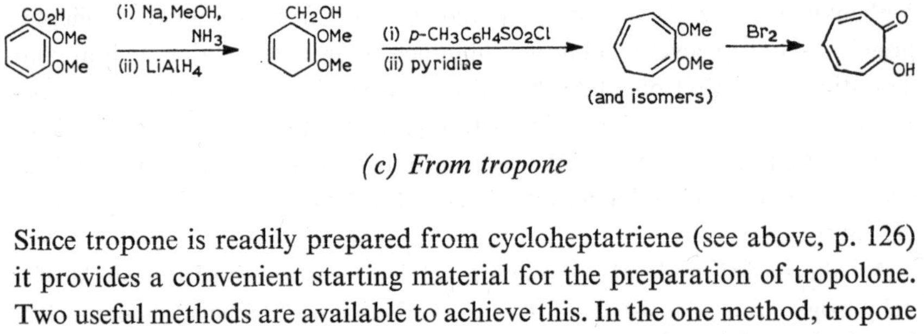

(c) From tropone

Since tropone is readily prepared from cycloheptatriene (see above, p. 126) it provides a convenient starting material for the preparation of tropolone. Two useful methods are available to achieve this. In the one method, tropone is first chlorinated and the chlorine atom then replaced by a hydroxyl group[52]. In the other 2-aminotropone is prepared by the action of hydrazine (or hydroxylamine) on tropone, and the amino-group is then replaced by a hydroxyl group[82,83].

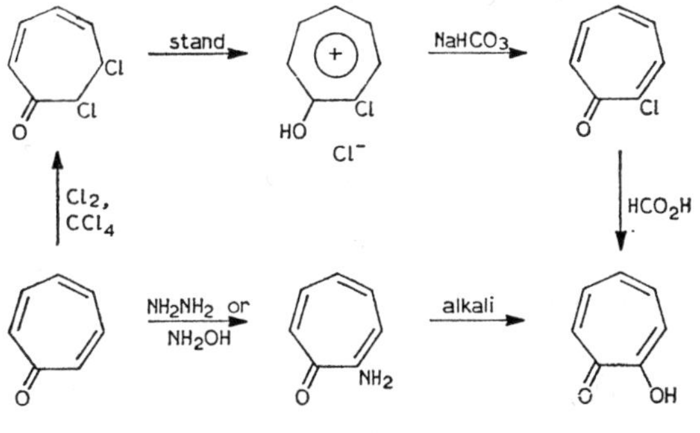

PREPARATION OF TROPOLONES 129

(d) From cyclopentadiene

Tetrafluoroethylene adds to cyclopentadiene to form a tetrafluorobicyclo-[3,2,0]heptene, which is isomerised on heating to a tetrafluorocycloheptadiene. The latter compound is converted into tropolone on treatment with potassium acetate and acetic acid, the overall yield from cyclopentadiene[84] being 20%.

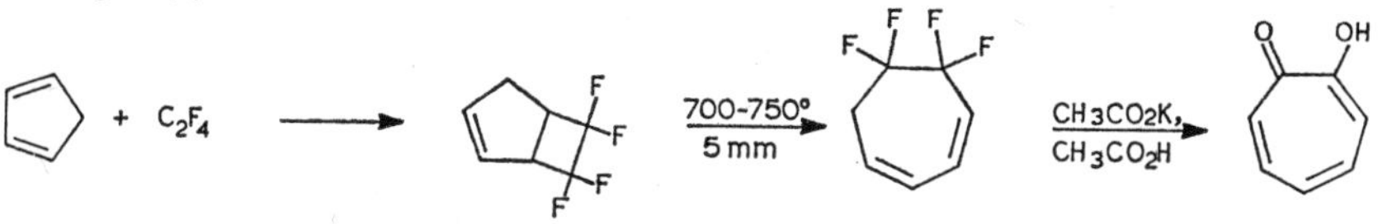

STRUCTURE OF TROPONE AND TROPOLONE

The dipole moment of tropone is 4.3 D[85-87]. This may be compared with the value of 3.04 D for cycloheptanone[88] and indicates a contribution from the dipolar form above, (see XII) although not a large one. The value for tropolone is 3.53 D[89]. The conjugation energies (based on heats of combustion) of tropone[60] and tropolone[90] are respectively ~29 and ~33–36 kcals/mole.

An X-ray crystallographic examination of the chelated copper salt of tropolone indicated that the molecule is a planar almost regular heptagon[91]. The carbon–carbon bond lengths average 1.40 Å; the values are as shown in (XV). Two of the most interesting facts to emerge are the non-equivalence of the two carbon–oxygen bonds and the length of the bond linking the two oxygen-bearing carbon atoms. On simple resonance theory this bond might be expected to be a "long" bond, similar to a normal single bond (see above, p. 119) although its hybrid length is in accord with molecular orbital calcu-

References p. 155

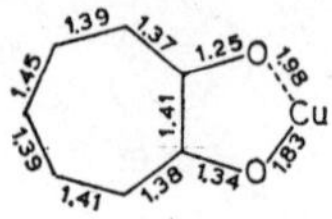

(XV)

lations. There is some evidence of alternating bond lengths in the ring, suggesting some measure of localisation of the double bonds. However the deviation from the mean cannot be said with certainty to exceed possible errors in the determination, and the alternation is so slight that very little significance can be attached to it. If it has any substance it is surprising that both –HC–CO– bonds are short.

X-ray crystallographic examinations have also been made of the sodium salt of tropolone (XVI)[92] and of the tropolonium ion (XVII)[93] formed from tropolone and concentrated hydrochloric acid.

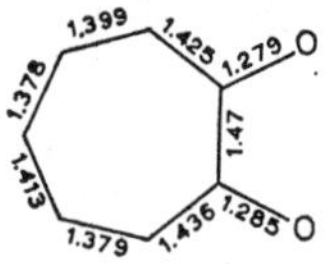

(XVI) 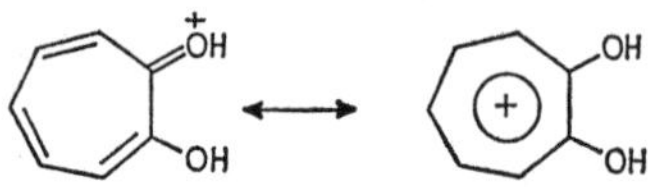 (XVII) 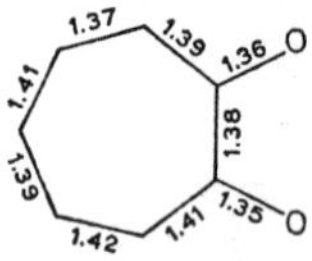 (XVII)

In both of these derivatives the two carbon–oxygen bonds are equivalent. The structure of the anion nearly corresponds to there being a plane of symmetry at right angles to the plane of the ring, and it is probable that greater reliance can be placed on mean values of pairs of corresponding bonds. It is certain that the OC–CO bond is longer than the remainder of the ring bonds and probable that the adjacent carbon–carbon bonds are also longer. In the tropolonium ion, however, the OC–CO bond is not longer than the other ring bonds. This difference is entirely reasonable for the tropolonium ion is a substituted tropylium ion. In view of the standard deviation in bond lengths in this determination it has been suggested that some of the differences in length in the ring bonds are possibly significant.

Electron diffraction measurements on tropolone[94] lead to similar conclusions, namely that the bond order of the ring bonds is intermediate between that of single and double bonds, the bond lengths being: C—C, 1.39; C—H, 1.10; C—O, 1.34; C=O, 1.26 Å.

Perhaps rather surprisingly similar measurements on tropone also show that the molecule consists of a plane regular heptagon, with C—C bond lengths = 1.405 Å intermediate between those of single and double bonds; the C=O and C—H bond lengths are 1.26 Å and 1.09 Å respectively[95].

In solutions of tropolone itself the two forms (A) and (B) (R=H) are thought to contribute equally to the equilibrium structure

On the other hand in unsymmetrically substituted tropolones this may no longer be the case. On the basis of their Raman spectra it has been suggested that discrete tautomers corresponding to (A) and (B) exist in the case of 3-isopropyl- and 4-isopropyl-tropolones[96], while n.m.r. spectroscopy suggests that in the case of 3-bromotropolone the equilibrium mixture of (A) and (B) (R=Br) contains more of form (B) than of form (A)[97].

SPECTRA

(a) Tropones

Tropones have two bands in their ultra-violet spectra, one in the region 200–280 mμ ($\varepsilon = 10^4$–10^5), and a simpler band in the region 290–400 mμ ($\varepsilon = 10^3$–10^4). Thus tropone itself has the following maxima in its spectra:

In water[19]		In isooctane[18]	
λ_{max}(mμ)	log ε	λ_{max}(mμ)	log ε
225	4.33	225	4.34
228	4.34		
231.5	4.34	297	3.74
239	4.10		
312.5	3.92	310	3.67

The band *ca.* 230 mμ is shifted to longer wavelengths by the presence of hydroxyl, carbonyl and aryl groups and by halogen atoms; it is almost unaffected by alkyl groups.

The infra-red spectrum shows C—H stretching bands of weak intensity in the region 3060–3010 cm^{-1} (3.27–3.32 μ), similar to benzenoid compounds. In addition there are bands at 1635 cm^{-1} (6.12 μ) and 1582 cm^{-1} (6.32 μ). Of these the band at 1635 cm^{-1} was initially ascribed to the carbonyl group and that at 1582 cm^{-1} to the carbon–carbon double bonds. Following some uncertainty about their assignation it now seems probable that the band at 1582 cm^{-1} is in fact associated with the carbonyl group. For further discussion of this point see refs. 87, 98–102. In any case the carbonyl peak appears at a markedly different position from that shown by saturated ketones (cf. cycloheptanone = 1702 cm^{-1})[103], presumably due to its highly polar character. It has been claimed that the intensities of the infra-red absorption spectrum of tropone suggest that the seven-membered ring has aromatic charcater[104].

The n.m.r. spectrum of tropone shows a single peak at approximately the same field strength as the peak in the spectrum of benzene[105].

(b) Tropolones

Like tropones, tropolones show an absorption band in their ultra-violet spectra in the region 200–280 mμ ($\varepsilon = 10^4$–10^5) which is almost unaltered by substitution of alkyl groups but is shifted to longer wavelengths by auxochromes or by additional conjugation, *i.e.* by substituted hydroxyl, carboxyl and carbonyl groups and halogen atoms. There are also two distinct series of bands in the region 290–400 mμ ($\varepsilon = 10^3$–10^4) having complex fine structure. Similar bands appear in the case of 2-aminotropone and they are thought to be connected with intramolecular hydrogen bonding. These bands vary markedly with solvent and with pH. The spectra of tropolone itself, in water and in cyclohexane, are as follows[60]:

In water		In cyclohexane	
λ_{max}(mμ)	log ε	λ_{max}(mμ)	log ε
228	4.36	222	4.37
		232	4.36
237	4.41	238	4.37
320	3.83	322	3.84
		340	3.64
351	3.76	356	3.73
		374	3.74

The bands at 300–340 mμ coalesce to one band in alkali and disappear when additional conjugation (*e.g.* a carboxyl group) is present. The bands at 340–375 mμ undergo a marked bathochromic shift in alkali but are not greatly changed by the introduction of additional conjugation.

The infra-red spectra of tropolone show the following peaks: (*inter alia*[77,107])

Solid: 3210, 1613, 1548, 1238 cm^{-1} (3.11, 6.20, 6.46, 8.08 μ)
Solution: 3140, 1620, 1555, 1252 cm^{-1} (3.18, 6.17, 6.43, 8.09 μ)

These peaks have been ascribed respectively to OH(stretching), C=O (stretching), C=C(stretching) and OH[108], but in view of the revision required in the assignment of peaks in the infra-red spectrum of tropone (see above) it is possible that this may require reconsideration. The C—H stretching peak appears as a shoulder overlapping the peak at 3210 or 3140 cm^{-1}. The OH and C=O stretching bands both appear at lower frequencies than normal hydroxyl and carbonyl bands; the C=C band corresponds to similar bands in benzenoid compounds. The band at 3140 cm^{-1} shifts to shorter wavelengths as the concentration of the solution increases and also in the solid; it has been suggested that under these circumstances intramolecular hydrogen bonding is weakened by intermolecular interaction[109].

The n.m.r. spectrum of tropolone (but not of substituted tropolones) consists of a single peak at slightly lower field than benzene, representing the ring hydrogens, and another peak resulting from the hydroxyl group, which appears at a position intermediate between that of the hydroxyl groups in phenol and acetic acid[105].

PROPERTIES OF TROPONES

Tropone is a colourless liquid completely miscible with water. It has a markedly higher boiling point (113°/15 mm.) than its isomer benzaldehyde (68°/15 mm.). It decolorises permanganate and is unstable to alkali. It can be catalytically reduced to cycloheptanone.

(a) Reactions of the carbonyl group

In strong acids a proton adds to the carbonyl group to give a hydroxytropylium ion:

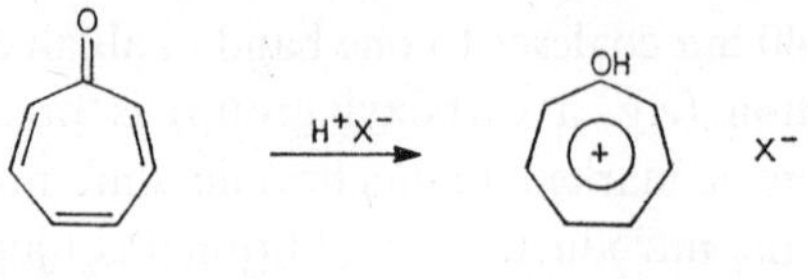

The solubility in acid is lowered by the presence of electron-withdrawing groups attached to the ring. Tropones form complexes with picric acid.

Tropone itself forms a semicarbazone and arylhydrazones but forcing conditions are required[38]. In contrast 2-phenyltropone hardly reacts with the normal ketonic reagents although a dinitrophenylhydrazone has been prepared[110]. With hydroxylamine tropones form not only oximes but also react by substitution to give 2-aminotropones[82] (see below).

(b) Addition reactions

Tropone reacts with bromine in carbon tetrachloride to give 1,2,5,6-tetrabromocyclohept-4-enone. This product slowly loses hydrogen bromide on standing, or rapidly on heating, forming 2,7-dibromotropone[38,111]. 2-Phenyltropone reacts similarly but addition takes place at the 4,5,6,7-positions[112,113]. If tropone is kept in a sealed tube with bromine for three weeks at room temperature hexabromocycloheptanone is formed; this is dehydrobrominated by sodium acetate to give 2,4,7-tribromotropone[111]. A mixture of the dibromo- and tribromo-tropones was obtained by allowing an acetic acid solution of bromine and tropone to stand and then heating the products[111].

A solution of chlorine in carbon tetrachloride adds to tropone to give 2,3-dichlorocycloheptadienone. On standing this loses hydrogen chloride forming 2-chlorohydroxytropylium chloride, which on treatment with aqueous bicarbonate produces 2-chlorotropone[52]. This has served as part of a route for converting tropone into tropolone.

Maleic anhydride reacts with tropone to give a normal adduct which has been formulated as an endo-*cis*-adduct[112]:

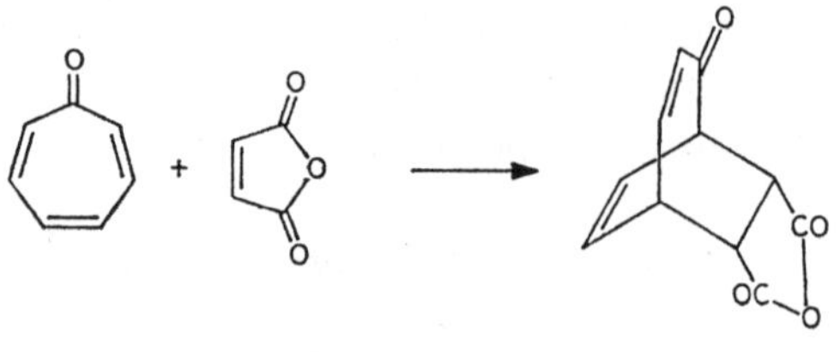

Tropone can be alkylated or arylated in the 2-position by reaction with a Grignard reagent or lithium derivative followed by dehydrogenation:

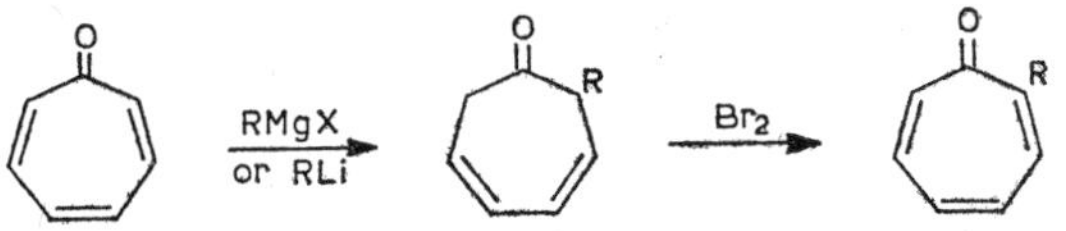

The initial reactions are probably examples of 1,8 conjugate addition followed by an enol→keto change[114,115]. A similar reaction occurs with lithium aluminium hydride.

It has been suggested[14] that the reactions of tropone with hydrazine or hydroxylamine (see below) which have usually been interpreted as substitution reactions could equally well be addition reactions proceeding by the following mechanism:

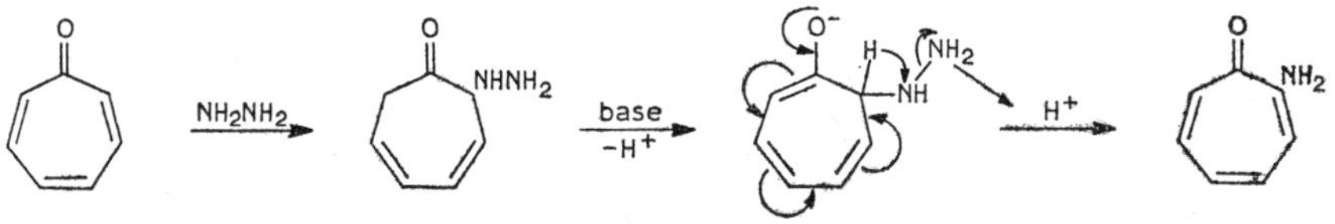

An additive mechanism for such reactions provides a more satisfactory explanation[14] for the course of the amination reaction of 4-bromo-2-phenyltropone[82] in which the bromine atom resists nucleophilic displacement.

(c) Substitution reactions

Tropone does not undergo direct electrophilic substitution, although its 2-amino- and 2-hydroxy- (tropolone) derivatives do.

It reacts with hydrazine or hydroxylamine to form 2-aminotropone[82]. This reaction was originally formulated as a nucleophilic substitution reaction:

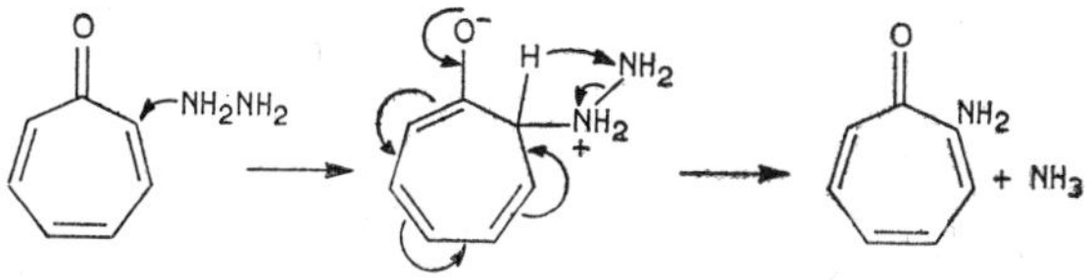

As discussed in the previous section it is possible to provide an alternative explanation in terms of an additive mechanism[14]. This reaction has served as an intermediate step in the conversion of tropone into tropolone. It has subsequently been shown that hydrazine hydrate in fact reacts to give a mixture consisting of 2-amino- and 2-hydrazinotropone[27]. The latter compound can be catalytically reduced to 2-aminotropone[27]. 2,7-Disubstituted tropones do not react with hydrazine.

(d) Replacement reactions of substituent groups

Substituted tropones react readily with nucleophiles if there is a substituent group attached to the ring which is easily removed as an anion, e.g. halogen atoms and alkoxyl groups are readily substituted by nucleophilic reagents. Halotropones and alkoxytropones can be regarded as "extended" acid chlorides and esters respectively, i.e. vinylogues of these compounds, see (XVIII), (XIX), (XX), and their reactions are very similar to those of simple acid halides and esters.

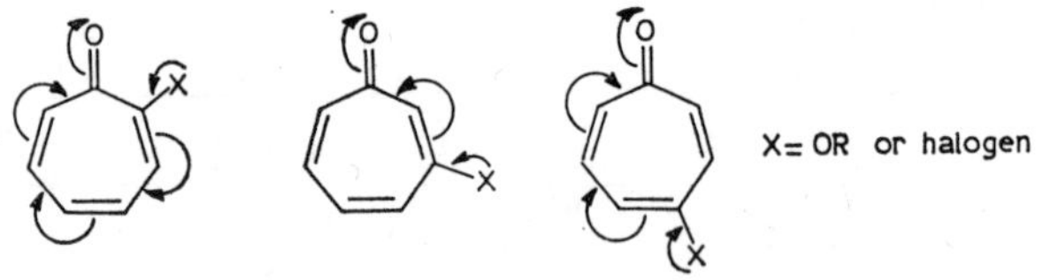

X = OR or halogen

(XVIII)   (XIX)   (XX)

Substitution takes place particularly readily in acid solution since under these conditions a conjugate acid is formed and the ring acquires a positive charge (see section (a) above) thereby facilitating nucleophilic attack. Thus halotropones undergo halogen exchange in acetic acid solution with, for example, hydrogen bromide or potassium iodide[116]. If water is present, the halogen atom is replaced by a hydroxyl group[117-119]. If halogen atoms are present at the 2 and 3 positions the one at the 3 position is replaced first[119], both being replaced under stronger conditions[120]. The hydrolysis of 2-chlorotropone by means of 80% formic acid serves as a convenient method for the preparation of tropolone[52]. Acid hydrolysis of methoxytropones similarly takes place readily[121,122].

Alkaline hydrolysis of 2-methoxytropone produces tropolone[77,118] but alkaline hydrolysis of substituted tropones is not infrequently accompanied by rearrangement reactions[123] (see also section (e) below). In polymethoxy-tropones the 2-methoxy-group is hydrolysed preferentially, more drastic treatment to remove the other groups normally causing concomitant rearrangement of the molecule[81,124,125]. With some more complicated tropones abnormal reactions may also occur involving migration of the carbonyl function[123]. This is believed to involve a series of reactions as follows:

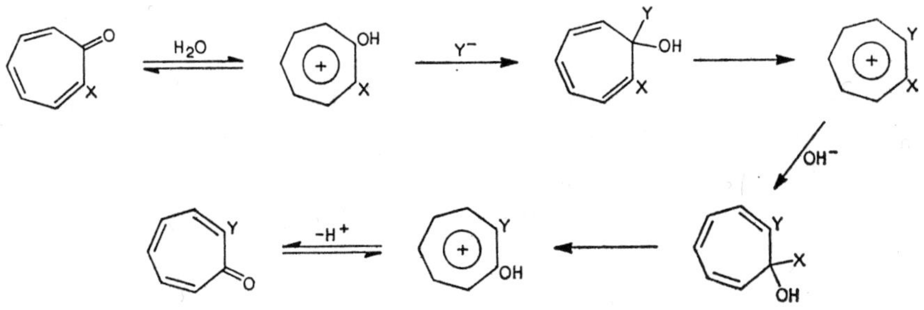

2-Bromotropone is converted into 2-cyanotropone by reaction with cuprous cyanide. 2-Halotropones react with Grignard reagents or lithium compounds with replacement of the halogen by a hydrocarbon function; for example with phenyllithium[126] or with benzyl magnesium bromide[127], 2-phenyl- and 2-benzyl-tropone, respectively, are obtained.

If methoxytropones are regarded as extended esters they would be expected to react with ammonia to produce the corresponding "amide". This does indeed happen, for example 2-methoxytropones react with ammonia to give 2-aminotropones[77,128–131]. Similarly 2-aminotropone as an extended amide is readily hydrolysed by alkali to tropolone and ammonia[82,112,127,132–135]. This reaction has also been utilised as a preparative route to tropolone[82,83]. 2-Aminotropones may also be converted into 2-anilino- and 2-hydrazino-tropones by heating with aniline or hydrazine hydrate respectively[82,129,130,136]. 2-Hydrazinotropone may also be prepared by reacting 2-methoxytropone with hydrazine; its hydrochloride on treatment with cupric sulphate is transformed into 2-chlorotropone[119].

2-Halo- and 2-methoxy-tropones react with various anionoid reagents such as acetoacetic ester or malonic ester to give heterocyclic analogues of azulene[126,133] *e.g.*:

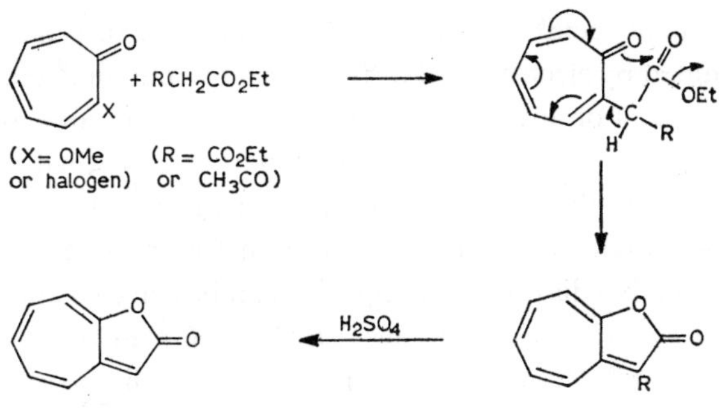

References p. 155

(e) Rearrangement reactions

It has been mentioned in the previous section that alkaline hydrolysis of 2-halotropones may lead to molecular rearrangement. This type of rearrangement is indeed very facile, the products being benzoic acid derivatives[20,125,137]. It has been proved that the carbon atom extruded from the ring is the 1-carbon atom[138].

2,7-Dihalotropones react with alkali at room temperature to give *o*-halobenzoic acids plus a small quantity of 3-halosalicylaldehydes as by-products[139] It was suggested that the aldehyde is formed by the following mechanism:

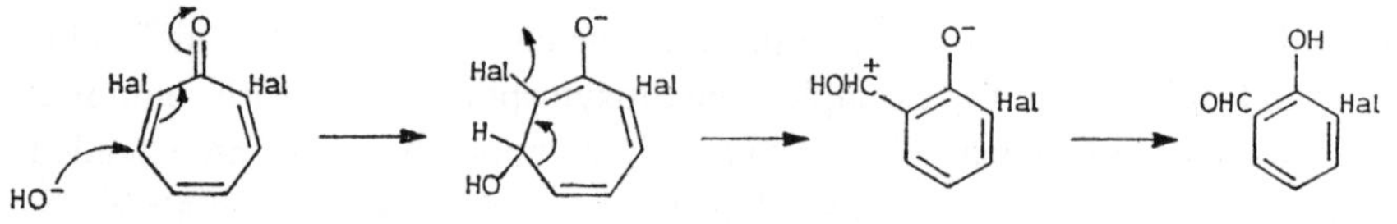

An alternative possibility differing only in detail is:

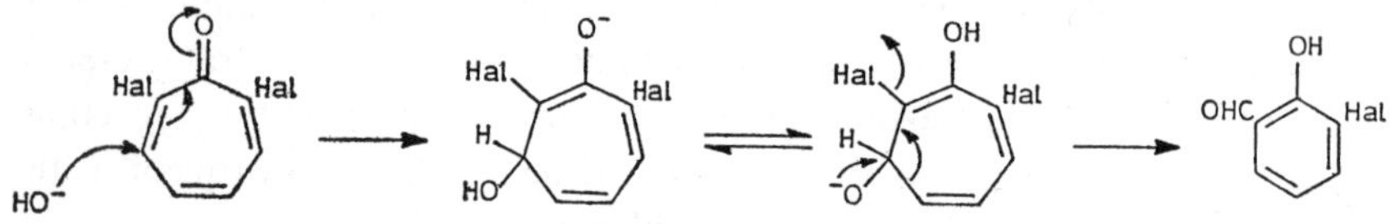

Alkaline hydrolysis of 2,3-dihalotropones leads to benzoic acid derivatives together with a small quantity of 3-hydroxy-2-halotropone[139].

Other tropone derivatives behave similarly, thus 2-phenyltropone and 4-carboxytropone both rearrange on treatment with alkali to give, respectively, 2-phenylbenzoic acid[82,132] and terephthalic acid[24]. In general rearrangement is facilitated by electron-withdrawing groups and hindered by electron-donating groups. For example, whereas 4-carboxytropone rearranges in aqueous alkali at room temperature tropolone[77,140] and 3-hydroxytropone[124] require fusion with alkali to bring about rearrangement.

2-Amino-4-methyltropone and 2-amino-6-methyltropone react vigorously with nitrous acid. No diazonium compound can be detected and the products are a mixture of *m*-toluic and either 2-hydroxy-6-methyl- or 2-hydroxy-4-methyl-benzaldehyde[130]. A possible mechanism for this reaction is:

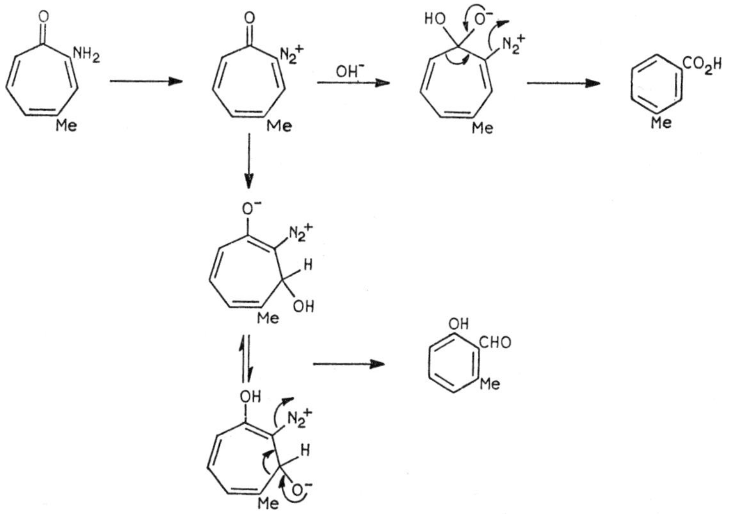

Examples of extrusion of the 3(6) and 4(5)carbon atoms from the ring in rearrangement reactions of tropones have also been observed[247].

PROPERTIES OF TROPOLONES

The tropolones are crystalline solids, tropolone itself melting at 50°. They dissolve more readily in hydroxylic solvents than in ether or hydrocarbons. They frequently sublime readily, a fact which has sometimes assisted in their purification.

(a) Oxidation

One method of preparation of tropolone involves permanganate oxidation of cycloheptatriene[59,77], but the yield is poor because the ring, although resistant to oxidation to some extent, nevertheless tends to undergo cleavage under these conditions. Drastic oxidation by permanganate or by chromium trioxide causes complete fission of the ring and has been used in the structural investigation of naturally occurring tropolones. Peracids in general do not affect tropolones; alkaline persulphate brings about hydroxylation at the 3- and 5-positions without rupture of the ring[39,130,141,142]. Alkaline hydrogen peroxide cleaves tropolone to produce *cis,cis*-muconic acid[60,143].

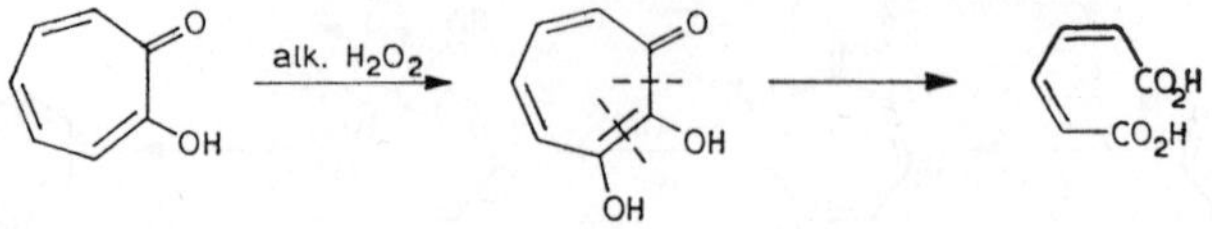

Other tropolones have been degraded similarly[6,7,143-146].

It is possible to oxidise side-chains without affecting the tropolone ring. For example double bonds in side chains have been oxidised to glycols by means of performic or perbenzoic acids[19,147-149] and the glycols cleaved by periodic acid without any oxidation of the ring[149]. Similarly 4-styryltropolone has been oxidised to 4-formyltropolone by means of periodate catalysed by osmium tetroxide[150]. Selenium dioxide has been used to oxidise side-chains attached to the tropolone ring, for example 4-methyltropolone methyl ethers are converted in this way to the corresponding 4-formyl derivatives[151,152]. Some tropolone derivatives, however, react with selenium dioxide to give diselenides[153], *viz.*

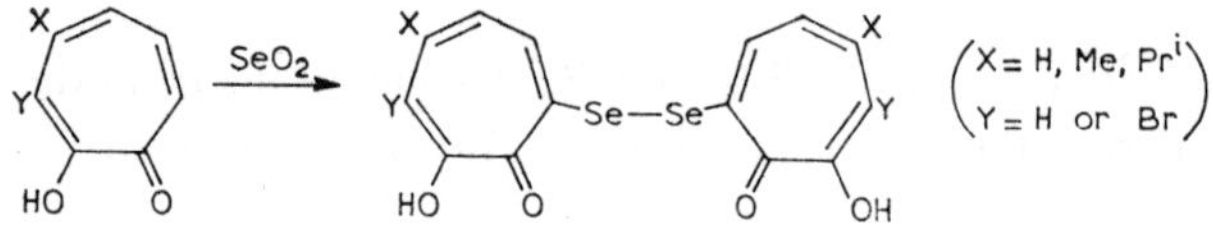

Hydroxymethyltropolones have been oxidised to formyltropolones by means of manganese dioxide[153]; further oxidation of the formyl group to a carboxyl group is achieved by using silver oxide[152].

(b) Reduction

Tropolones are fairly resistant to reducing agents. Catalytic hydrogenation does not proceed in the presence of palladium, but in the presence of Raney nickel or Adams' catalyst a variety of products is produced including cycloheptanone, cycloheptanol and cycloheptane-1,2-diol derivatives.

Tropolone is reduced by lithium aluminium hydride to cycloheptane-1,2-dione[155]. Under the same conditions its methyl ether rearranges to give benzaldehyde[155]; alkyl substituted tropolone methyl ethers rearrange similarly[130].

The methyl ether of 4-isopropyltropolone has been reduced by the Bouveault-Blanc method to the corresponding cycloheptanediol but the unmethylated tropolone is unattacked[154]. Clemmensen reduction of this tropolone produces what is believed to be a mixture of menthadienes[154].

Nitro and azo groups attached to a tropolone ring can be reduced to amines without the ring being affected[156,157].

(c) Reactions of the carbonyl group

In strong acids the carbonyl group is protonated to form the tropolonium ion. A crystalline hydrochloride has been isolated[77].

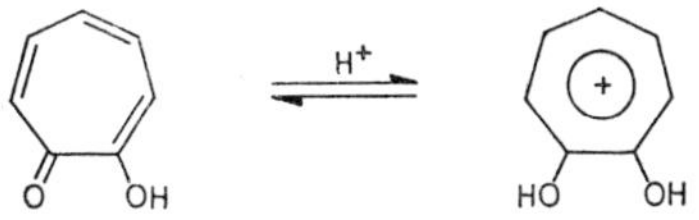

Tropolones form complexes such as picrates. The presence of electron-withdrawing substituents lower the solubilities of tropolones in acid.

Tropolone itself does not react with the normal ketonic reagents, but some tropolones having electron-withdrawing substituents such as nitro or nitroso groups do show some carbonyl reactivity[60]. Some even react in the form of α-diketones. Amines generally only form salts but the presence of electron-withdrawing groups sometimes leads to complex formation.

(d) Reactions of the hydroxyl group

The hydroxyl group is acidic, pK = 7[77,158]. This is intermediate between the values for phenol (pK = 10.0) and acetic acid (pK = 4.8). Simpler tropolones are soluble in sodium bicarbonate solution. As enols tropolones give coloured products with ferric chloride. They form chelated complexes with various metals. The cupric and ferric complexes are soluble in organic solvents and are easily purified crystalline compounds[77]; they have been used for the separation and purification of tropolones.

As already mentioned earlier in the chapter tropolone may be regarded as an "extended" acid, since interaction of its keto and enol groups through a conjugated chain is possible:

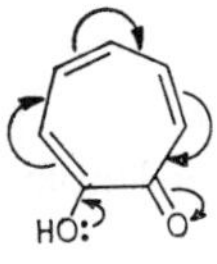

Conversion of a tropolone into its alkyl ether thus corresponds to an esterification reaction and it may be brought about by reaction of tropolones with the normal esterification reagents, e.g., with alcohols plus mineral acid[159], alkyl sulphates[160,161] or diazomethane, *inter alia*[77,129,152,160,162,163] or by converting the tropolone to its sodium or silver salt and reacting this with an alkyl halide[77,121]. *O*-Alkylation of unsymmetrically substituted tropolones leads to two isomeric alkoxytropones as follows:

In the O-alkylated products tautomerism is excluded and two products may thus be isolated.

Similarly these O-alkyltropones are hydrolysed more easily than ethers, the conditions required being analogous to those used to hydrolyse esters, *viz.* by refluxing them with aqueous sodium hydroxide or hydrochloric acid;[77] ammonia converts them to 2-aminotropones[77] (see also p. 137). When more than one alkoxide group is present, that in the 2-position is hydrolysed first.

Tropolone methyl ethers form hydrates and are more soluble in water and have higher boiling points than the parent tropolones[60,121].

On the basis of this analogy with carboxylic acids, the acylated products obtained by the action of acid chlorides and anhydrides on tropolones may be likened to acid anhydrides; like acid anhydrides they are readily hydrolysed[60,77,154,164,165].

Tropolones react with thionyl chloride to give 2-chlorotropones[78,116,122,137,166]. In the case of tropolone itself, *o*-chlorobenzaldehyde is formed as a by-product[137,167]. Two isomeric chlorotropones are obtained from unsymmetrically substituted tropolones[78,166]. Tropolones have been converted into 2-bromotropones and 2-fluorotropones by the action of phosphorus tribromide[137] and sulphur tetrafluoride[168] respectively.

The reactions of the hydroxyl group in tropolones may be contrasted with those of the same group in phenols, despite the fact that formally tropolone might be regarded as a tropone-phenol. The ease of alkylation and the difficulty of acylation of tropolones, the ease of hydrolysis of O-alkyltropolones, and the ready replacement of the hydroxyl group by a halogen atom are all in marked contrast to the behaviour of phenols and phenol ethers. These reactions are however strictly comparable to those of the carboxyl group.

(e) Addition reactions

Halogens normally react with tropolone by substitution but coloured addition complexes are sometimes obtained by reaction with bromine in carbon tetrachloride or chloroform. These complexes decompose in water or ethanol to tropolone and a mixture of bromotropolones[58,140,152,169].

Tropolone and some substituted tropolones form normal Diels–Alder adducts with maleic anhydride, when the compounds are heated together for several hours either in xylene or without solvent[170-172].

(f) Electrophilic substitution reactions

Unlike tropones, tropolones readily undergo electrophilic substitution reactions, substitution taking place at the 3, 5 and 7 positions. The reasons for this enhanced reactivity are discussed earlier in the chapter (p. 121). Reaction may be inhibited in strongly acidic media, however, owing to formation of the tropolonium cation (XVII) (p. 130) which, as a tropylium derivative is very resistant to electrophilic attack. A number of typical electrophilic substitution reactions will now be considered.

(i) Halogenation

The bromination of many tropolones, including tropolone itself, has been recorded[58,77,140,173,248]. Substitution takes place at the 3, 5 and 7 positions, predominantly at the latter two positions. The bromination of the cupric complex of tropolone produces 5-bromotropolone, however[140]. By the action of excess bromine in acetic acid 3,5,7-tribromocyclohepta-3,5-diene-1,2-dione is produced[145]. Mention has been made in the previous section of the scarlet addition complex produced in chloroform or carbon tetrachloride. Bromination of 4-isopropenyltropolone results in addition to the side-chain if the reaction if effected in carbon tetrachloride and both side-chain addition and substitution in the ring if it is carried out in acetic acid and in the presence of sodium acetate[106].

Chlorine also reacts readily with tropolone by substitution, entering the 5-position more readily than does bromine[169,174,175] *(inter alia)*. Another halogen atom may be replaced in the course of the reaction[175,176].

Direct iodination may be carried out by reacting the alkali metal salts of tropolones with iodine[177,178,248]. The product is usually the 3-iodotropolone but because of the alkalinity of the reaction mixture, concurrent rearrangement reactions tend to take place, giving rise to benzoic acid derivatives as by-products[177]. Reaction of tropolone with iodine monochloride produces a mixture of 5-chloro- and 3-iodo-tropolone[177].

(ii) Nitration

Tropolone is readily nitrated by nitric acid in glacial acetic acid, giving mainly the 5-nitro compound plus a little of the 3-nitro compound[74,156,163],

[179,248]. Other nitrating agents which have been used include dilute nitric acid[77], cupric nitrate in acetic anhydride[180], and nitrogen tetroxide[156]. Nitration does not occur if sulphuric acid is present, owing to formation of the tropolonium cation. The 3- and 5-nitrotropolones may be reduced to the corresponding amino-compounds by means of sodium dithionite[74,156]. On heating with nitrous acid 5-aminotropolone gives 5-hydroxytropolone but in the case of 3-aminotropolone the reaction is complex and among the products is some salicylic acid[74,156].

(iii) Sulphonation

Because of tropolonium ion formation, tropolone remains unaffected by fuming sulphuric acid at 100–150°, (ref. 181). It may, however, be sulphonated by heating with sulphamic acid[181]. In the case of tropolone itself the main product is the ammonium salt of tropolone-5-sulphonic acid; a small amount of the 3,5-disulphonic acid is also produced. Other tropolones may give the 3- or 5-isomers.

(iv) Azo-coupling

Coupling with diazonium salts takes place at the 5-position[58,77,156,182,248]. 5-Alkyltropolones do not undergo azo-coupling.

(v) Alkylation and acylation

Alkylation and acylation of tropolones by such means as the Friedel–Crafts, Gattermann–Hoesch and Fries reactions do not take place, since tropolone forms a complex in the presence of metal salts and a tropolonium cation in the presence of strong acids[155,183].

(vi) Nitrosation

Tropolone may be converted into its 5-nitroso derivative in good yield using barium nitrite/sulphuric acid[77], nitrous acid or sodium nitrite in acetic acid[165,184] or nitrosyl chloride/acetic anhydride[184]. 5-Substituted tropolones usually resist nitrosation.

(vii) Reimer–Tiemann reaction

Tropolone reacts with chloroform or carbon tetrachloride in the presence of alkali to give in low yield, respectively, 5-formyl or 5-carboxy-tropolone[155].

(viii) Hydroxymethylation and aminomethylation

When tropolone is treated with one equivalent of formaldehyde and alkali, a mixture of 3- and 5-hydroxymethyltropolones together with polysubstituted products are obtained. With three molar equivalents of formaldehyde 3,5,7-trishydroxymethyltropolone is obtained in good yield[37,185].

With morpholine and formaldehyde tropolone reacts in a Mannich-type reaction to give 3,5,7-trimorpholinomethyltropolone[186]. Other substituted tropolones react similarly[187].

(g) Replacement reactions of substituent groups

Tropolones are less prone to take part in reactions of this sort than are tropones, since as enols, they form anions in the presence of bases. None the less, tropolones having substituent groups which can readily be transformed into anions will undergo nucleophilic displacement reactions, the reactivity being comparable to that of similar groups in benzenoid compounds which are activated by nitro-groups. Frequently, however, nucleophilic attack is accompanied by rearrangement reactions leading to benzenoid products.

Halotropolones undergo both halogen exchange and hydrolysis in acid conditions[122]; nitro-groups may be replaced by halogen under similar conditions[188]. 3-Bromotropolone reacts with cuprous cyanide to give 3-cyanotropolone[137,189] but if potassium cyanide is used 4-cyanotropolone is formed[190]

Tropolone or its methyl ether or copper complex react with Grignard reagents or organo-lithium compounds to produce 2-alkyl- or 2-aryl-tropones[112,116,117,191], *e.g.*

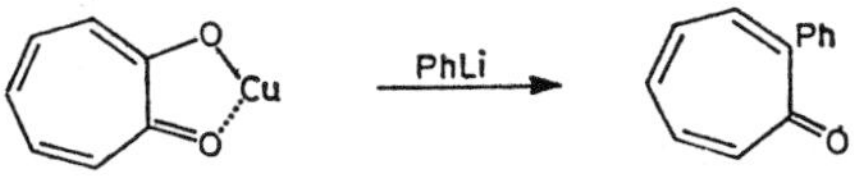

Reaction may not be straightforward, however, for the 4-methyltropolone methyl ethers give rearranged products with phenyl magnesium bromide as follows[191]:

Such reactions may involve 1,8-addition:

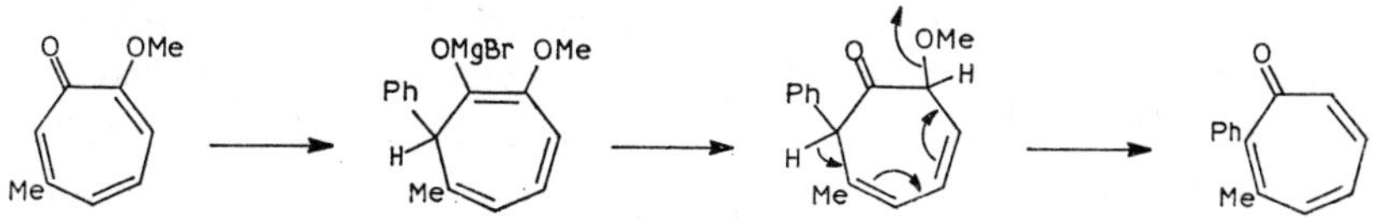

References p. 155

(h) Rearrangement reactions

Examples have already been mentioned occasionally in preceding sections of rearrangement reactions accompanying other reactions. Such rearrangements are particularly prone to take place under alkaline conditions. Thus iodination of tropolone in the presence of sodium bicarbonate results in the formation of 2,4-diiodo- and 2,4,6-triiodo-benzoic acids[177]. Similarly treatment of tropolone by hypobromite[77] or hypoiodite[77,140] produces 2,4,6-trihalophenols. In the reaction with hypoiodite the formation of iodoform was also noted[140].

Tropolones rearrange with varying degrees of ease on treatment with alkali to give benzoic acid derivatives. The presence of electron withdrawing groups greatly facilitates rearrangement; in their absence rearrangement does not usually take place under gentle conditions. For example 4-methyl-5-nitrotropolone is converted into 4-nitro-*m*-toluic acid on warming with 8% sodium hydroxide[182], while 5,7-dinitrotropolones give benzoic acid derivatives merely on heating with water, alcohol or acetic acid, without even addition of base[192,193]. In contrast tropolone itself requires fusion with alkali to above 200° (ref. 77, 140), and 4-methyltropolone requires heating to 300° (ref. 152) to bring about a rearrangement reaction. The ease of rearrangement of nitrotropolones is indeed such that rearrangement may take place during the nitration reaction. Thus nitration of 4-methyltropolone leads directly to some 4,6-dinitro-*m*-toluic acid[140].

Tropolone methyl ethers rearrange more easily than the parent tropolones, for example tropolone methyl ether itself is converted into methyl benzoate by boiling methanolic sodium hydroxide[77].

The mechanism of these rearrangements is probably analogous to those of halotropones and resembles a benzil-benzilic acid type rearrangement.

Diazotisation of 3-aminotropolones also leads to rearrangement reactions, among the products being salicylic acid derivatives[74,156,159,194,195]. 5-Aminotropolones behave normally on diazotisation and can be converted into 5-halo- or 5-cyanotropolones by means of a Sandmeyer reaction[77,156,196]. The action of nitrous acid on 4-aminotropolones gives the corresponding hydroxytropolones and no intermediate diazocompound could be detected, even at −20°, (ref. 141).

Tropolone and its methyl ether also undergo rearrangement on irradiation with ultra-violet light[197-199].

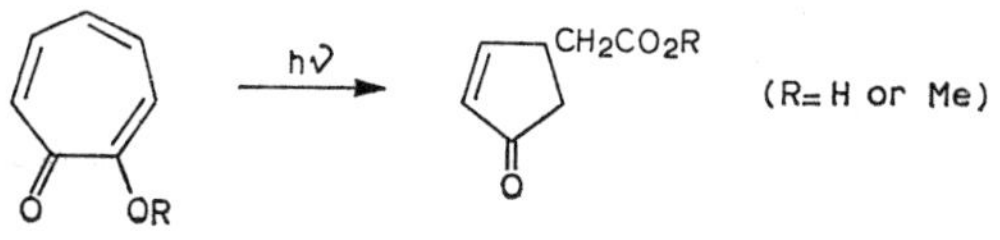

$$\underset{\text{OR}}{\text{tropolone}} \xrightarrow{h\nu} \underset{\text{O}}{\text{cyclopentenone-CH}_2\text{CO}_2\text{R}} \quad (R = H \text{ or } Me)$$

Allyl ethers of tropolones undergo Claisen rearrangements[200], the allyl group usually migrating to the 3- or 7-positions; occasionally migration takes place to the 5-position, e.g.

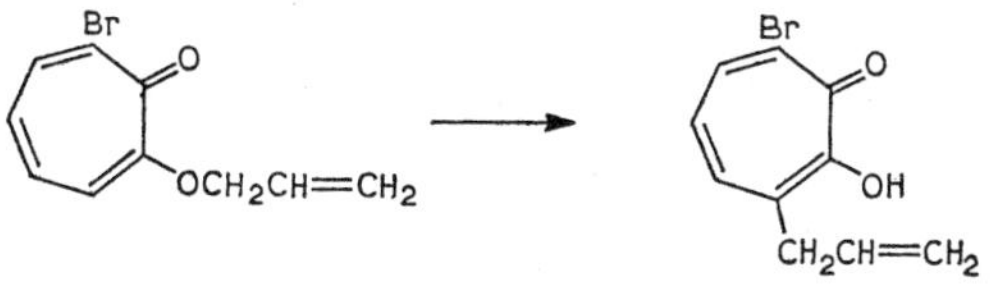

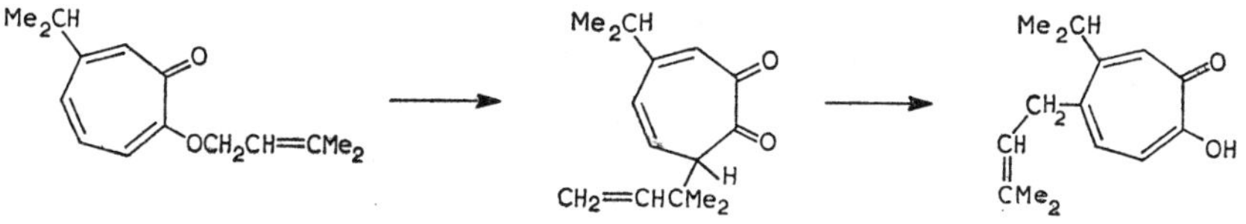

3- AND 4-HYDROXYTROPONES

3- and 4-Hydroxytropones, which are positional isomers of tropolone, have been prepared by methods similar to those used for the preparation of tropolones. Thus ring expansion of resorcinol dimethyl ether or quinol dimethyl ether with diazoacetic ester followed by decarboxylation and bromination/dehydrobromination leads to 3- and 4-hydroxytropone respectively[124,201-204].

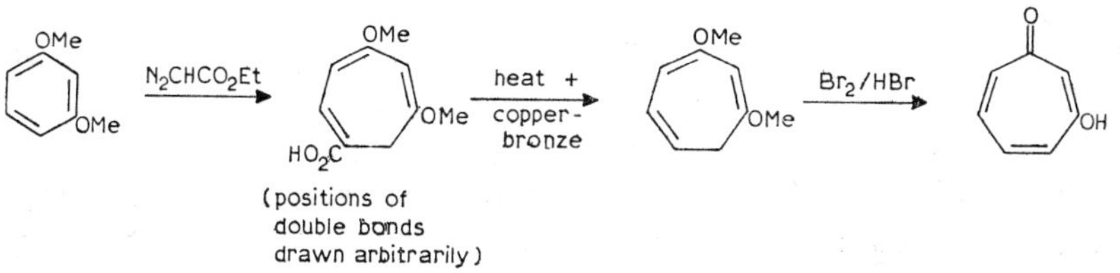

(positions of double bonds drawn arbitrarily)

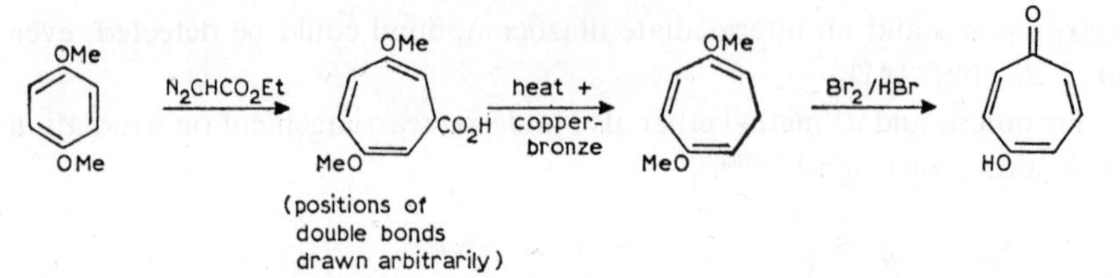

148   TROPONES, TROPOLONES, ETC.

[Reaction scheme: 1,2-dimethoxybenzene → (with N₂CHCO₂Et) → cycloheptatriene with OMe, CO₂H, MeO (positions of double bonds drawn arbitrarily) → (heat + copper-bronze) → methoxytropone → (Br₂/HBr) → 3-hydroxytropone]

An alternative route for the preparation of 3-hydroxytropone starts from catechol dimethyl ether[30]:

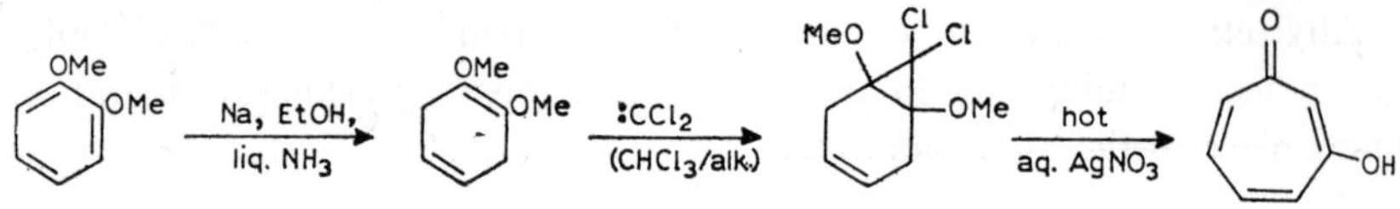

[Reaction scheme: catechol dimethyl ether → (Na, EtOH, liq. NH₃) → dihydro compound → (:CCl₂, CHCl₃/alk.) → bicyclic dichloride with OMe → (hot aq. AgNO₃) → 3-hydroxytropone]

3- and 4-Hydroxytropones may also be conveniently prepared from dimethoxybenzoic acids[32]:

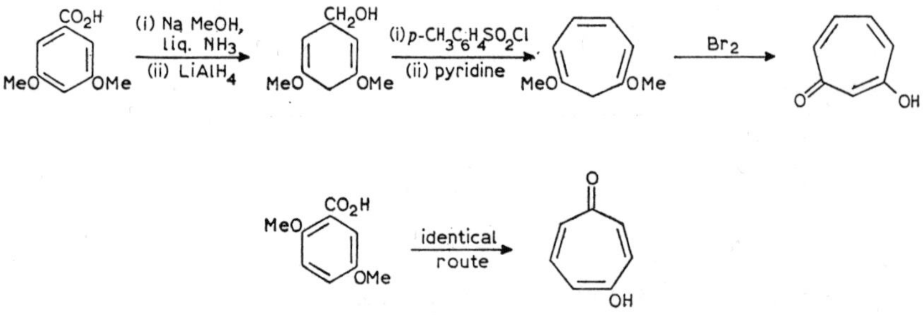

[Reaction scheme: 3,4-dimethoxybenzoic acid → (i) Na MeOH, liq. NH₃ (ii) LiAlH₄ → hydroxymethyl compound → (i) p-CH₃C₆H₄SO₂Cl (ii) pyridine → dimethoxycycloheptatriene → Br₂ → 3-hydroxytropone]

[Reaction scheme: 3,5-dimethoxybenzoic acid → identical route → 4-hydroxytropone]

4-Hydroxytropone has also been obtained by acid hydrolysis of 4-bromotropone, a by-product in the bromination of cycloheptanone[118], and by treating teloidinone methobromide or its dimethyl ether with alkali[205,206]:

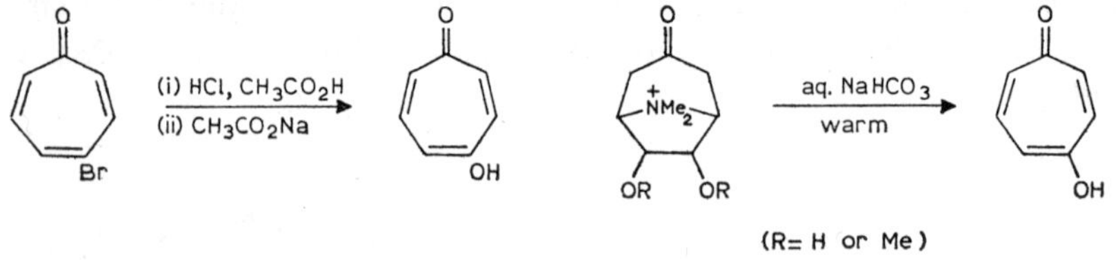

[Reaction scheme: 4-bromotropone → (i) HCl, CH₃CO₂H (ii) CH₃CO₂Na → 4-hydroxytropone]

[Reaction scheme: teloidinone methobromide (NMe₂⁺, OR, OR) → (aq. NaHCO₃, warm) → 4-hydroxytropone (R = H or Me)]

The bicyclic ketone (XXI) undergoes rearrangement to 4-hydroxytropone, the suggested mechanism being as follows[207]:

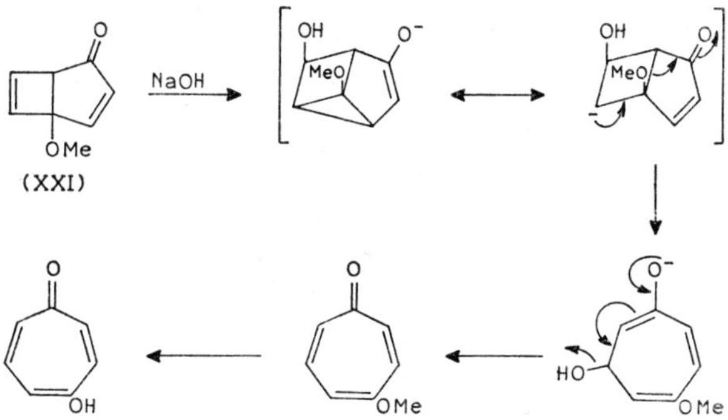

In chemical properties both 3- and 4-hydroxytropones resemble tropolone closely. Thus neither reacts with ketonic reagents[118,120,204], both react with diazomethane to form methyl esters, which can be readily hydrolysed by acid or alkali[25,118,203] and both react with thionyl chloride to give chlorotropones[142,202]. Both undergo electrophilic substitution. This takes place in the 2-position of 3-hydroxytropone; it reacts thus to give chloro, bromo, iodo, nitro and phenylazo derivatives[120]. The 2-nitro group can be catalytically reduced and the resultant amine diazotised and made to undergo Sandmeyer reactions[120]. The halogen atoms in 2-halo-3-hydroxytropones appear to be rather unreactive[120]. 4-Hydroxytropone gives a dibromo-derivative at room temperature[204] and the 2,5,7-tribromo-derivative under forcing conditions[118]; it also couples with diazonium compounds[118].

Both hydroxycompounds are amphoteric and form salts with bases and strong acids. They are somewhat stronger acids than tropolone with pK values of 5.4 (3-hydroxy)[124] and 5.64 (4-hydroxy)[118,158]. Unlike tropolone they do not give colours with ferric chloride.

The intramolecular hydrogen-bonding between carbonyl and hydroxyl groups which is typical of tropolone is not possible in its isomers. This is reflected in differences in physical properties, comparable to the differences between *o*-nitrophenol and its *m*- and *p*-isomers. Thus whereas tropolone sublimes readily, the 3- and 4-hydroxytropones do not. They have much higher melting points (179–180° (3-OH); 212° (4-OH)) than tropolone (49°) and unlike tropolone are insoluble in non-polar solvents. For a detailed comparison see ref. 204.

THIOTROPOLONES, AMINOTHIOTROPONES AND AMINOIMINO-CYCLOHEPTATRIENES

Thiotropolone (XXII) has been prepared by the reaction between a halotropone and sodium hydrogen sulphide[208]. Physical measurements such as dissociation constant and ultra-violet, infra-red and n.m.r. spectra suggest that the compound exists predominantly in form (A)[209].

(R= H, alkyl, aryl)

(XXII A)   (XXII B) 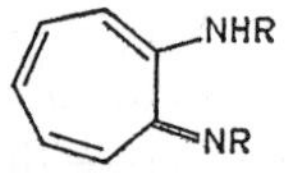   (XXIII) 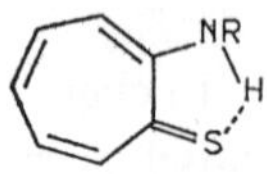   (XXIV)

Aminoiminocycloheptatrienes (XXIII) have been prepared by the action of ammonia or amines on 5,5,6,6-tetrafluorocyclohepta-1,3-diene[210]. The N,N-dialkyl and diaryl compounds are stable and highly coloured, and undergo electrophilic substitution at the 5-position. They react with hydrogen sulphide to give aminothiotropones (XXIV); ultra-violet and n.m.r. spectra show that the latter exist predominantly in the tautomeric form shown. Compound (XXIII; R=H) also undergoes bromination and azo-coupling at the 5-position. It forms stable complexes with Cu^{++}, Ni^{++} and Co^{++} and its ultra-violet and n.m.r. spectra show that the two nitrogen atoms are equivalent.

A study of the dipole moments and infra-red and n.m.r. spectra of the aminoiminocycloheptatrienes has led to the suggestion[211,212] that the aromaticity of these compounds is fundamentally different from that of tropolones and that they are better represented by formula (XXV).

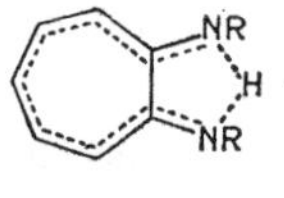

(XXV)    (XXVI)   (XXVII)

In this formulation ten electrons are shared among the atoms linked by a dotted line, producing a complete peripheral conjugation (cf. azulene, Chapter VIII). On the other hand the structure of these aminoiminocycloheptatrienes can equally well be explained by assuming a rapid tautomeric change between (XXVI) and (XXVII) which would result in a symmetrical time average structure akin to (XXV); hydrogen/deuterium exchange could not however be detected even at $-80°$.

BENZOTROPONES AND BENZOTROPOLONES

Preparation

4,5-Benzotropones were first prepared almost sixty years ago by a condensation reaction between phthalaldehyde and acetonedicarboxylic ester[21-23]. The failure of 4,5-benzotropone to form an oxime or phenylhydrazone was observed at the time.

It is interesting to note that this preparative method involved the use of a condensation reaction leading directly to a tropone ring system, for no simple monocyclic tropone or tropolone has yet been prepared by a condensation reaction of this sort.

The same method has also been used more recently to prepare benzotropolones by reaction of phthalaldehyde with hydroxyacetone and its derivatives[145,161,213-215,218]. The reaction of phthalaldehyde with ketones has also been used to prepare benzotropones[21,216-218].

The first synthesis of a benzotropolone (or indeed of any tropolone) was carried out in 1949 by dehydrogenation of 1,2-benzocycloheptene-3,4-dione to 3,4-benzotropolone[56,144]. This method has been used to prepare a number of 3,4-benzotropolones[176,219-222]. The benzocycloheptenedione is obtained from benzocycloheptene-3-one; the latter compound also serves as a convenient starting point for the preparation of 2,3-benzotropone[14]. Among other methods which have been used to prepare benzotropones and benzotropolones, mention might be made of the ring expansion of methoxynaphthalenes with dichlorocarbene, which results in the formation of mixtures of benzotropolones and chlorobenzotropones[223].

Properties of benzotropones

Hydrogenation data on 4,5-benzotropones suggest that they have greater stability than open chain ketones such as dibenzylideneacetone[224]. The double bonds appear to be greatly deactivated towards reduction but once one double bond has been reduced the second one is reduced normally[224]. 4,5-Benzotropone also has a larger dipole moment (4.25 D) than dibenzylideneacetone (3.3 D) and an abnormal carbonyl absorption frequency in its Raman spectrum. It does not form an oxime or phenylhydrazone[22].

2,3-Benzotropone forms a 2,4-dinitrophenylhydrazone, reacts with bromine by addition, and in general behaves as a conjugated dienone[14]. None the less the carbonyl absorption in its infra-red spectrum occurs at an unusually low frequency[14].

2-Chloro-4,5-benzotropone and 7-chloro-2,3-benzotropone are both resistant to acid hydrolysis and cannot be converted into the corresponding benzotropolones; both form oximes[223]. The former cannot be nitrated by nitric acid/acetic acid nor does it undergo skeletal rearrangement with alkali to naphthoic acid derivatives[223].

2,3,6,7-Dibenzotropone shows no aromatic character in the seven-membered ring[225-227]; both it and its 2,3,4,5-dibenzo isomer form 2,4-dinitrophenylhydrazones and oximes[227-232].

Tribenzotropone has also been prepared[233]. Measurements of its ultra-violet spectrum in various solvents and at different pH values indicate little aromatic stabilisation in the seven-membered ring. The position of the carbonyl absorption in the infra-red spectrum differs little from that of the dibenzotropones or of normal diaryl ketones.

A most interesting series of 2,7-polymethylene-4,5-benzotropones (XXVIII) has been prepared[98] and the variation in their physical properties with length of the polymethylene chain has been closely investigated[98,234,235].

(XXVIII)

If n is small the seven-membered ring must be non-planar and electronic delocalisation in it is inhibited. This is not the case where n is large for here the seven-membered ring can still be planar. Measurements of the infra-red spectra[98], dipole moments[235] and enthalpy of formation[234] indicate that planarity of the tropone ring is lost if $n<7$. Convincing evidence of delocalisation of the electrons in the planar form of the tropone ring comes from measurements of the heats of combustion, for (XXVIII, $n = 12$), wherein this ring is planar, has a resonance energy of 82.7 kcal/mole, whereas (XXVIII, $n = 5$), wherein the ring is buckled, has a resonance energy of only 48.7 kcal/mole[234]. The corresponding value for benzotropone itself is 84.6 kcal/mole.

Properties of benzotropolones

Neither 3,4- nor 4,5-benzotropolones react with ketonic reagents but their methyl ethers both form dinitrophenylhydrazones[144,145,161,213,214]. Unlike monocyclic tropolones they do not react with thionyl chloride or phosphorus pentachloride to give chlorotropones[223]. On the other hand, like their monocyclic analogues, 3,4-benzotropolones react with chlorine or bromine by substitution (giving 7-halo derivatives)[144,176] while 4,5-benzotropolones rearrange to naphthoic acid derivatives on treatment with alkali[236].

3,4,5,6-Dibenzotropolone behaves both as a 1,2-ketoenol and as an α-diketone[237,238]. Thus its infra-red spectrum shows the presence of a hydroxyl group, it forms an acetyl derivative, gives an enol colour with ferric chloride and forms a chelate complex with cupric ions[238]. On the other hand it forms a monoxime and dioxime and a monodinitrophenylhydrazone[238], and reacts with o-phenylenediamine to give a quinoxaline derivative[237,238]. It rearranges readily to phenanthrene-9-carboxylic acid[237]. 3,4,6,7-Dibenzotropolone appears to react as a diketone[239]. Methyl ethers of dibenzotropolones are unreactive and resemble phenol ethers rather than tropolone ethers.

HEPTAFULVENES

Heptafulvene (XXIX) or methylenecycloheptatriene is the methylene homologue of tropone and is a vinylogue of fulvene (XXX) (see also Chapter IV). It is an extremely labile red oil, which polymerises even at −80°. It was prepared as shown, its vapour being trapped by a cold solvent[240]. Its structure was shown by its reduction to methylcycloheptane. It reacts with methyl acetylenedicarboxylate to give methyl azulene-1,2-dicarboxylate, and with concentrated hydrobromic acid to give the salt (XXXI), presumably *via* the paths shown[240].

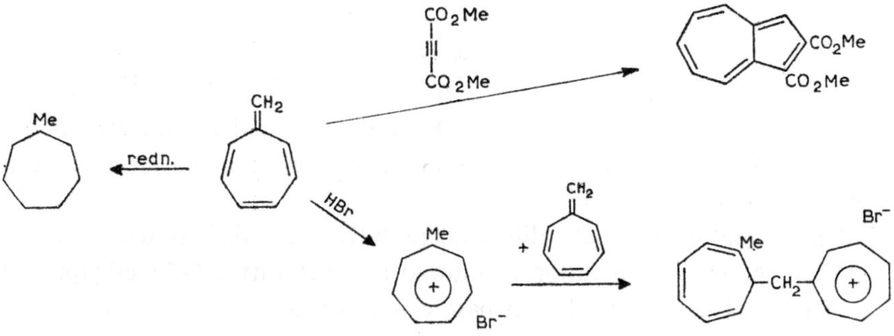

(XXXI)

References p. 155

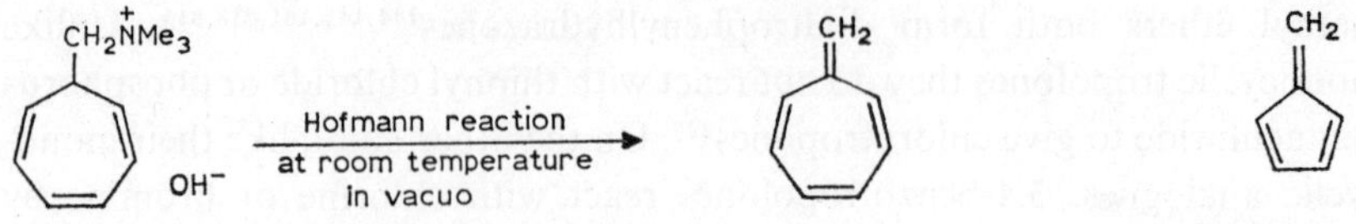

TROPONES, TROPOLONES, ETC.

(XXIX) (XXX)

Heptafulvene has also been formed in solution (but not isolated) by the action of triethylamine on a methyltropylium salt at −70° (ref. 241):

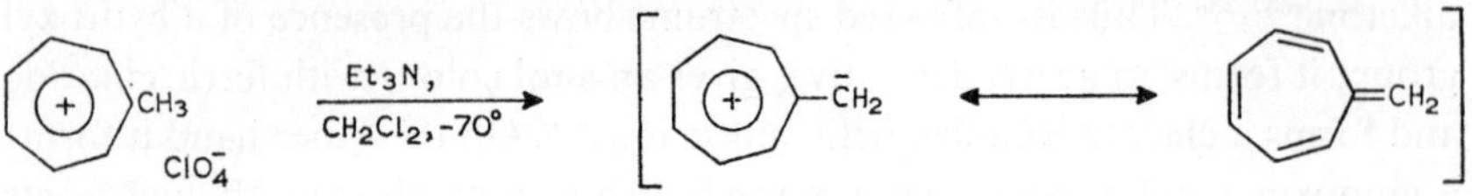

This reaction proceeds by elimination of a proton from the exocyclic methyl group; the dipolar ion which results represents one canonical form of the heptafulvene.

Heptafulvene is also known to be formed in minute amount by thermal rearrangement of 6-methylenebicyclo[2,2,1]hept-2-ene or of methylenecycloheptadienes[242].

Heptafulvenes are stabilised by the presence of electron-withdrawing groups attached to the exocyclic methylene group. These groups will tend to stabilise the dipolar form by assisting delocalisation of the negative charge. Examples of such heptafulvene derivatives have been isolated as follows[243,244]

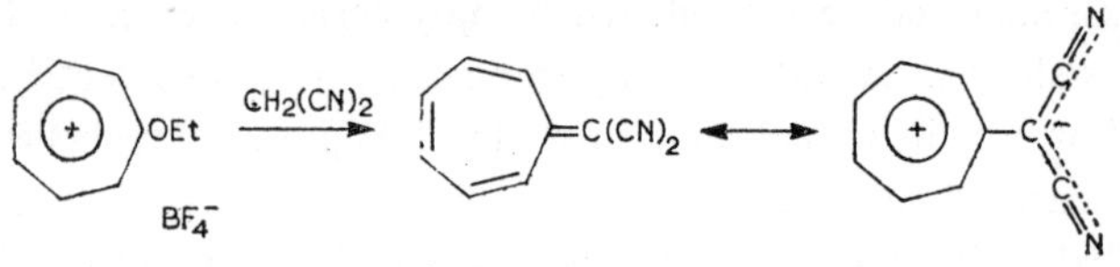

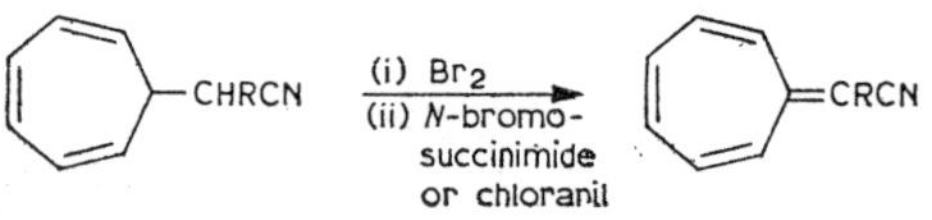

(R= CN or CO$_2$Et)

These compounds are stable up to 300° and react neither with dienophiles nor with electrophilic reagents. With bases the dicyanoheptafulvene is converted into dicyanostyrene.

Benzoheptafulvenes and a dibenzoheptafulvene are also known[245,250].

When the tropylium cation is allowed to react with 2,6-dimethylphenol and the product is treated with base a quinocycloheptatriene is formed[249] (for details see Chapter V, p. 111):

This compound is another example of a heptafulvene derivative whose stability is increased by the presence of an exocyclic electron withdrawing group.

The *sesquifulvalenes* or cyclopentadienylidenecycloheptatrienes are special examples of heptafulvene derivatives, wherein both positive and negative charges may be delocalised over carbocyclic rings:

The parent compound has not yet been isolated but some derivatives of sesquifulvalene are known. They are discussed in Chapter VIII, p. 203.

REFERENCES

1 M. J. S. DEWAR, *Nature*, 155 (1945) 50.
2 For a more complete account see: T. NOZOE, in L. ZECHMEISTER (Herausgeber) *Fortschritte der Chemie organischer Naturstoffe, Bd XIII*, Springer, Vienna, 1956, p. 232.
3 J. GRIPENBERG, *Acta Chem. Scand.*, 2 (1948) 639.
4 A. B. ANDERSON AND J. GRIPENBERG, *Acta Chem. Scand.*, 2 (1948) 644.
5 H. ERDTMAN AND J. GRIPENBERG, *Acta Chem. Scand.*, 2 (1948) 625.
6 R. E. CORBETT, C. H. HASSALL, A. W. JOHNSON AND A. R. TODD, *Chem. and Ind.*, (1949) 626.
7 R. E. CORBETT, A. W. JOHNSON AND A. R. TODD, *J. Chem. Soc.*, (1950) 6.
8 R. E. CORBETT, A. W. JOHNSON AND A. R. TODD, *J. Chem. Soc.*, (1950) 147.
9 M. J. S. DEWAR, *Nature*, 155 (1945) 141, 479.
10 For details see: J. W. COOK AND J. D. LOUDON, in R. H. F. MANSKE AND H. L. HOLMES (Editors) *The Alkaloids, Vol. II*, Academic Press, New York, 1952, p. 261.
11 T. W. EVANS AND W. M. DEHN, *J. Am. Chem. Soc.*, 52 (1930) 3647.
12 J. A. BARLTROP AND J. S. NICHOLSON, *J. Chem. Soc.*, (1948) 116.
13 R. D. HAWORTH, B. P. MOORE AND P. L. PAUSON, *J. Chem. Soc.*, (1948) 1045.
14 G. L. BUCHANAN AND D. R. LOCKHART, *J. Chem. Soc.*, (1959) 3586.
15 O. TOUSTER, *Org. Reactions*, VII, (1953) 327.
16 N. KORNBLUM, *Org. Reactions*, XII, (1962) 101.
17 S. M. PARMENTER, *Org. Reactions*, X, (1959) 1.
18 W. VON E. DOERING AND F. L. DETERT, *J. Am. Chem. Soc.*, 73 (1951) 876.
19 H. J. DAUBEN AND H. J. RINGOLD, *J. Am. Chem. Soc.*, 73 (1951) 876.
20 T. NOZOE, Y. KITAHARA, T. ANDO AND S. MASAMUNE, *Proc. Japan Acad.*, 27 (1951) 415.
21 J. THIELE AND J. SCHNEIDER, *Ann.*, 369 (1909) 287.
22 J. THIELE AND E. WEITZ, *Ann.*, 377 (1910) 1.

23 J. THIELE AND K. G. FALK, *Ann.*, 347 (1906) 112.
24 J. R. BARTELS-KEITH, A. W. JOHNSON AND A. LANGEMANN, *J. Chem. Soc.*, (1952) 4461.
25 R. B. JOHNS, A. W. JOHNSON, A. LANGEMANN AND J. MURRAY, *J. Chem. Soc.*, (1955) 309.
26 W. VON E. DOERING AND H. KRAUCH, *Angew. Chem.*, 68 (1956) 661.
27 E. ZBIRAL, F. TAKACS AND F. WESSELY, *Monatsh.*, 95 (1964) 402.
28 E. ZBIRAL, J. JAZ AND F. WESSELY, *Monatsh.*, 92 (1961) 1155.
29 G. L. CLOSS AND L. E. CLOSS, *J. Am. Chem. Soc.*, 83 (1961) 599.
30 A. J. BIRCH, J. M. H. GRAVES AND F. STANSFIELD, *Proc. Chem. Soc.*, (1962) 282.
31 O. L. CHAPMAN AND P. FITTON, *J. Am. Chem. Soc.*, 83 (1961) 1005.
32 O. L. CHAPMAN AND P. FITTON, *J. Am. Chem. Soc.*, 85 (1963) 41.
33 E. E. VAN TAMELEN AND G. T. HILDAHL, *J. Am. Chem. Soc.*, 75 (1953) 5451.
34 E. E. VAN TAMELEN AND G. T. HILDAHL, *J. Am. Chem. Soc.*, 78 (1956) 4405.
35 J. SCHREIBER, M. PESARO, W. LEIMGRUBER AND A. ESCHENMOSER, *Helv. Chim. Acta*, 41 (1958) 2103.
36 T. NOZOE, Y. KITAHARA, T. ANDO, S. MASAMUNE AND H. ABE, *Sci. Reports Tôhoku Univ.*, I 36 (1952) 166.
37 T. NOZOE, T. MUKAI AND K. TAKASE, *Sci. Reports Tôkohu Univ.*, I 36 (1952) 40.
38 T. NOZOE, T. MUKAI ,K. TAKASE AND T. NAGASE, *Proc. Japan Acad.*, 28 (1952) 477.
39 T. NOZOE, S. ITO AND K. SONOBE, *Proc. Japan Aacd.*, 29 (1953) 101; *Sci. Reports Tôhoku Univ.*, I 38 (1954) 141.
40 N. J. LEONARD, L. A. MILLER AND J. W. BERRY, *J. Am. Chem. Soc.*, 79 (1957) 1482.
41 M. MÜHLSTÄDT, *Naturwiss.*, 45 (1958) 240.
42 W. TREIBS AND P. GROSSMAN, *Chem. Ber.*, 92 (1959) 273.
43 A. EINHORN, *Ber.*, 20 (1887) 1227.
44 G. BUCHI, N. C. YANG, S. L. EMERMAN AND J. MEINWALD, *Chem. and Ind.*, (1953) 1063.
45 J. MEINWALD, S. L. EMERMAN, N. C. YANG AND G. BUCHI, *J. Am. Chem. Soc.*, 77 (1955) 4401.
46 D. ELAD AND D. GINSBURG, *J. Chem. Soc.*, (1954) 471.
47 E. E. VAN TAMELEN, P. BARTH AND F. LORNITZO, *J. Am. Chem. Soc.*, 78 (1956) 5442.
48 T. NOZOE, T. MUKAI, T. TEZUKU AND K. OSAKA; G. SUNAGAWA, N. SAMA AND H. NAKAO; quoted in ref. 49.
49 T. NOZOE, in J. W. COOK (Editor), *Progress in organic chemistry*, Vol. 5, Butterworth, London, 1961, p. 132.
50 M. E. VOL'PIN, I. S. AKHREM AND D. N. KURSANOV, *Izv. Akad. Nauk S.S.S.R., Otdel. khim. Nauk*, (1957) 760.
51 H. J. DAUBEN, F. A. GADECKI, K. M. HARMON AND D. L. PEARSON, *J. Am. Chem. Soc.*, 79 (1957) 4557.
52 A. P. TER BORG, R. VAN HELDEN AND A. F. BICKEL, *Rec. Trav. chim.*, 81 (1962) 177.
53 A. P. TER BORG, R. VAN HELDEN, A. F. BICKEL ,W. RENOLD AND A. S. DREIDING, *Helv. Chim. Acta*, 43 (1960) 457.
54 T. IKEMI, T. NOZOE AND H. SUGIYAMA, *Chem. and Ind.*, (1960) 932.
55 Z. N. PARNES, G. D. MUR, R. V. KUDRIAVTSEV AND D. N. KURSANOV, *Dokl. Akad. Nauk S.S.S.R.*, 155 (1964) 6.
56 J. W. COOK AND A. R. SOMERVILLE, *Nature*, 163 (1949) 410.
57 J. W. COOK, A. R. GIBB, R. A. RAPHAEL AND A. R. SOMERVILLE, *Chem. and Ind.*, (1950) 427.
58 T. NOZOE, S SETO, Y. KITAHARA, M. KUNORI AND Y. NAKAYAMA, *Proc. Japan. Acad.*, 26(7) (1950) 38.

59 W. VON E. DOERING AND L. H. KNOX, *J. Am. Chem. Soc.*, 72 (1950) 2305.
60 J. W. COOK, A. R. GIBB, R. A. RAPHAEL AND A. R. SOMERVILLE, *J. Chem. Soc.*, (1951) 503.
61 T. NOZOE, S. SETO, K. KIKUCHI, T. MUKAI, S. MATSUMOTO AND M. MURASE, *Proc. Japan. Acad.*, 26(7) (1950) 43.
62 T. NOZOE, Y. KITAHARA AND S. ITO, *Proc. Japan. Acad.*, 26(7) (1950) 47.
63 J. W. COOK, R. A. RAPHAEL AND A. I. SCOTT, *J. Chem. Soc.*, (1951) 695.
64 T. NOZOE, T. MUKAI AND S. MATSUMOTO, *Proc. Japan. Acad.*, 27 (1951) 110.
65 T. NOZOE, S. SETO, K. KIKUCHI AND H. TAKEDA, *Proc. Japan. Acad.*, 27 (1951) 146.
66 T. NOZOE, H. KISHI AND A. YOSHIKOSHI, *Proc. Japan. Acad.*, 27 (1951) 149.
67 T. MUKAI, M. KUNORI, H. KISHI, T. MUROI AND K. MATSUI, *Proc. Japan. Acad.*, 27 (1951) 410.
68 T. NOZOE, T. MUKAI AND K. MATSUI, *Proc. Japan. Acad.*, 27 (1951) 646.
69 T. NOZOE, T. MUKAI, M. KUNORI, T. MUROI AND K. MATSUI, *Sci. Reports Tôhoku Univ.*, I 35 (1951) 242.
70 T. NOZOE, *Nature*, 167 (1951) 1055.
71 B. E. BRYANT AND W. C. FERNELIUS, *J. Am. Chem. Soc.*, 76 (1954) 1696.
72 W. VON E. DOERING AND A. A-R. SAYIGH, *J. Org. Chem.*, 26 (1961) 1365.
73 J. D. KNIGHT AND D. J. CRAM, *J. Am. Chem. Soc.*, 73 (1951) 4136.
74 J. W. COOK, J. D. LOUDON AND D. K. V. STEEL, *J. Chem. Soc.*, (1954) 530.
75 N. J. LEONARD AND G. C. ROBINSON, *J. Am. Chem. Soc.*, 75 (1953) 2143.
76 N. J. LEONARD AND J. W. BERRY, *J. Am. Chem. Soc.*, 75 (1953) 4989.
77 W. VON E. DOERING AND L. H. KNOX, *J. Am. Chem. Soc.*, 73 (1951) 828.
78 W. VON E. DOERING AND L. H. KNOX, *J. Am. Chem. Soc.*, 75 (1953) 297.
79 J. R. BARTELS-KEITH AND A. W. JOHNSON, *Chem. and Ind.*, (1950) 677.
80 J. R. BARTELS-KEITH, A. W. JOHNSON AND W. I. TAYLOR, *Chem. and Ind.*, (1951) 337.
81 J. R. BARTELS-KEITH, A. W. JOHNSON AND W. I. TAYLOR, *J. Chem. Soc.*, (1951) 2352.
82 T. NOZOE, T. MUKAI, J. MINEGISHI AND T. FUJISAWA, *Sci. Reports Tôhoku Univ.*, I 37 (1953) 388.
83 T. NOZOE, T. MUKAI AND K. TAKASE, *Sci. Reports Tôhoku Univ.*, I 39 (1955) 164.
84 J. J. DRYSDALE, W. W. GILBERT, H. K. SINCLAIR AND W. H. SHARKEY, *J. Am. Chem. Soc.*, 80 (1958) 245, 3672.
85 A. DI GIACOMO AND C. P. SMYTH, *J. Am. Chem. Soc.*, 74 (1952) 4411.
86 Y. KURITA, S. SETO, T. NOZOE AND M. KUBO, *Bull. Chem. Soc. Japan*, 26 (1953) 272.
87 Y. G. BOROD'KO AND Y. R. SYRKIN, *Dokl. Akad. Nauk S.S.S.R.*, 134 (1960) 1272.
88 H. H. GUNTHARD AND T. GAUMANN, *Helv. Chim. Acta*, 34 (1951) 39.
89 Y. KURITA, *Sci. Reports Tôhoku Univ.*, I 38 (1954) 85.
90 W. N. HUBBARD, C. KATZ, G. B. GUTHRIE AND G. WADDINGTON, *J. Am. Chem. Soc.*, 74 (1952) 4456.
91 J. M. ROBERTSON, *J. Chem. Soc.*, (1951) 1222.
92 Y. SASADA, K. OSAKI AND I. NITTA, *Acta Cryst.*, 7 (1954) 113; Y. SASADA AND I. NITTA, *Acta Cryst.*, 9 (1956) 205.
93 Y. SASADA AND I. NITTA, *Bull. Chem. Soc. Japan*, 30 (1957) 62.
94 E. HEILBRONNER AND K. HEDBERG, *J. Am. Chem. Soc.*, 73 (1951) 1386; M. KIMURA AND M. KUBO, *Bull. Chem. Soc. Japan*, 26 (1953) 250.
95 K. KIMURA, S. SAZUKI, M. KIMURA AND M. KUBO, *J. Chem. Phys.*, 27 (1957) 320; *Bull. Chem. Soc. Japan*, 31 (1958) 1051.
96 Y. IKEGAMI, *Bull. Chem. Soc. Japan*, 36 (1963) 1118.
97 H. SUGIYAMA, S. ITO AND T. NOZOE, *Tetrahedron Letters*, (1965) 179.
98 E. KLOSTER-JENSEN, N. TARKÖY, A. ESCHENMOSER AND E. HEILBRONNER, *Helv. Chim. Acta*, 39 (1956) 786.

99 N. J. LEONARD, L. A. MILLER AND J. W. BERRY, *J. Am. Chem. Soc.*, 79 (1957) 1485.
100 H. H. RENNHARD, G. DI MODICA, J. SIMON, A. ESCHENMOSER AND E. HEILBRONNER, *Helv. Chim. Acta*, 40 (1957) 957.
101 H. GOETZ, E. HEILBRONNER, A. R. KATRITZKY AND R. A. JONES, *Helv. Chim. Acta*, 44 (1961) 387.
102 B. E. ZAITSEV, Y. N. SHEINKER AND Y. D. KORESHKOV, *Dokl. Akad. Nauk S.S.S.R.*, 136 (1961) 1090.
103 D. LLOYD, D. R. MARSHALL AND M. RANDALL, *Chem. and Ind.*, (1960) 1132.
104 V. D. ZAITSEV, Y. N. SHEINKER, Y. D. KORESHKOV AND M. E. VOL'PIN, *Fiz. Probl. Spektroscopii, Akad. Nauk S.S.S.R., Materialy 13-go [Trinadtsatogo]Soveshch. Leningrad*, 1 (1960) 442; *Chem. Abs.*, 59 (1963) 12303.
105 T. ISOBE AND K. TAKAHASHI, quoted in ref. 89.
106 T. NOZOE, in D. GINSBURG (Editor), *Non-benzenoid Aromatic Compounds*, Interscience, New York, 1959, p. 339.
107 H. P. KOCH, *J. Chem. Soc.*, (1951) 512.
108 J. KINUMAKI, K. AIDA AND Y. IKEGAMI, *Sci. Reports Res. Inst., Tôhoku Univ.*, A 8 (1956) 263.
109 Y. IKEGAMI, *Bull. Chem. Soc. Japan*, 34 (1961) 91.
110 T. NOZOE, T. MUKAI AND J. MINEGISHI, *Proc. Japan. Acad.*, 27 (1951) 419.
111 T. MUKAI, *Bull. Chem. Soc., Japan*, 31 (1958) 846.
112 T. NOZOE, T. MUKAI AND J. MINEGISHI, *Proc. Japan. Acad.*, 28 (1952) 287.
113 T. MUKAI, *Sci. Reports Tôhoku Univ.*, I 38 (1954) 280.
114 O. L. CHAPMAN, A. A. GRISWOLD AND D. J. PASTO, *J. Am. Chem. Soc.*, 84 (1962) 1213.
115 A. F. BICKEL, A. P. TER BORG AND R. VAN HELDEN, *Rec. Trav. chim.*, 81 (1962) 591.
116 W. VON E. DOERING AND C. F. HISKEY, *J. Am. Chem. Soc.*, 74 (1952) 5688.
117 W. VON E. DOERING AND J. R. MAYER, *J. Am. Chem. Soc.*, 75 (1953) 2387.
118 T. NOZOE, T. MUKAI, Y. IKEGAMI AND T. TODA, *Chem. and Ind.*, (1955) 66.
119 S. SETO, *Sci. Reports Tôhoku Univ.*, I 37 (1953) 275, 297.
120 A. W. JOHNSON AND M. TISLER, *J. Chem. Soc.*, (1955) 1841.
121 T. NOZOE, S. SETO, T. IKEMI AND T. ARAI, *Proc. Japan. Acad.*, 27 (1951) 102.
122 W. VON E. DOERING AND L. H. KNOX, *J. Am. Chem. Soc.*, 74 (1952) 5683.
123 *E.g.* T. NOZOE, Y. KITAHARA AND S. MASAMUNE, *Proc. Japan. Acad.*, 27 (1951) 649.
124 R. B. JOHNS, A. W. JOHNSON AND M. TISLER, *J. Chem. Soc.*, (1954) 4605.
125 Y. KITAHARA, *Sci. Reports Tôhoku Univ.*, I 39 (1956) 258.
126 T. NOZOE, S. SETO AND S. MATSUMURA, *Proc. Japan. Acad.*, 28 (1952) 483.
127 T. NOZOE, T. MUKAI AND I. MURATA, *Proc. Japan. Acad.*, 29 (1953) 169.
128 R. E. CORBETT, C. H. HASSALL, A. W. JOHNSON AND A. R. TODD, *J. Chem. Soc.*, (1950) 1.
129 P. AKROYD, R. D. HAWORTH AND J. D. HOBSON, *J. Chem. Soc.*, (1951) 3427.
130 P. AKROYD, R. D. HAWORTH AND P. R. JEFFERIES, *J. Chem. Soc.*, (1954) 286.
131 T. NOZOE, S. SETO, H. TAKEDA, S. MOROSAWA AND K. MATSUMOTO, *Proc. Japan. Acad.*, 27 (1951) 556; 28 (1952) 192; *Sci. Reports Tôhoku Univ.*, 136 (1952) 126.
132 T. NOZOE, S. SETO, T. IKEMI, T. SATO AND K. WATANABE, *Proc. Japan. Acad.*, 28 (1952) 413.
133 S. SETO, *Sci. Reports Tôhoku Univ.*, I 37 (1953) 367.
134 K. KIKUCHI, *J. Chem. Soc. Japan*, 77 (1956) 1439.
135 H. AKINO, *Sci. Reports Tôhoku Univ.*, I 40 (1956) 92.
136 S. SETO, *Sci. Reports Tôhoku Univ.*, I 37 (1953) 286.
137 B. J. ABADIR, J. W. COOK, J. D. LOUDON AND D. K. V. STEEL, *J. Chem. Soc.*, (1952) 2350.
138 W. VON E. DOERING AND D. B. DENNY, *J. Am. Chem. Soc.*, 77 (1955) 4619.

REFERENCES

139 S. Seto, *Sci. Reports Tôhoku Univ.*, I 37 (1953) 377.
140 J. W. Cook, A. R. Gibb and R. A. Raphael, *J. Chem. Soc.*, (1951) 2244.
141 W. D. Crow, R. D. Haworth and P. R. Jefferies, *J. Chem. Soc.*, (1952) 3705.
142 T. Nozoe, S. Seto, S. Ito, M. Sato and T. Katono, *Proc. Japan. Acad.*, 28 (1952) 488; *Sci. Reports Tôhoku Univ.*, I 37 (1953) 191.
143 T. Nozoe, S. Ito, K. Matsui and T. Ozeki, *Proc. Japan. Acad.*, 30 (1954) 604.
144 J. W. Cook, A. R. Gibb, R. A. Raphael and A. R. Somerville, *J. Chem. Soc.*, (1952) 603.
145 H. Fernholz, E. Hartwig and J. C. Salfeld, *Ann.*, 576 (1952) 131.
146 T. Nozoe, M. Sato, S. Ito, K. Matsui and T. Ozeki, *Proc. Japan. Acad.*, 30 (1954) 599.
147 H. Erdtman and W. E. Harvey, *Chem. and Ind.*, (1952) 1267.
148 R. E. Corbett and D. E. Wright, *Chem. and Ind.*, (1953) 1258.
149 T. Nozoe, K. Takase and M. Ogata, *Chem. and Ind.*, (1957) 1070.
150 D. S. Tarbell, K. I. H. Williams and E. J. Sehm, *J. Am. Chem. Soc.*, 81 (1959) 3443.
151 R. D. Haworth and J. D. Hobson, *Chem. and Ind.*, 149 (1950) 441.
152 R. D. Haworth and J. D. Hobson, *J. Chem. Soc.*, (1951) 561.
153 E. Sebe and S. Matsumoto, *Sci. Reports Tôhoku Univ.*, I 38 (1954) 308.
154 T. Nozoe, *Sci. Reports Tôhoku Univ.*, I 34 (1950) 199.
155 J. W. Cook, R. A. Raphael and A. I. Scott, *J. Chem. Soc.*, (1952) 4416.
156 J. W. Cook, J. D. Loudon and D. K. V. Steel, *Chem. and Ind.*, (1951) 669.
157 T. Nozoe, E. Sebe and S. Ebine, *Proc. Japan. Acad.*, 26(8) (1950) 24; T. Nozoe, Y. Kitahara, E. Kunioka and K. Doi, *Proc. Japan. Acad.*, 26(9) (1950) 38; T. Nozoe, Y. Kitahara and K. Doi, *Proc. Japan. Acad.*, 26(10) (1950) 25; 27(10) (1951) 156.
158 N. Yui, *Sci. Reports Tôhoku Univ.*, I 40 (1956) 102.
159 J. W. Cook, A. R. Gibb and R. A. Raphael, *J. Chem. Soc.*, (1951) 2067.
160 R. M. Horowitz and G. E. Ullyat, *J. Am. Chem. Soc.*, 74 (1952) 587.
161 D. S. Tarbell and J. C. Bill, *J. Am. Chem. Soc.*, 74 (1952) 1234.
162 T. Nozoe, Y. Kitahara, K. Yamane and T. Ikemi, *Proc. Japan. Acad.*, 27 (1951) 193.
163 T. Nozoe, Y. Kitahara, T. Ando and E. Kunioka, *Proc. Japan. Acad.*, 27 (1951) 231.
164 T. Nozoe, S. Ebine, S. Ito and A. Konishi, *Proc. Japan. Acad.*, 27 (1951) 10.
165 T. Nozoe and S. Seto, *Proc. Japan. Acad.*, 27 (1951) 188.
166 Y. Kitahara, *Sci. Reports Tôhoku Univ.*, I 39 (1956) 265.
167 Y. Kitahara, *Sci. Reports Tôkohu Univ.*, I 39 (1956) 250.
168 W. R. Hasek, W. C. Smith and V. A. Engelhardt, *J. Am. Chem. Soc.*, 82 (1960) 543.
169 T. Nozoe, S. Seto, T. Mukai, K. Yamane and A. Matsukuma, *Proc. Japan. Acad.*, 27 (1951) 224.
170 T. Nozoe, S. Seto and T. Ikemi, *Proc. Japan Acad.*, 27 (1951) 655.
171 E. Sebe and Y. Itsuno, *Proc. Japan. Acad.*, 29 (1953) 107.
172 E. Sebe, C. Osako and Y. Itsuno, *Kumamoto Med. J.*, 6 (1953) 9.
173 T. Nozoe, Y. Kitahara, K. Yamane and A. Yoshikoshi, *Proc. Japan. Acad.*, 27 (1951) 18.
174 T. Nozoe, K. Kikuchi and T. Ando, *Proc. Japan. Acad.*, 26(10) (1950) 32.
175 E. Sebe, T. Nozoe, P. Y. Yen and S. Iwamoto, *Sci. Reports Tôhoku Univ.*, I 36 (1952) 307.
176 T. Nozoe, Y. Kitahara and T. Ando, *Proc. Japan. Acad.*, 27 (1951) 107.
177 Y. Kitahara and T. Arai, *Proc. Japan. Acad.*, 27 (1951) 423.
178 T. Nozoe, E. Sebe, L. S. Cheng, S. Mayama and T. J. Hsü, *Sci. Reports Tôhoku Univ.*, I 36 (1952) 299.
179 T. Nozoe, Y. Kitahara, E. Kunioka and T. Ando, *Proc. Japan. Acad.*, 27 (1951) 190.
180 K. Yamane and S. Morosawa, *Bull. Chem. Soc. Japan*, 27 (1954) 18.

181 T. Nozoe, S. Seto, T. Ikemi and T. Arai, *Proc. Japan. Acad.*, 26(9) (1950) 50; 27 (1951) 24.
182 R. D. Haworth and P. R. Jefferies, *Chem. and Ind.*, (1950) 841.
183 T. Nozoe, *Sci. Reports Tôhoku Univ.*, I 36 (1952) 82.
184 T. Nozoe, S. Seto, H. Takeda and T. Sato, *Sci. Reports Tôhoku Univ.*, I 35 (1951) 274.
185 T. Nozoe, T. Mukai and K. Takase, *Proc. Japan. Acad.*, 27 (1951) 561.
186 E. Hartwig, *Angew. Chem.*, 66 (1954) 605.
187 S. Seto and K. Ogura, *Bull. Chem. Soc. Japan*, 32 (1959) 493.
188 K. Yamane, *J. Chem. Soc. Japan*, 76 (1955) 787.
189 Y. Kitahara, *Sci. Reports Tôhoku Univ.*, I 40 (1956) 74.
190 T. Nozoe and Y. Kitahara, *Proc. Japan. Acad.*, 30 (1954) 204.
191 R. D. Haworth and P. B. Tinker, *J. Chem. Soc.*, (1955) 911.
192 T. Nozoe, Y. Kitahara, K. Yamane and K. Yamaki, *Proc. Japan. Acad.*, 26(8) (1950) 14.
193 T. Muroi, *Bull. Yamagata Univ. (Nat. Sci.)*, 3 (1954) 155; *Chem. Abs.*, 50 (1956) 287.
194 T. Nozoe, H. Akino and K. Sato, *Proc. Japan. Acad.*, 27 (1951) 565.
195 T. Nozoe, Y. Kitahara and K. Doi, *Proc. Japan. Acad.*, 27 (1951) 156, 282; *J. Am. Chem. Soc.*, 73 (1951) 1895.
196 T. Nozoe, S. Seto, S. Ebine and S. Ito, *Proc. Japan. Acad.*, 26(9) (1950) 45.
197 E. J. Forbes and R. A. Pipley, *J. Chem. Soc.*, (1959) 2770.
198 W. G. Dauben, K. Koch and W. E. Thiessen, *J. Am. Chem. Soc.*, 81 (1959) 6087.
199 W. G. Dauben, K. Koch, S. L. Smith and O. L. Chapman, *J. Am. Chem. Soc.*, 85 (1963) 2616.
200 E. Sebe and S. Matsumoto, *Proc. Japan. Acad.*, 29 (1953) 207.
201 R. B. Johns and A. W. Johnson, *Chem. and Ind.*, (1954) 192.
202 R. B. Johns, A. W. Johnson and J. Murray, *J. Chem. Soc.*, (1954) 198.
203 R. S. Coffey, R. B. Johns and A. W. Johnson, *Chem. and Ind.*, (1955) 658.
204 R. S. Coffey and A. W. Johnson, *J. Chem. Soc.*, (1958) 1741.
205 J. Meinwald and O. L. Chapman, *J. Am. Chem. Soc.*, 78 (1956) 4816.
206 J. Meinwald and O. L. Chapman, *J. Am. Chem. Soc.*, 80 (1958) 633.
207 O. L. Chapman and D. J. Pasto, *J. Am. Chem. Soc.*, 81 (1959) 5510.
208 T. Nozoe, M. Sato and K. Matsui, *Proc. Japan. Acad.*, 28 (1952) 407; 29 (1953) 22.
209 T. Nozoe and K. Matsui, *Bull. Chem. Soc. Japan*, 34 (1961) 616.
210 W. R. Brasen, H. E. Holmquist and R. E. Benson, *J. Am. Chem. Soc.*, 82 (1960) 995.
211 R. E. Benson, *J. Am. Chem. Soc.*, 82 (1960) 5948.
212 W. R. Brasen, H. E. Holmquist and R. E. Benson, *J. Am. Chem. Soc.*, 83 (1961) 3125.
213 D. S. Tarbell, G. P. Scott and A. D. Kemp, *J. Am. Chem. Soc.*, 72 (1950) 379.
214 G. A. Nicholls and D. S. Tarbell, *J. Am. Chem. Soc.*, 74 (1952) 4935.
215 M. Kerfanto and J. P. Quentin, *Compt. rend.*, 257 (1963) 2660.
216 W. Ried and H. J. Schwenecke, *Chem. Ber.*, 91 (1958) 566.
217 W. Treibs and E. Lippman, *Chem. Ber.*, 91 (1958) 1999.
218 W. Davey and H. Gottfried, *J. Org. Chem.*, 26 (1961) 3699, 3705.
219 D. Caunt, W. D. Crow, R. D. Haworth and C. A. Vodoz, *Chem. and Ind.*, (1950) 149; *J. Chem. Soc.*, (1950) 1631.
220 J. A. Barltrop, J. A. Johnson and G. D. Meakin, *J. Chem. Soc.*, (1951) 181.
221 D. Caunt, W. D. Crow and R. D. Haworth, *J. Chem. Soc.*, (1951) 1313.
222 T. Moroe, T. Matsushima and O. Miyamato, *J. Pharm. Soc. Japan*, 72 (1952) 1238.
223 W. E. Parham, D. A. Bolon and E. E. Schweizer, *J. Am. Chem. Soc.*, 83 (1961) 603.
224 M. E. Vol'pin and A. F. Plate, *Dokl. Akad. Nauk S.S.S.R.*, 70 (1950) 843.

REFERENCES

225 W. Treibs and H. J. Klinkhammer, *Chem. Ber.*, 84 (1951) 671.
226 A. C. Cope and S. W. Fenton, *J. Am. Chem. Soc.*, 73 (1951) 1673.
227 E. D. Bergmann, E. Fischer, D. Ginsburg, Y. Hirshberg, D. Lavie, M. Mayot, A. Pullman and B. Pullman, *Bull. Soc. chim. France*, (1951) 684.
228 G. L. Buchanan, J. W. Cook and J. D. Loudon, *J. Chem. Soc.*, (1944) 325.
229 J. W. Cook, G. T. Dickson and J. D. Loudon, *J. Chem. Soc.*, (1947) 746.
230 J. W. Cook, J. Jack, J. D. Loudon, G. L. Buchanan and J. MacMillan, *J. Chem. Soc.*, (1951) 1397.
231 G. L. Buchanan, *Chem. and Ind.*, (1952) 855.
232 E. D. Bergmann and D. Ginsburg, *Chem. and Ind.*, (1954) 45.
233 M. Stiles and A. J. Libbey, *J. Org. Chem.*, (1957) 1243.
234 R. W. Schmid, E. Kloster-Jensen, E. Kovats and E. Heilbronner, *Helv. Chim. Acta*, 39 (1956) 806.
235 T. Gäumann, R. W. Schmid and E. Heilbronner, *Helv. Chim. Acta*, 39 (1956) 1985.
236 W. Treibs and G. Herdmann, *Ann.*, 609 (1957) 70.
237 T. Sakan and M. Nakazaki, *J. Inst. Polytechnics, Osaka City Univ.*, 1 (1950) 23; *Chem. Abs.*, 46 (1952) 5036.
238 B. Eistert and H. Minas, *Tetrahedron Letters*, (1964) 1361.
239 J. Rigandy and L. Nedelec, *Compt. rend.*, 236 (1953) 1287.
240 W. von E. Doering and D. W. Wiley, *Tetrahedron*, 11 (1960) 183.
241 C. Jutz, *Chem. Ber.*, 97 (1964) 2050.
242 D. S. Matteson, J. J. Drysdale and W. H. Sharkey, *J. Am. Chem. Soc.*, 82 (1960) 2853.
243 K. Hafner, H. W. Riedel and M. Danielisz, *Angew. Chem.*, 75 (1963) 344.
244 T. Mukai, T. Nozoe, K. Osaka and N. Shishido, *Bull. Chem. Soc. Japan*, 34 (1961) 1384.
245 A. C. Cope and R. D. Smith, *J. Am. Chem. Soc.*, 77 (1955) 4596.
246 W. E. Parham, R. W. Soeder, J. R. Throckmorton, K. Kuncl and R. M. Dodson, *J. Am. Chem. Soc.*, 87 (1965) 321.
247 E. J. Forbes and D. C. Warrell, *Chem. and Ind.*, (1964) 2056; T. Nozoe, T. Mukai and K. Sakai, *Tetrahedron Letters*, (1965) 1041.
248 T. Nozoe, K. Takase and K. Umino, *Bull. Chem. Soc. Japan*, 38 (1965) 358.
249 A. F. Bickel, A. P. ter Borg and R. van Helden, *Rec. Trav. Chim.*, 81 (1962) 599.
250 D. J. Bertelli and C. C. Ong, *J. Am. Chem. Soc.*, 87 (1965) 3719.

CHAPTER VII

Medium and large ring compounds

CYCLOOCTATETRAENIDES

Cyclooctatetraene was first prepared in 1911[1]. The total lack of any "aromatic" properties surprised most organic chemists, conditioned by the properties of a cyclic alternating system of double and single bonds as found in benzene[2]. Now that cyclooctatetraene is recognised as a non-planar hydrocarbon having 8 π-electrons its properties are unexceptionable.

Electron-diffraction measurements[3] indicate a system of alternating double and single bonds of length 1.334 and 1.462 Å, while its infra-red spectrum[4] is typically that of a non-conjugated aliphatic olefin. It has been shown to be less stable than isomeric styrene by 34 kcal/mole[5]. Chemically it behaves as a very reactive olefin, readily undergoing oxidation, reduction, addition of halogens and participating in the Diels–Alder reaction[6].

The carbon–carbon bond lengths in cyclooctatetraene exclude any extensive delocalisation of the double bonds. In order to attain a strain-free structure the molecule takes up a buckled tub shape in which the planes of adjacent pairs of double bonds are almost exactly at right angles to each other. Molecular orbital calculations (cf. Chapter I) show that even were the molecule forced into a planar form, the stabilisation energy gained thereby would be very small, and the increase in angle strain would greatly outweigh it, so that the molecule would take up the buckled shape again in order to relieve this strain.

It was known that cyclooctatetraene formed a dilithium derivative, $C_8H_8Li_2$, and in 1956 the suggestion was made[7], based on magnetic susceptibility measurements, that this might be a dilithium salt of the cyclooctatetraenide dianion, $(C_8H_8)^{2-}$.

In 1960 the dipotassium salt was isolated as very pale yellow crystals which exploded on exposure to air[8].

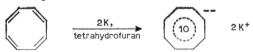

This salt has also been prepared by the reaction between cyclooctatetraene and potassium in ether[9] or liquid ammonia[10]. It is stable in solution but the crystalline compound explodes on the least contact with oxygen or moisture and turns brown even in an atmosphere of nitrogen or argon[11]. Its melting-point is above 250° (ref. 11).

Such an anion has a decet of π-electrons and thus according to Hückel's rule should be aromatic in character. This has been shown to be the case from a study of its spectral properties. Thus its n.m.r. spectrum[8], which shows a single sharp peak at 4.3 τ, and its ultra-violet and infra-red spectra[9,11] and magnetic susceptibility[11] are in accord with its having a symmetrical planar monocyclic structure with aromatic character.

The action of acetyl and benzoyl chlorides on the cyclooctatetraenide dianion has been investigated[12]. A mixture of bicyclic and open-chain products was obtained; the ion (I) being suggested as a first intermediate.

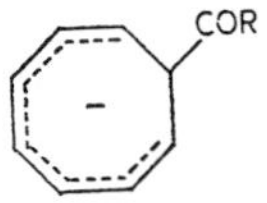

(I)

The cyclooctatetraenide dianion reacts with aldehydes and ketones to give 1,2 and 1,4 addition products[50,51]. The 1,2 adducts may undergo rearrangement, *e.g.* [51]

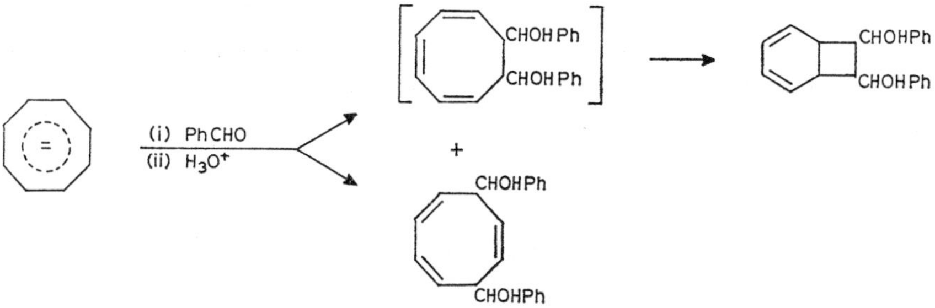

Ring-opened products may also be obtained in some cases[51].

The reduction of cyclooctatetraene by means of alkali metals in the presence of proton donors (*e.g.* liquid ammonia or methanol)[13] presumably proceeds *via* the cyclooctatetraenide dianion.

CYCLONONATETRAENIDES

The cyclononatetraenide anion $(C_9H_9)^-$ is a vinylogue of the cyclopentadi-

References p. 179

enide anion, and, having a decet of π-electrons fits Hückel's conditions for aromaticity. It was first prepared in 1963, independently by two sets of workers[14,15,65], by the methods shown in the following chart.

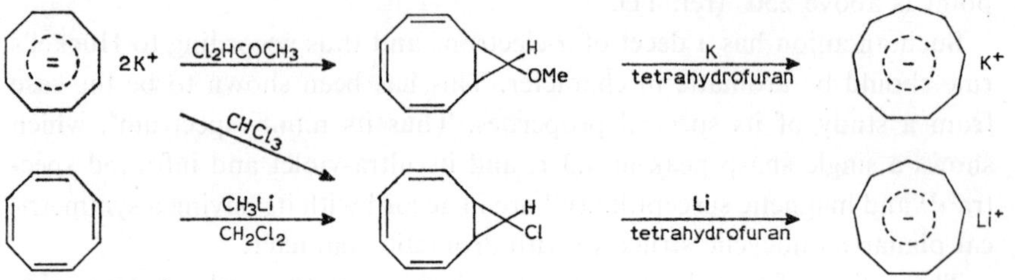

It can also be obtained in lower yield by proton abstraction from bicyclo-[6,1,0]nona-2,4,6-triene[65].

Despite having ten π-electrons the cyclononatetraenide anion might not be aromatic owing to excessive energy being required to distort the valency angles in order to make the molecule planar. Furthermore it might valence tautomerise to give a bicyclic structure. However the n.m.r. spectrum of the anion (one sharp peak ca. 3 τ)[14,15,65] clearly shows it to have a planar monocyclic structure and to be aromatic in character. The simplicity of the spectrum precludes the possibility of a bicyclic structure since this would give several peaks, while the value of the chemical shift rules out the possibility of the ion existing as a mixture of rapidly interconverting valency tautomers. Furthermore the presence of only one peak suggests that the anion cannot have a shape involving re-entrant angles:

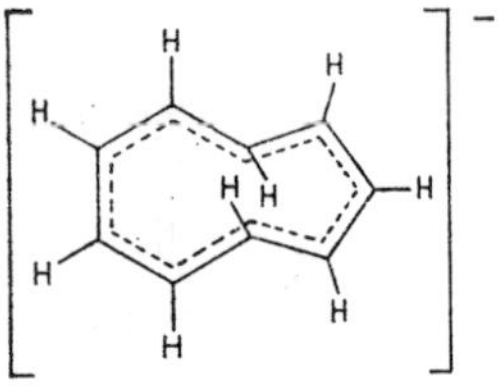

If the anion had this shape the exocyclic and endocyclic hydrogen atoms should give rise to separate peaks. The spectrum shows no line-broadening over a range from $-40°$ to $+60°$. (cf. p. 172 and ref. 57). The difference between the τ-values for this anion and the cyclooctatetraenide dianion accords well with theory. Assuming that there is a similar ring current in each ion, the chemical shifts should vary with the charge densities on each carbon atom; were it not for these differences in charge density the chemical shifts for the two ions would be the same. (*cf.* Chapter IV, p. 58). ^{13}C n.m.r. studies are also in accord with the assigned structure[65].

As might be expected[66] for this highly symmetrical ion the ultra-violet and infra-red spectra are very simple. It absorbs ultra-violet light intensely (λ_{max} 252, 318, 325; log ε_{max} 5.0, 3.9, 3.9)[14,15,65].

The lithium salt is very sensitive to moisture and to oxygen but is stable in an inert atmosphere. Thus after being heated for 30 hours in an inert solvent at 163° over 70% of the salt remained unchanged[65].

The lithium salt is converted into the tetraethylammonium salt by interaction with tetraethylammonium chloride[15,65]. Tetraethylammonium cyclononatetraenide is a white solid, stable in an inert atmosphere, and having n.m.r. and ultra-violet spectra the same as those of the alkali metal salts. Like the lithium salt it is sensitive to moisture but it is apparently stable in dry oxygen. In solution, however, it is fairly readily oxidised.

The cyclononatetraenide anion reacts with water to form a mixture from which indene and 8,9-dihydroindene have been isolated[14,65]. There is evidence that cyclononatetraene is also formed[65]. It reacts with deuterium oxide to give 1-deutero-8,9-dihydroindene and with carbon dioxide to give 1-carboxy-8,9-dihydroindene[14]. Catalytic reduction leads to a mixture of products, including cyclononane, thus demonstrating the presence of the nine-membered ring[65].

The cyclononatetraenide anion extracts a proton from cyclopentadiene to give the cyclopentadienide anion[15,65]. Equilibration of the cyclononatetraenide anion with cyclopentadiene suggests that the $C_9H_9^-$ ion is more thermodynamically stable than the $C_5H_5^-$ ion[65]. No proton exchange takes place between the cyclononatetraenide anion and indene[65]. These results suggest that the pK of cyclononatetraene lies between 16 and 21, (ref. 65).

Methyl iodide reacts with the cyclononatetraenide anion to give 1-methyl-8,9-dihydroindene, but no reaction takes place between the anion and methyl chloride[14].

ANNULENES

The name "annulenes" has been given to completely conjugated cyclic polyolefins. Individual compounds are described by adding a number in square brackets indicating the number of atoms in the ring. Thus cyclodecapentaene and cyclododecahexaene are known, respectively, as [10]annulene and [12]annulene.

References p. 179

Alternate members of the annulene series have $(4n + 2)$ π-electrons and might therefore possess aromatic character. It is possible to draw rings which have the correct valency angles, *e.g.*

[10] Annulene   [12] Annulene   [14] Annulene   [18] Annulene

On the other hand such planar structures are not possible for the medium-sized rings because there would be overlap of hydrogen atoms in the centre of the rings:

(o= H atom)

A calculation made from scale drawings[16] suggested that there would be overlap in all annulenes from $C_{10}H_{10}$ to $C_{28}H_{28}$ and that such molecules would therefore probably be buckled, but that [30]annulene would probably be planar since the overlap of the internal hydrogen atoms would be small.

It cannot be deduced with certainty from this that all annulenes with less than thirty carbon atoms would not be aromatic, for aromatic compounds are known which have buckled rings (*e.g.* di-*p*-xylylene) or rings with angles distorted appreciably from the preferred values (*e.g.* tropylium salts). On the other hand lack of planarity would necessarily lower the conjugation energy.*

Whereas calculations[16,17] predicted that [18]annulene should be non-planar, other considerations[18,52] suggested that it would be almost if not completely planar. In fact experimental work has confirmed the latter prediction.

The stabilisation energy, measured from the heat of combustion, of [18]-annulene is 100 ± 6 kcal/mole[53], which is close to the calculated value[54] and

---

* The possibility of annulenes twisted into a Möbius strip conformation has also been considered theoretically[49]. It was deduced that annulenes having $(4n+2)$ atomic orbitals would be destabilised in this conformation, but that for an annulene having $4n$ atomic orbitals a closed shell configuration could be obtained without loss of π-electron energy.

suggests that the molecule is nearly planar and not seriously deformed. This is confirmed by n.m.r. spectrum determinations and by X-ray crystallographic investigation (see below).

Another possible reason for lack of aromaticity in larger rings arises from the suggestion that above a certain ring size, alternating single and double bonds may be expected rather than hybridised bonds[19]. Unfortunately the exact value of the limiting size is uncertain. In general the delocalisation energy per double bond in annulenes decreases as the ring size increases.

Preparation of annulenes

The first examples of annulenes to be prepared were benzo-derivatives, *e.g.* (II)[20] and (III)[21].

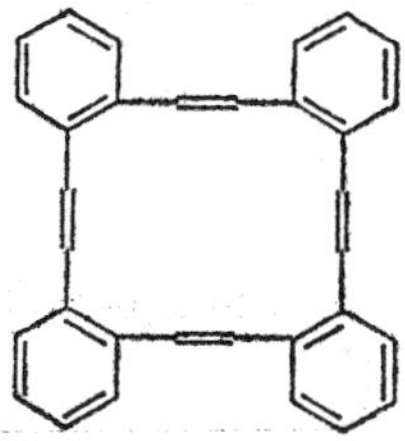

(II) 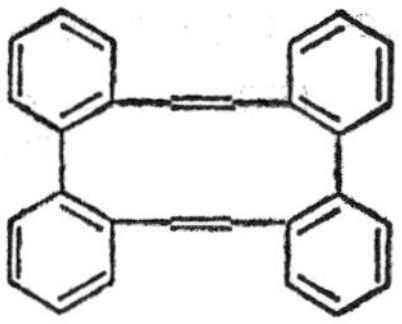 (III)

Both of these molecules have buckled but unstrained rings. Tetra-*o*-phenylene, hexa-*o*-phenylene and octa-*o*-phenylene have also been prepared[22], but these molecules cannot in any case be planar owing to overlap between adjacent *o*-hydrogen atoms. The n.m.r. and ultra-violet spectra of hexa-*m*-phenylene show that there is no annular conjugation in the central eighteen-membered ring[23].

Attempts have been made to prepare annulenes by catalytic dehydrogenation of cycloalkanes[24]. No cyclic polyolefins were obtained. Instead polycylic essentially benzenoid compounds were formed by transannular bridging, *e.g.* cyclodecane was converted to naphthalene and azulene, cyclotetradecane to phenanthrene and anthracene, and cyclooctadecane to triphenylene. In view of the conditions under which dehydrogenation was carried out (over palladium/charcoal at 400°) it is possible that annulenes might have been formed first but then reacted further to produce the products actually isolated.

An attempt to prepare [10]annulene by isomerisation of 9,10-dihydronaphthalene, either by heating the latter compound under nitrogen or by irradiation, also failed to produce any of the desired product[25]. Similarly methyl 9,10-dihydronaphthalene-9,10-dicarboxylate could not be isomerised to give the [10]annulene system[55].

The successful syntheses of annulenes have all used an oxidative coupling of diacetylenes as the ring-closure step. Two methods are available to do this.

When α,ω-diacetylenes (IV) ($n = 3, 4$, or 5) are oxidised in aqueous ethanolic solution by oxygen in the presence of cuprous chloride and ammonium chloride, some cyclic dimer (V) is produced as well as linear condensation products which, however form the major part of the yield[26]. When $n = 2$ or 6 only polymeric material was obtained.

$HC{\equiv}C{-}(CH_2)_n{-}C{\equiv}CH$
(IV)

$\xrightarrow{O_2, CuCl, NH_4Cl, aq. EtOH}$

(V) cyclic dimer with $(CH_2)_n$ bridges

$\xrightarrow{Cu(OAc)_2, pyridine}$

$\left[C{\equiv}C{-}(CH_2)_n{-}C{\equiv}C\right]_x$ (VI)

An alternative method involves cyclisation by means of cupric acetate in pyridine; cyclic products (VI) are obtained, with $x = 1,2,3,4,5,6$, etc.[27]. By this means cyclic tri-, tetra- and penta-mers have been obtained from hepta-1,6-diyne, octa-1,7-diyne, nona-1,8-diyne and deca-1,9-diyne[28].

Utilising one or other of these modes of ring-closure as the first step, the 12-, 14-, 16-, 18-, 20-, 24- and 30-membered ring annulenes have been prepared.

As an example, the preparation of [18]annulene will be considered in some detail. Treatment of hexa-1,5-diyne [(IV), $n = 2$] with cupric acetate in pyridine at 55° for 4 hours gave a complex mixture which was separated by chromatography on alumina to give 6% of the trimer [(VI), $x = 3$], 6% of the tetramer [(VI), $x = 4$] and 6% and 3% of the pentamer and hexamer [(VI), $x = 5, 6$] respectively[28,29]. On treatment of the trimer (VII) with potassium tert. butoxide in tert. butanol a prototropic rearrangement took place giving the completely conjugated polyenyne (VIII), 1,7,13-trisdehydro[18]-annulene[29,30,67]. Partial hydrogenation of this compound over palladium/charcoal produced [18]annulene (IX)[18,31].

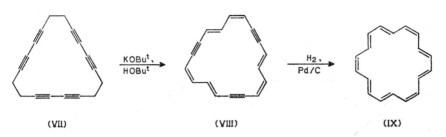

(VII)   (VIII)   (IX)

It is interesting to note that an attempt to make hexamethyl[18]annulene by an analogous route resulted in only very low yields in the reduction step[32]. [24]Annulene[18,24,33] and [30]annulene[18,29,34] were prepared by an identical method starting from the tetramer [(VI), $x = 4$] and pentamer [(VI), $x = 5$] respectively.

Hexa-1,5-diyne was also the starting material for the preparation of [12]annulene[35]. Ring-closure by the oxygen/cuprous chloride/ammonium chloride method in a large amount of benzene gave a solution of cyclododeca-1,3,7,9-tetrayne [(V), $n = 2$]. This tetrayne polymerised very rapidly but if treated at once with potassium tert. butoxide in tert. butanol under mild conditions gave in small yield, two isomers of bisdehydro[12]annulene. The main product isolated, in 7% yield, was biphenylene. Catalytic hydrogenation of the bisdehydro[12]annulenes resulted in the formation of a complex mixture, probably containing [12]annulene, but this annulene appears to undergo transannular reactions on standing. The methods used to obtain [14]annulene[36], [16]annulene[37], and [20]annulene[38] are shown in the following charts.

[14]annulene

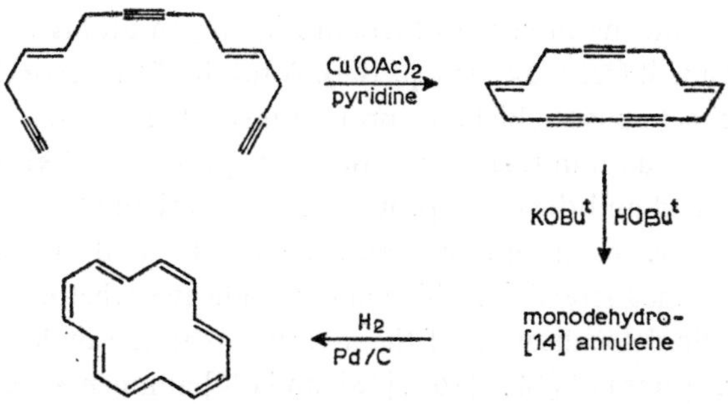

[16]annulene

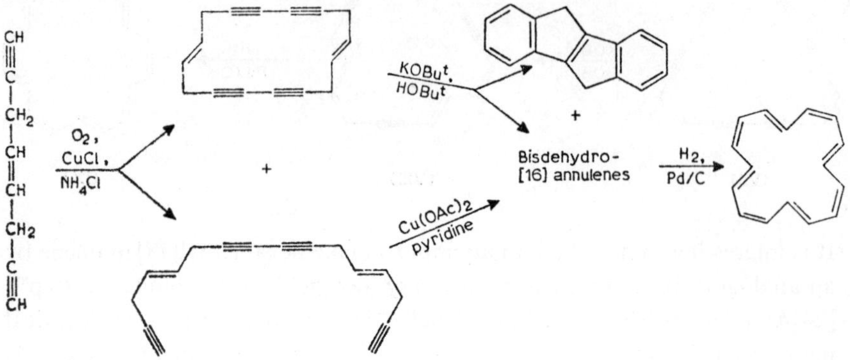

[16]annulene

[20]annulene

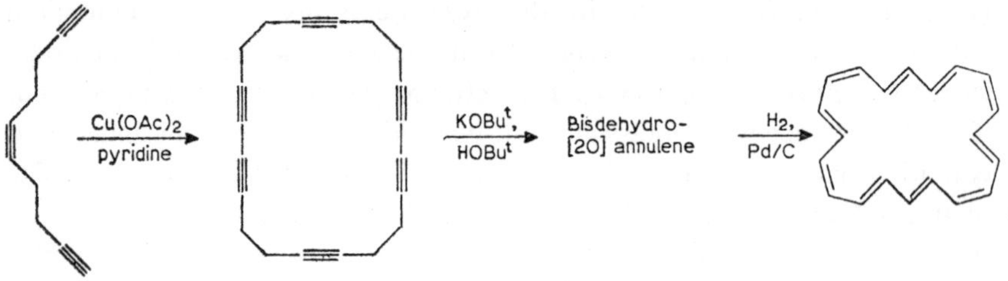

[20]annulene

N.m.r. spectra and aromaticity of annulenes

A consequence of π-electron delocalisation in a completely conjugated cyclic polyene is that the molecules have the ability to sustain a magnetically induced ring current; associated with this ring current is a secondary magnetic field[39,40]. In annulene molecules there are hydrogen atoms both within the ring and outside it (see diagrams, p. 166). A result of the secondary magnetic field which is set up is to shield the protons inside the ring and deshield those outside it, *i.e.* in an n.m.r. spectrum the inner protons will absorb at an unusually high field and the outer protons at an unusually low field. In consequence the n.m.r. spectrum of an annulene may be used to discover if there is any induced magnetic field, which in turn indicates whether or not there is complete cyclic delocalisation of the π-electrons in the molecule.

The n.m.r. spectra of [14]-, [16]-, [18]- and [24]-annulenes have been determined[40,41]. Of these four annulenes, [16]annulene and [24]annulene should be non-aromatic as they do not fit Hückel's requirements for aromaticity,

[14]annulene, although meeting these requirements, almost certainly could not be fully planar, and complete cyclic delocalisation might be impaired. [18]Annulene has a suitable number of π-electrons and could be planar.

In keeping with this the n.m.r. spectrum of [18]annulene has two broad bands, one at very low field (1.1 τ) and one at very high field (11.8 τ). This is exactly what would be expected for an aromatic molecule having a cyclic delocalised system of π-electrons. Furthermore the areas beneath these peaks are in the ratio 2:1, suggesting that there are 12 exocyclic and 6 endocyclic hydrogen atoms which is in complete accord with the proposed structure of [18]annulene (see above, p. 166). On the other hand the n.m.r. spectra of [16]annulene and of [24]annulene gave no indication of any induced ring current or aromaticity. The spectra of both of these annulenes consisted of one band, at 3.27 τ and 3.16 τ respectively, which is close to the position of absorption of linear conjugated polyenes.

The n.m.r. spectrum of [14]annulene in solution at room temperature consists of two bands, at 3.93 τ and 4.42 τ, in the ratio 1:6. These bands have been ascribed to the presence of two conformers differing in the spatial arrangement of the endocyclic hydrogen atoms, which form an equilibrium mixture in solution, *viz.* [56]

(A) ⇌ (B)

These conformers have been separated by subjecting [14]annulene to thin-layer chromatography on Kieselguhr coated with silver nitrate. Two unequal neighbouring spots were formed[56].

On the other hand, immediate examination of fresh solutions of crystalline [14]annulene showed essentially only the peak at 4.42 τ with only a very small peak at 3.93 τ. This suggests that the crystalline compound consists almost entirely of the one conformer. In solution equilibration of the two conformers takes place, the equilibrium ratio of 1:6 being achieved in about 30 minutes[56].

Since X-ray crystallographic examination of crystalline [14]annulene show that the molecule has a centre of symmetry this preponderant conformer must have structure (A)[56]. At $-60°$ the n.m.r. spectrum of this major constituent is completely different and consists of two peaks at 2.4 τ and 10.0 τ [57]. This is the characteristic pattern for an aromatic annulene and suggests that

[14]annulene has a complete cyclic conjugated system even if it cannot achieve complete coplanarity. The areas beneath the two peaks at $-60°$ are in the ratio *ca.* 5:2, in accord with the expected numbers of exocylic and endocyclic hydrogen atoms. The change in the n.m.r. spectrum at room temperature has been explained by assuming that at this temperature the protons change position at such a rate that an average value results[57]. This interpretation is supported by the observation that on heating a solution of [18]annulene to about 110°, the two peaks in the n.m.r. spectrum disappear in this case also, and are replaced by one peak at 4.55 τ [57].

The existence of a single n.m.r. band in the non-aromatic region cannot therefore be considered as evidence for lack of aromaticity in an annulene, unless it is shown that there is no change in the spectrum on cooling[57]. It appears that the single "average" peaks given by [14]annulene and [18]-annulene occur at higher τ-values than the single peaks given by annulenes which should not be aromatic according to Hückel's rule.

U.V. spectra

All the annulenes investigated have a series of intense absorption maxima. The positions of the main peaks and their intensities are shown in the following table.

Ring-size (no. of ring atoms)	14[36]	16[37]	18[18]	20[38]	24[18]	30[18]
λ_{max}	317	285	349	284	363	432
ε_{max}	4.84	4.86	5.48		5.30	5.15

X-ray crystallographic analysis

An X-ray crystallographic investigation of [18]annulene entirely confirms the predicted shape of the molecule[42]. The molecule has a centre of symmetry and in consequence cannot be made up from alternating single and double bonds. It is planar or almost so, confirming one set of predictions[18,52] and contradicting another[17]. The cisoid bonds, shown by thick lines in formula (X) are somewhat longer (1.419±0.004 Å) than the transoid bonds (1.382± 0.003 Å) as shown by thin lines in (X). The reasons for these differing bond lengths are not certain. They serve to emphasise the absence of an alternating single-double bond system, such as is found in cyclooctatetraene where alternate bonds have lengths of 1.462 and 1.334 Å. An X-ray examination

(X)

has also been made of [14]annulene[43]. Results so far available confirm the general shape of the ring (see formula, p. 171) but do not yet provide quantitative information about the bond lengths, bond angles or planarity of the ring.

Chemical properties

[14]-, [18]-, [24]- and [30]-annulenes are crystalline compounds, [24]annulene being very dark blue and the other three brown. [16]Annulene and [20]annulene are yellow oils. Irrespective of their aromaticity, as shown by n.m.r. spectra, all of these annulenes are relatively unstable and decompose in a short time on standing. [18]Annulene is the most stable; its decomposition is accelerated by light[18]. Its hexamethyl derivative is also less stable[32].

Annulenes can be reduced catalytically to cycloalkanes; this reaction has been used in proof of their structures[18,36,38]. [18]Annulene reacts with bromine and with maleic anhydride to give addition products[18]; [14]annulene also reacts with maleic anhydride[44]. Attempted acylation by means of a Friedel–Crafts reaction, and sulphonation of [18]annulene led only to desstruction of the starting material[18]. It also proved impossible to get substitution derivatives from [14]annulene using conditions under which dehydro-[14]annulene (see below) reacted[44]. On the other hand it has been reported that annulenes, which from their n.m.r. spectra are aromatic, can be nitrated under carefully defined conditions and form complexes with compounds such as trinitrobenzene[41].

It thus appears that, although those annulenes which have $(4n + 2)\pi$-electrons have n.m.r. spectra which indicate that they are aromatic compounds, they neither have the unusual stability nor take part in the typical reactions of classical aromatic compounds. This does not, however, invalidate their classification as aromatic compounds; many benzenoid compounds are unstable and many others undergo typical aromatic substitution reactions only under specified conditions yet no-one regards them as other than aromatic compounds (see also Chapter I).

Dehydroannulenes

Dehydroannulenes are fully conjugated cyclic polyenynes. Many have been prepared as intermediates in the preparation of annulenes (see above, pp. 168–170).

They are, however, of considerable interest in their own rights. It was suggested in 1948[45] that they should be aromatic compounds, provided that there were the correct number of electrons to form a delocalised π-orbital and that the rings were planar. Thus trisdehydro[18]annulene can have a continuous molecular orbital involving 18 π-electrons, as shown in (XI) and should therefore have aromatic character.

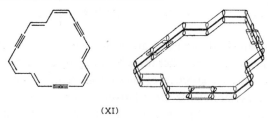

(XI)

Similarly pentadehydro[30]annulene should be aromatic in character. On the other hand tetradehydro[24]annulene, like [24]annulene should not have aromatic properties. "Benzyne" may be considered as dehydro[6]annulene, but owing to the distortion of bond angles involved, it is unlikely to be isollable as a stable compound.

Of those dehydroannulenes which are not expected to be aromatic, bisdehydro[12]annulenes[35] and bisdehydro[16]annulenes[37] (in each case more than one isomer) have been shown by their n.m.r. spectra to be non-planar and non-aromatic; tetradehydro[24]annulene[29,33] was similarly shown to be non-aromatic[40].

On the other hand monodehydro[14]annulene[36,44], bisdehydro[14]annulene[44,46], trisdehydro[18]annulene[29,30,40,47] all have n.m.r. spectra which show that they are able to sustain an induced ring current, *i.e.* that they are aromatic[40].

Of particular interest is the 1,8-bisdehydro[14]annulene since it is impossible to write a conventional formula for it with a conjugated system of single and double bonds linking the two formal triple bonds, for there is an odd number of carbon atoms joining these triple bonds on either side. It is possible, in fact, to write Kekulé type formulae only by representing one of the acetylenic links as a cumulene as in (XII).

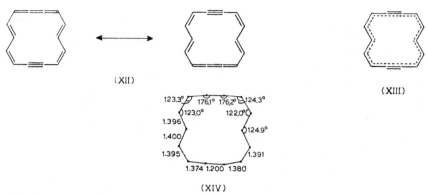

(XII)   (XIII)

(XIV)

It is more satisfactory to represent this molecule by the formula (XIII) showing a delocalised group of 14 π-electrons with two extra dehydro groups.

Models show that this structure can be planar and accommodate the two internal hydrogen atoms. X-ray crystallographic analysis confirms this structure[48]. The molecule is planar and has a centre of symmetry. The bond-lengths and bond-angles are as shown in (XIV). The bond-lengths, apart from the triple bonds, correspond to a bond order almost identical with that in benzene. The shortening of the bonds next to the triple bonds arises from the change of atomic radius of carbon on rehybridisation from an sp^2 to an sp state rather than from increased π-bonding, and the π-electron delocalisation in the molecule is complete. It is interesting to note that the acetylenic sides of the molecule are slightly bent. All the hydrogen atoms have been located by difference Fourier analysis, and another point of interest is that the separation of the two internal hydrogen atoms is only 1.85 Å, suggesting a lower value for the van der Waals radius than the usually accepted figure of 1.25 Å.

As mentioned previously the n.m.r. spectrum also demonstrates the aromaticity of bisdehydro[14]annulene, clearly showing the internal hydrogen atoms at very high field and the external ones at very low field.

Additionally 1,8-bisdehydro[14]annulene undergoes normal aromatic substitution reactions[44], as shown in the following chart:

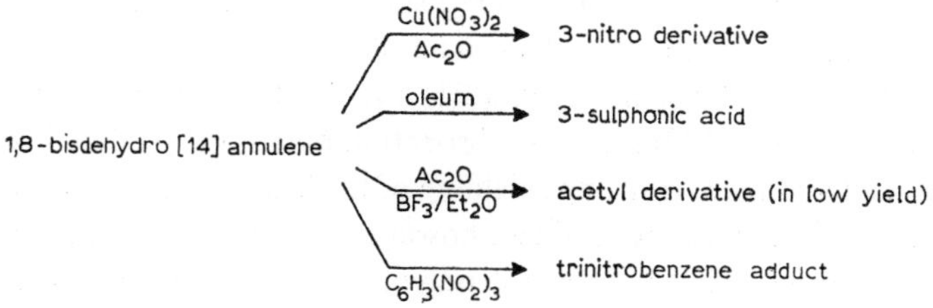

1,8-bisdehydro [14] annulene
- $Cu(NO_3)_2$ / Ac_2O → 3-nitro derivative
- oleum → 3-sulphonic acid
- Ac_2O / BF_3/Et_2O → acetyl derivative (in low yield)
- $C_6H_3(NO_2)_3$ → trinitrobenzene adduct

References p. 179

Under comparable conditions monodehydro[14]annulene also forms nitro and sulphonic acid derivatives and a trinitrobenzene adduct, yet [14]annulene itself does not form substitution products under these conditions[44]. The reactivity of [14]annulene and its monodehydro and bisdehydro derivatives towards maleic anhydride in boiling benzene — a diagnostic reagent for non-aromatic conjugated double bonds — was also investigated[44]. It was found that after one hour the bisdehydro compound was recovered unchanged, there was 85% recovery of the monodehydro compound but less than 10% unreacted [14]annulene remained.

The 1,8-bisdehydro[14]annulene is obtained[46] in the prototropic rearrangement of cyclotetradeca-4,12-dien-1,7,9-triyne, together with monodehydro-[14]annulene (see p. 169). It is the first member of a potential series of aromatic compounds of considerable interest.

Bridged ring annulenes

It is possible to have bridged ring compounds such as (XV, X = O, CH_2, etc.) with a completely conjugated peripheral system of double bonds and with a bridge linking two points of this peripheral system. Such compounds are known as bridged ring annulenes. Like annulenes, they may be aromatic if there is the correct number of π-electrons in the peripheral conjugated system and if the central group X does not interfere with this system electronically or sterically.

(XV)   (XVI)

A number of bridged derivatives of [10]annulene (see p. 166) have been prepared very recently and do indeed show aromatic character.

An example is 1,6-methano[10]annulene (XVI, X = CH_2)[58,59]. The ultraviolet spectrum of this compound suggests an extensive conjugated system while its n.m.r. spectrum [an A_2B_2 system (8 protons) in the range 2.5–3.2 τ (centred at 2.8 τ) and a sharp signal (2H) at 10.5 τ] indicates the presence of a ring current (see p. 170) since the absorption due to the olefinic hydrogens is at low field and there is strong shielding of the bridging methylene group[58]. The molecule is apparently strained, however, and exhibits olefinic reactivity[58]. On the other hand it does not polymerise nor react in a Diels–Alder reaction with maleic anhydride, but does undergo substitution reactions with electrophilic reagents[59]. Thus it can be acetylated with acetic anhydride in the presence of stannic chloride to give a 2-acetylderivative and is chlorinated or brominated in the 2-position. With cupric nitrate in acetic anhydride it gives a mixture of the 2-nitro- and 2,7-dinitro-derivatives. Similarly N-bromo succinimide gives a mixture of 2,7-dibromo- and 2-bromo-1,6-methano[10]-annulenes. The latter compound has been converted into the corresponding carboxylic acid via formation of a Grignard reagent.

1,6-Oxido[10]annulene (XVI, X = O)[60,61] has a similar ultra-violet spectrum to its methano-analogue. Its n.m.r. spectrum (A_2B_2 pattern in region 2.23–2.81 τ) is not only similar to the methano analogue (less the peak due to the methylene group) but also to naphthalene (similar pattern at 2.05–2.71 τ). The low field at which this band appears again indicates aromatic character. 1,6-Oxido[10]annulene also undergoes electrophilic substitution reactions, five minutes treatment with cupric nitrate in acetic anhydride giving rise to the 2-nitrocompound together with some isomer, probably the 1-nitro compound[60]. Acids convert this oxidoannulene into β-naphthol[61].

The corresponding amido- and acetamido-annulenes (XVI, X = NH, NAc) have also been prepared, their structures being confirmed by their ultra-violet and n.m.r. spectra[61,68]. The amidoannulene may be protonated, acetylated and methylated at the nitrogen atom[68].

The successful preparation of the bridged ring [10]annulenes is in contrast to the hitherto unsuccessful efforts to prepare [10]annulene itself. Whereas in the latter compound the hydrogen atoms on the 1,6-position would overlap one another in the planar structure required for aromaticity and such a planar structure is therefore impossible, in the case of the bridged ring annulenes not only is this overlap eliminated but the bridge actually assists in maintaining the planar form of the remainder of the molecule. It may, however, introduce extra angle strain into the molecule.

1,6-Methano[10]annulene reacts with diazomethane in the presence of cuprous chloride to give a bicyclo[5,4,1]dodecapentaene, from which a proton may be abstracted by means of triphenylmethyl fluoroborate producing a bridged ion having a peripheral delocalised system of 10 π-electrons[64]:

The resultant salt exists as yellow-orange needles. It must have the same carbon skeleton as the intermediate bicyclic olefin since both compounds are reduced to the same saturated hydrocarbon. The ultra-violet spectrum of the salt resembles those of benzotropylium salts. The aromaticity of the ion is deduced from its n.m.r. spectrum, since the ring protons and the bridge methylene protons show the expected large shifts to lower (0.4–1.7 τ) and higher (10.3–11.8 τ) fields respectively, relative to the spectrum of the intermediate bicyclododecapentaene[64].

Other examples of bridged ring annulenes which have been prepared recently are 15,16-*trans*-dimethyl-15,16-dihydropyrene (XVII, X = H) and its 2,7-diacetoxy derivative (XVII, X = CH_3CO_2)[62]. These compounds have a peripheral conjugated system of 14 π-electrons and their aromaticity is evident from their visible and ultra-violet and n.m.r. spectra. The diacetoxy compound is stable to heat, light and air, but the hydrocarbon is transformed into the isomeric metacyclophane (XVIII) by the action of light[63]. In the dark the cyclophane reverts to the more stable dihydropyrene form.

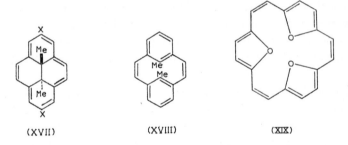

(XVII)   (XVIII)   (XIX)

[18]Annulene-1,4;7,10;13,16-trioxide[69] (XIX) and the corresponding trisulphide[70], oxide-disulphide[69] and dioxide-sulphide[71] have been prepared. Models suggest that the trioxide should be planar, the trisulphide and oxide-disulphide non-planar, and that the dioxide-sulphide should be nearly planar[71]. In fact the trioxide and dioxide-sulphide, which are red crystalline compounds, both have n.m.r. spectra which indicate the presence of an appreciable peripheral ring current[69,71]. Their u.v. spectra are very similar to those of [18]annulene itself. In contrast the n.m.r. spectra of the trisulphide and oxide-disulphide show that there is no appreciable ring current in these cases. The u.v. spectrum of the trisulphide is consistent with its formulation as a compound having three thiophen rings linked by three essentially olefinic vinylic groups; the u.v. spectrum of the oxide-disulphide is similar[69,70,72].

REFERENCES

1 R. WILLSTÄTTER AND E. WASER, *Ber.*, 44 (1911) 3423; R. WILLSTÄTTER AND M. HEIDELBERGER, *Ber.*, 46 (1913) 517.
2 For an interesting summary see W. BAKER, *J. Chem. Soc.*, (1945) 258.
3 I. L. KARLE, *J. Chem. Phys.*, 20 (1952) 65.
4 E. R. LIPPINCOTT, R. C. LORD AND R. S. MCDONALD, *J. Am. Chem. Soc.*, 73 (1951) 3370.
5 E. J. PROSEN, W. H. JOHNSON AND F. D. ROSSINI, *J. Am. Chem. Soc.*, 69 (1947) 2068; 72 (1950) 626.
6 For further details see R. A. RAPHAEL in D. GINSBURG (Editor), *Non-Benzenoid Aromatic Compounds*, Chapter VIII, Interscience, New York, 1959, p. 465.
7 See A. R. UBBELOHDE, *Chem. and Ind.*, (1956) 153.
8 T. J. KATZ, *J. Am. Chem. Soc.*, 82 (1960) 3784.
9 P. G. FARRELL AND S. F. MASON, *Z. Naturforsch.*, 16b (1961) 848.
10 H. P. FRITZ AND H. KELLER, *Chem. Ber.*, 95 (1962) 158.
11 H. P. FRITZ AND H. KELLER, *Z. Naturforsch.*, 16b (1961) 231.
12 T. S. CANTRELL AND H. SCHECHTER, *J. Am. Chem. Soc.*, 85 (1963) 3300.
13 W. REPPE, O. SCHLICHTING, K. KLAGER AND T. TOEPEL, *Ann.*, 560 (1948) 1; A. C. COPE AND F. A. HOCHSTEIN, *J. Am. Chem. Soc.*, 72 (1950) 2515; A. C. COPE, A. C. HAVEN, F. L. RAMP AND E. R. TURNBULL, *J. Am. Chem. Soc.*, 74 (1952) 4867; L. E. CRAIG, R. M. ELOFSON AND I. J. RESSA, *J. Am. Chem. Soc.*, 75 (1953) 480.
14 T. J. KATZ AND P. J. GARRATT, *J. Am. Chem. Soc.*, 85 (1963) 2852; 86 (1964) 5194.
15 E. A. LALANCETTE AND R. E. BENSON, *J. Am. Chem. Soc.*, 85 (1963) 2853.
16 K. MISLOW, *J. Chem. Phys.*, 20 (1952) 1489.
17 C. A. COULSON AND A. GOLEBIEWSKI, *Tetrahedron*, 11 (1960) 125.
18 F. SONDHEIMER, R. WOLOVSKY AND Y. AMIEL, *J. Am. Chem. Soc.*, 84 (1962) 274.
19 H. C. LONGUET-HIGGINS AND L. SALEM, *Proc. Roy. Soc.*, A 251 (1959) 172; A 257 (1960) 445.
20 E. D. BERGMANN AND Z. PELCHOWICZ, *J. Am. Chem. Soc.*, 75 (1953) 4281.
21 G. WITTIG, G. KOENIG AND K. CLAUSZ, *Ann.*, 593 (1955) 127.
22 G. WITTIG AND G. LEHMANN, *Chem. Ber.*, 90 (1957) 875.
23 H. A. STAAB AND F. BINNIG, *Tetrahedron Letters*, (1964) 319.
24 V. PRELOG, M. P. V. BOARLAND AND S. POLYÁK, *Helv. Chim. Acta.*, 38 (1955) 434.
25 E. E. VAN TAMELEN AND B. PAPPONS, *J. Am. Chem. Soc.*, 85 (1963) 3296.
26 F. SONDHEIMER AND Y. AMIEL, *J. Am. Chem. Soc.*, 78 (1956) 4178; 79 (1957) 5817; Y. AMIEL, F. SONDHEIMER AND R. WOLOVSKY, *J. Am. Chem. Soc.*, 79 (1957) 6263.
27 G. EGLINTON AND A. R. GALBRAITH, *Chem. and Ind.*, (1956) 737; *J. Chem. Soc.*, (1959) 889.
28 F. SONDHEIMER, Y. AMIEL AND R. WOLOVSKY, *J. Am. Chem. Soc.*, 79 (1957) 4247; 81 (1959) 4600.
29 F. SONDHEIMER AND R. WOLOVSKY, *J. Am. Chem. Soc.*, 84 (1962) 260.
30 F. SONDHEIMER AND R. WOLOVSKY, *J. Am. Chem. Soc.*, 81 (1959) 1771.
31 F. SONDHEIMER AND R. WOLOVSKY, *Tetrahedron Letters*, No. 3 (1959) 3.
32 F. SONDHEIMER AND D. A. BEN-EFRAIM, *J. Am. Chem. Soc.*, 85 (1963) 52.
33 F. SONDHEIMER AND R. WOLOVSKY, *J. Am. Chem. Soc.*, 81 (1959) 4755.
34 F. SONDHEIMER, R. WOLOVSKY AND Y. GAONI, *J. Am. Chem. Soc.*, 82 (1960) 754, 755.
35 R. WOLOVSKY AND F. SONDHEIMER, *J. Am. Chem. Soc.*, 84 (1962) 2844.
36 F. SONDHEIMER AND Y. GAONI, *J. Am. Chem. Soc.*, 82 (1960) 5765.
37 F. SONDHEIMER AND Y. GAONI, *J. Am. Chem. Soc.*, 83 (1961) 4863.
38 F. SONDHEIMER AND Y. GAONI, *J. Am. Chem. Soc.*, 83 (1961) 1259; 84 (1962) 3520.
39 J. A. ELVIDGE AND L. M. JACKMAN, *J. Chem. Soc.*, (1961) 859.

40 L. M. JACKMAN, F. SONDHEIMER, Y. AMIEL, D. A. BEN-EFRAIM, Y. GAONI, R. WOLOVSKY AND A. A. BOTHNER-BY, *J. Am. Chem. Soc.*, 84 (1962) 4307.
41 F. SONDHEIMER, *Pure Appl. Chem.*, 7 (1963) 363.
42 J. BREGMAN, F. L. HIRSHFELD AND D. RABINOVICH. Private communication quoted in ref. 41.
43 J. BREGMAN, *Nature*, 194 (1962) 679.
44 Y. GAONI AND F. SONDHEIMER, *J. Am. Chem. Soc.*, 86 (1964) 521.
45 T. J. SWORSKI, *J. Chem. Phys.*, 16 (1948) 550.
46 F. SONDHEIMER, Y. GAONI, L. M. JACKMAN, N. A. BAILEY AND R. MASON, *J. Am. Chem. Soc.*, 84 (1962) 4595.
47 F. SONDHEIMER, Y. AMIEL AND Y. GAONI, *J. Am. Chem. Soc.*, 84 (1962) 270.
48 N. A. BAILEY AND R. MASON, *Proc. Chem. Soc.*, (1963) 180.
49 E. HEILBRONNER, *Tetrahedron Letters* (1964) 1923.
50 V. D. AZATYAN AND R. S. GYULI-KVKHYAN, *Izv. Akad. Nauk Arm. S.S.R. khim. Nauk*, 14 (1961) 451.
51 T. S. CANTRELL AND H. SHECHTER, *J. Am. Chem. Soc.*, 87 (1965) 136.
52 W. BAKER AND J. F. W. MCOMIE, in D. GINSBURG (Editor) *Non-benzenoid Aromatic Compounds*, Interscience, New York, 1959, p. 480.
53 A. E. BEEZER, C. T. MORTIMER, H. D. SPRINGALL, F. SONDHEIMER AND R. WOLOVSKY, *J. Chem. Soc.*, (1965) 216.
54 D. W. DAVIES, *Tetrahedron Letters*, No. 8 (1959) 4.
55 E. VOGEL, W. MECKEL AND W. GRIMME, *Angew. Chem.*, 76 (1964) 786.
56 Y. GAONI AND F. SONDHEIMER, *Proc. Chem. Soc.*, (1964) 299.
57 Y. GAONI, A. MELERA, F. SONDHEIMER AND R. WOLOVSKY, *Proc. Chem. Soc.*, (1964) 397.
58 E. VOGEL AND H. D. ROTH, *Angew. Chem.*, 76 (1964) 145.
59 E. VOGEL AND W. A. BÖLL, *Angew. Chem.*, 76 (1964) 784.
60 F. SONDHEIMER AND A. SHANI, *J. Am. Chem. Soc.*, 86 (1964) 3168.
61 E. VOGEL, M. BISKUP, W. PRETZER AND W. A. BÖLL, *Angew. Chem.*, 76 (1964) 785.
62 V. BOEKELHEIDE AND J. B. PHILLIPS, *J. Am. Chem. Soc.*, 85 (1963) 1545; *Proc. Nat. Acad. Sci. U.S.*, 51 (1964) 1963; F. GERSON, E. HEILBRONNER AND V. BOEKELHEIDE, *Helv. Chim. Acta*, 47 (1964) 1123.
63 H.-R. BLATTMANN, D. MEUCHE, E. HEILBRONNER, R. J. MOLYNEUX AND V. BOEKELHEIDE, *J. Am. Chem. Soc.*, 87 (1965) 130.
64 W. GRIMME, H. HOFFMANN AND E. VOGEL, *Angew. Chem.*, 77 (1965) 348.
65 E. A. LALANCETTE AND R. E. BENSON, *J. Am. Chem. Soc.*, 87 (1965) 1941.
66 H. E. SIMMONS, D. B. CHESNUT AND E. A. LALANCETTE, *J. Am. Chem. Soc.*, 87 (1965) 982.
67 R. WOLOVSKY, *J. Am. Chem. Soc.*, 87 (1965) 3638.
68 E. VOGEL, W. PRETZER AND W. A. BÖLL, *Tetrahedron Letters*, (1965) 3613.
69 G. M. BADGER, J. A. ELIX, G. E. LEWIS, U. P. SINGH AND T. M. SPOTSWOOD, *Chem. Comm.*, (1965) 269.
70 G. M. BADGER, J. A. ELIX AND G. E. LEWIS, *Proc. Chem. Soc.*, (1964) 82; *Austral. J. Chem.*, 18 (1965) 70.
71 G. M. BADGER, G. E. LEWIS, U. P. SINGH AND T. M. SPOTSWOOD, *Chem. Comm.*, (1965) 492.
72 C. A. COULSON AND M. D. POOLE, *Proc. Chem. Soc.*, (1964) 220.

CHAPTER VIII

Polycyclic compounds

INTRODUCTION

In addition to monocyclic non-benzenoid aromatic compounds it is possible to have compounds in which two or more rings participate in an aromatic delocalised system of π-electrons. A large number of possible bicyclic and polycyclic systems have been considered from time to time (see, *e.g.* ref. 137). Recently increased attention has been paid to polycyclic non-benzenoid aromatic hydrocarbons and a number have been prepared and their properties compared with those predicted by molecular orbital calculations.

Bicyclic compounds can be divided into four groups depending on the mode of junction of the rings, namely (1) compounds with discrete rings joined either directly or by a chain of carbon atoms and in which *no* carbon atom is common to the two rings, (2) spirans, in which *one* carbon atom is common to two rings, (3) condensed ring systems, in which *two* carbon atoms are common to the two rings, and (4) bridged ring compounds, in which two rings share *three or more* carbon atoms. Polycyclic compounds may fall into one or other of these groups or may belong to more than one of them.

The term "*polycyclic non-benzenoid aromatic hydrocarbons*" usually refers to condensed ring systems and the greater part of this chapter deals with this type of compounds. Some compounds having two discrete rings are also considered, in every case the two rings being linked together by a formal olefinic bond thus permitting electronic interaction between the rings.

Some examples of bridged ring compounds are considered at the end of the previous chapter (p. 176). These compounds have delocalised π-electron systems around their periphery only, not involving the bridge atoms, and are thus treated as special examples of annulenes. Spirans cannot be involved in aromatic systems since the atom common to both rings must be an sp^3 hybridised carbon atom.

Benzo derivatives of monocyclic aromatic ring systems are not discussed in this chapter but in the chapters dealing with the related monocyclic systems.

Representatives of one group of polycyclic non-benzenoid aromatic compounds, the azulenes, have been known for a long time and have been intensively studied. They will be considered first in this chapter and then shorter discussions of some other groups of compounds follow.

AZULENES

Of polycyclic non-benzenoid hydrocarbons by far the most numerous and the most investigated series consists of the azulenes. The parent compound is azulene itself, bicyclo[5,3,0]decapentaene (I).

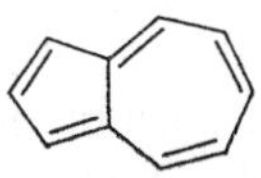

(I) 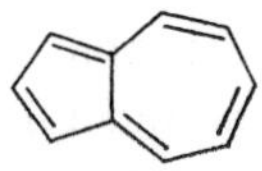   (II)

The azulene molecule has a peripheral conjugated system involving 10 π-electrons and hence fits Hückel's rule. It shows many properties typical of an aromatic compound. As in the case of benzene it is possible to represent it by two alternative Kekulé forms, (I) and (II).

Azulene has an intense blue colour; it is perceptible at dilutions as great as 1 part in 10000 of a solvent such as light petroleum.

The development of blue colours in certain essential oils, e.g. oil of camomile, after such simple operations as distillation, steam distillation or treatment with acids or oxidising agents has been noted at various times since the fifteenth century. Many such oils contain hydrogenated azulene derivatives (i.e. derivatives of bicyclo[5,3,0]decane) in their higher boiling fractions and the development of blue colours stems from dehydrogenation of these compounds. Thus, for example, an intensely blue coloured fraction, b.p. 165–170°/10 mm. was isolated in 1894[1] from the distillation of geranium oil. A treatise on oils[2] published in 1906 listed about twenty different oils showing this kind of behaviour.

The name "azulenes" was first used for these blue oils in 1864[3]. When the chemical structure of azulenes was elucidated the name "azulene" was applied to the parent hydrocarbon of the series[4]. Although it is now customary to name substituted azulenes systematically, many which were first obtained from natural sources were named according to the oil from which they derived, *e.g.* vetivazulene from oil of vetiver. Many of these azulenes from natural sources have a molecular formula $C_{15}H_{18}$ and are dimethylisopropylazulenes. In the systematic nomenclature the various positions in the azulene ring system are numbered as follows:

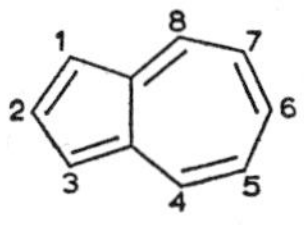

An important step in the elucidation of the structures of azulenes was the discovery[5] in 1915 that azulenes can be extracted into concentrated aqueous mineral acids, in which they give pale brown solutions, and can be recovered unchanged from these solutions by diluting them with water. It thus became possible to isolate azulenes from the other organic material accompanying them when they were obtained from natural sources.

Early attempts to determine their structures usually represented them as benzene derivatives but this was soon shown to be unlikely[6] since even at low temperatures they are oxidised by potassium permanganate to give only small fragments. It was suggested, therefore[6], that azulenes must contain a hitherto unknown bicyclic system not involving benzene rings.

Another most important step was the discovery[4] of the value of chromatography on alumina and the formation of trinitrobenzene adducts for the purification and identification of azulenes. The use of these techniques led to the final elucidation of the fundamental ring-system of azulenes[4].

A more complete account of the history of the developments in early azulene chemistry may be found in a review article on these compounds[7].

Syntheses of azulenes

The first azulenes were obtained by dehydrogenation of naturally occurring hydroazulenes. The first intentional synthesis of an azulene was carried out in 1936, starting from 9,10-octalin[4]. The method, together with later modifications[8,9] is shown in the following chart.

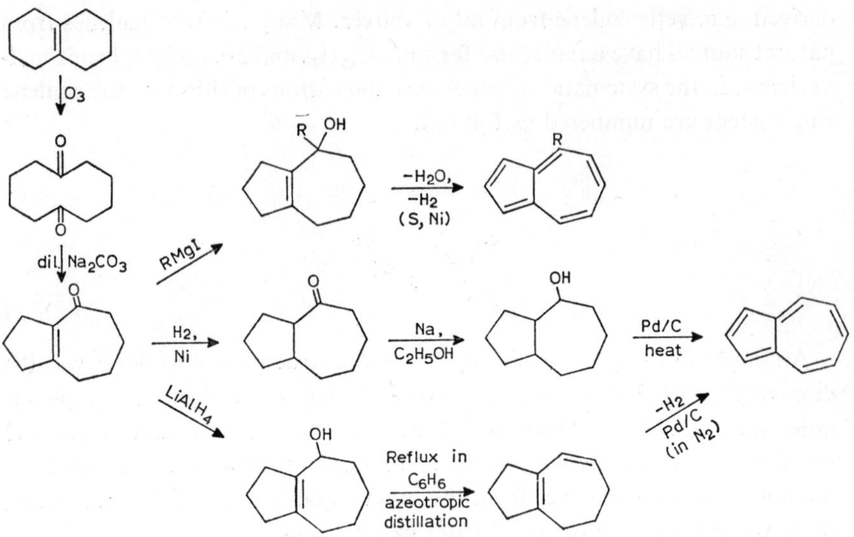

It is interesting to note in passing that azulene was in fact prepared in 1893 as a by-product in the preparation of cyclopentanone by distillation of calcium adipate[22]. A mechanism which has been suggested[8] for its formation is as follows:

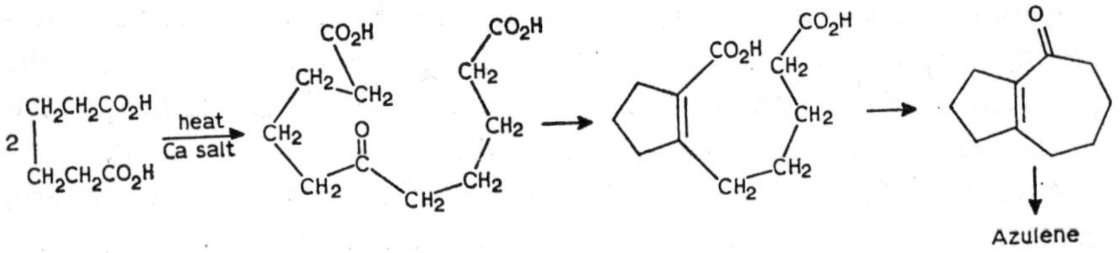

In a more recent investigation azulene was shown to be formed when adipic acid was heated in the presence of various catalysts; using a mixture of Cr_2O_3/Al_2O_3 at 475° it is formed to the extent of 3%[111].

All the earlier methods of preparation of azulenes involved a high temperature dehydrogenation stage and great losses were frequently encountered in this process. On the other hand by careful adjustment of the dehydrogenation catalyst and of the reaction conditions reasonable yields have been obtained[10], e.g.

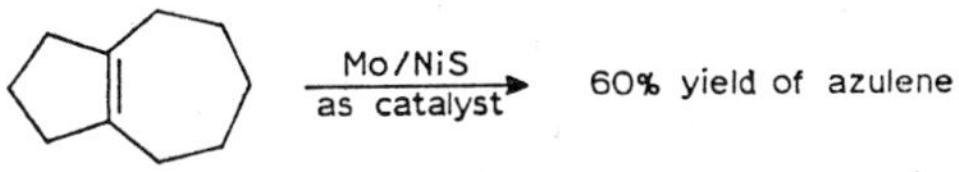

60% yield of azulene (Mo/NiS as catalyst)

In general these methods have been superceded by synthetic routes which do not involve dehydrogenation. A few examples of the earlier methods are now given before the more modern methods are discussed.

One approach involved the ring-enlargement of the six-membered ring of indane derivatives by means of diazoacetic ester[11], diazomethane[12,13], or a Demjanov rearrangement[14], *viz.*

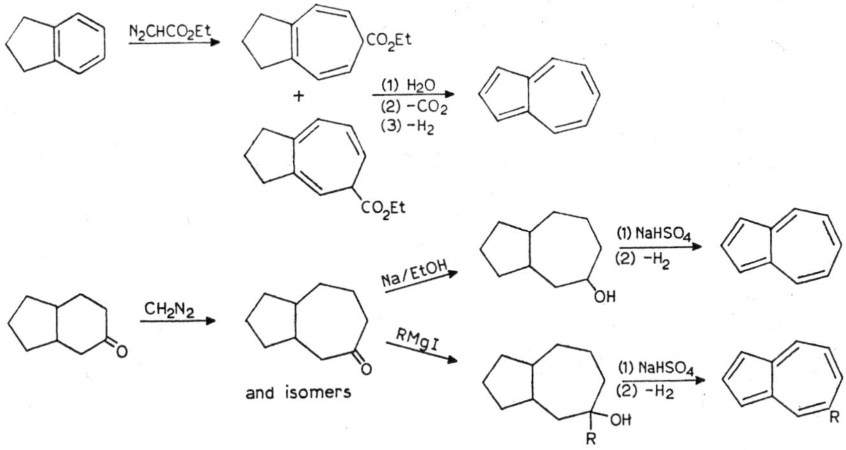

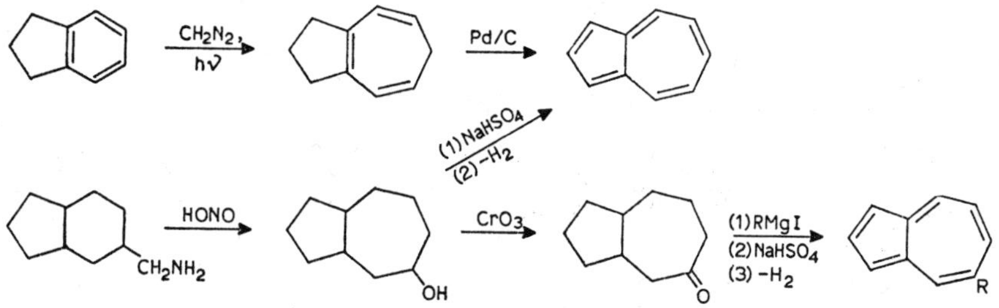

An alternative approach involved the closure of a five-membered ring on to a pre-formed seven-membered ring or (less frequently) *vice versa*. Some examples of these preparative methods are shown in the charts (a) and (b)[15-20]

References p. 212

(a) Closure of five-membered ring onto a cycloheptane derivative.

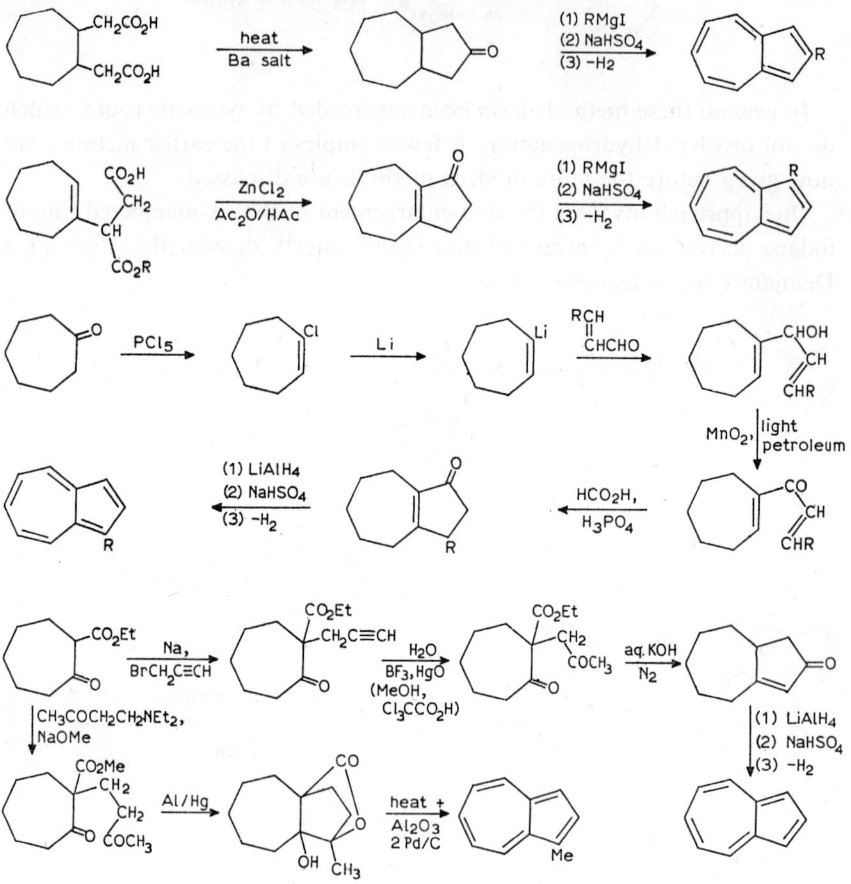

(b) Closure of seven-membered ring onto a cyclopentane derivative.

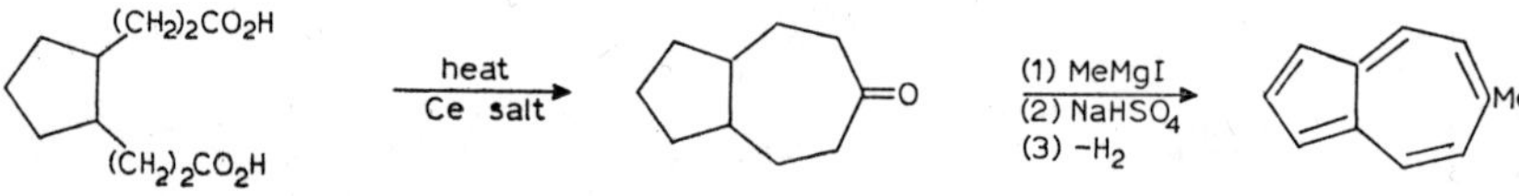

Yet another approach employed the catalytic dehydrogenation of cyclodecane derivatives, dehydrogenation being accompanied by transannular ring-closure, *e.g.* ref. 21, 101.

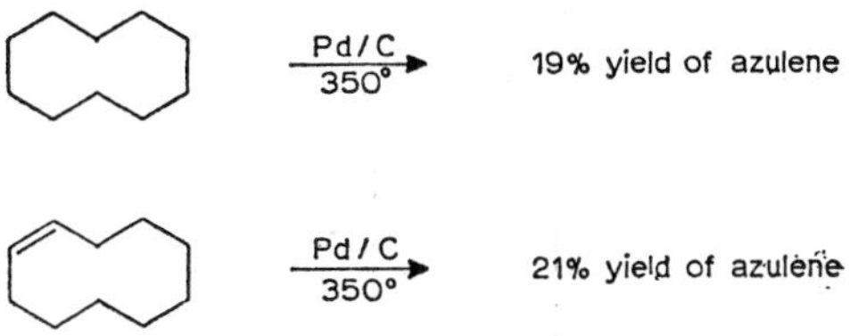

AZULENES

[Structure: decalin] → Pd/C, 350° → 19% yield of azulene

[Structure: octalin] → Pd/C, 350° → 21% yield of azulene

Syntheses not involving dehydrogenation

The first recorded synthesis[23] of an azulene not involving a dehydrogenation step utilised the dimerisation of diphenylacetylene in the presence of 2,4-dinitrophenylsulphenyl chloride and aluminium chloride:

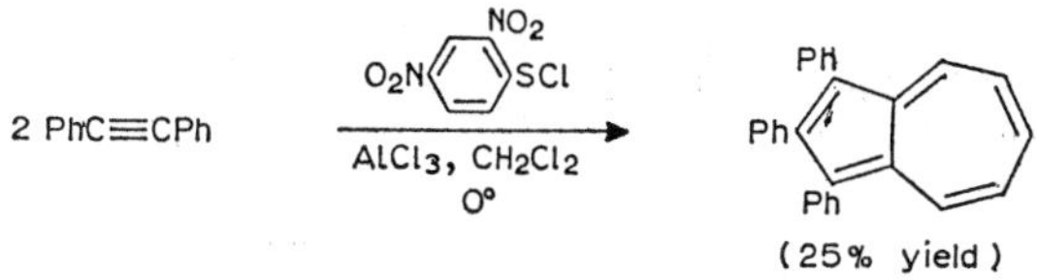

2 PhC≡CPh → (with 2,4-dinitrophenylsulphenyl chloride, $AlCl_3$, CH_2Cl_2, 0°) → 1,3-diphenylazulene (25% yield)

A suggested mechanism for this reaction, which must also involve ring-expansion of a benzene ring is as follows[24]:

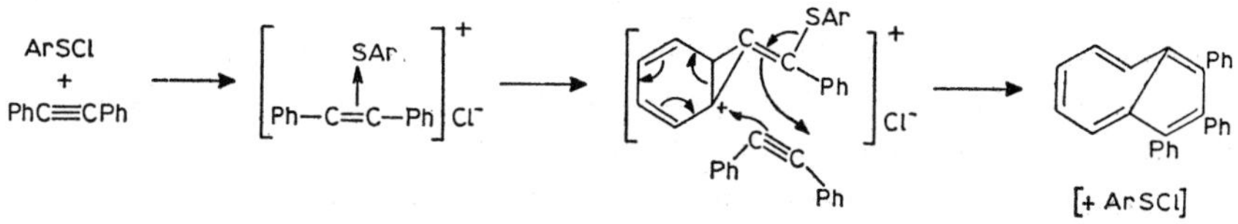

ArSCl + PhC≡CPh → [Ph—C=C—Ph, SAr]⁺ Cl⁻ → [intermediate with SAr, Ph groups]⁺ Cl⁻ → 1,3-diphenylazulene [+ ArSCl]

The most valuable series of methods of preparation of azulenes involves elegant adaptations of the alkali catalysed condensation of glutacondialdehyde with cyclopentadiene. This may be represented as:

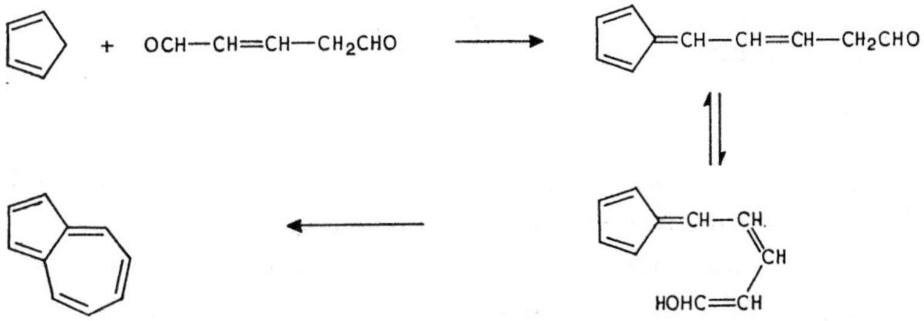

cyclopentadiene + $OCH-CH=CH-CH_2CHO$ → cyclopentadienyl=CH—CH=CH—CH_2CHO

⇅

azulene ← cyclopentadienyl=CH—CH, CH, HOHC=CH

Glutacondialdehyde is itself unstable, however, but a di-N-methylaniline derivative is readily obtained from pyridine:

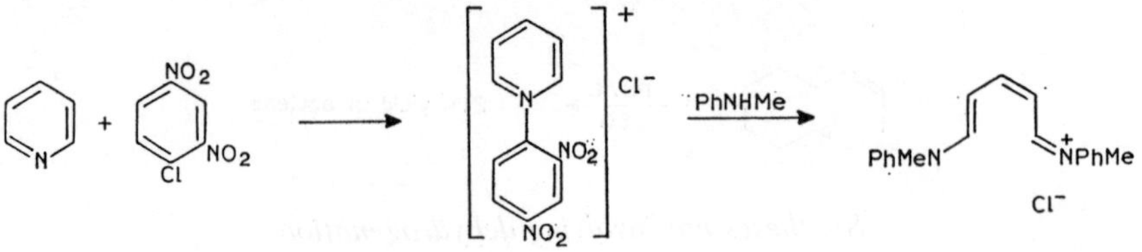

This derivative was then treated with alkali to give a monoaldehyde which condensed with cyclopentadiene producing a fulvene (III), which on heating was converted into azulene[25-27]:

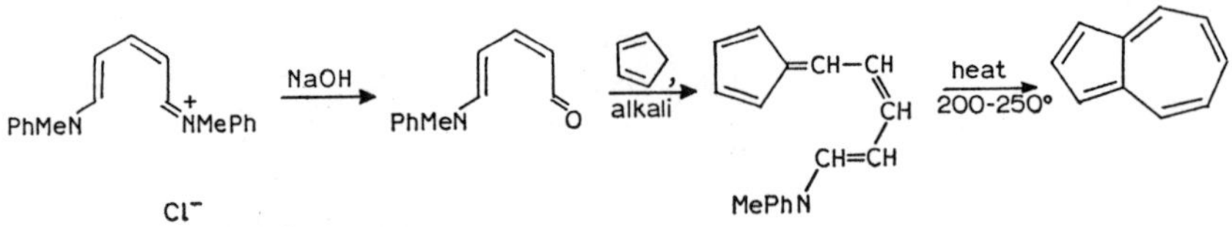

A maximum yield of 70% was obtained if ring-closure was carried out in a high-boiling basic organic solvent, e.g. benzidine, and by a continuous process in which only small amounts of the intermediate fulvene (III) were present at any time. Alkylcyclopentadienes react similarly to give 1-substituted azulenes, while substituted pyridines may also be used (giving products substituted in the seven-membered ring).

A simple modification of this reaction[28] involved the treatment of an N-alkylpyridinium salt with sodium cyclopentadienide. A not very stable N-alkyl-2-cyclopentadienyl-1,2-dihydropyridine is formed, which on heating to ca. 200° in benzidine is converted directly into azulene.

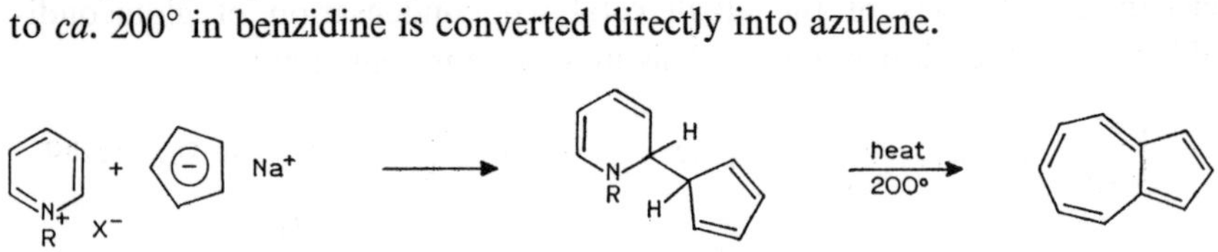

This method too has been used to obtain azulene derivatives with substituent groups in either the five- or seven-membered rings.

A further modification[29] uses a pyrylium salt instead of a pyridinium salt as starting material. For example 4,6,8-trimethylazulene was prepared in 65%

yield from cyclopentadiene and 2,4,6-trimethylpyrylium perchlorate as follows:

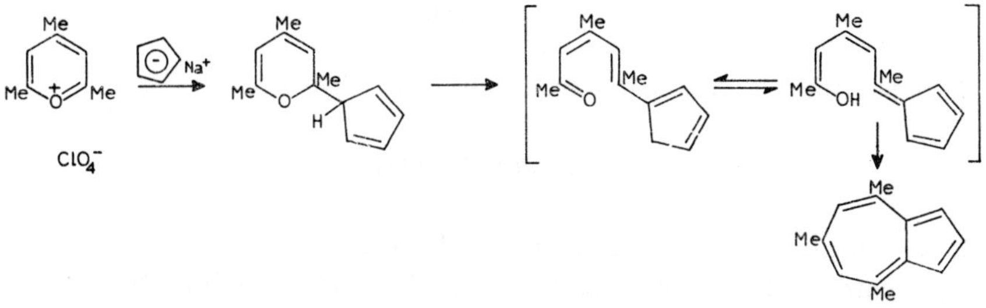

The reaction proceeds readily at room temperature and intermediates can only be isolated if the reaction is carried out below $-20°$. The reaction is not successful starting from the unsubstituted pyrylium perchlorate. Pyrylium salts have also been converted into azulenes by opening the ring with methylaniline and reacting the resultant product with sodium cyclopentadienide[157].

Using one or other of these methods starting from pyridinium or pyrylium salts a large number of azulene derivatives has been readily prepared (see *inter al.*, ref. 24).

A further azulene synthesis[30] starting from pyrylium salts makes use of their reaction with triphenylphosphinemethylene in methylene chloride or acetonitrile:

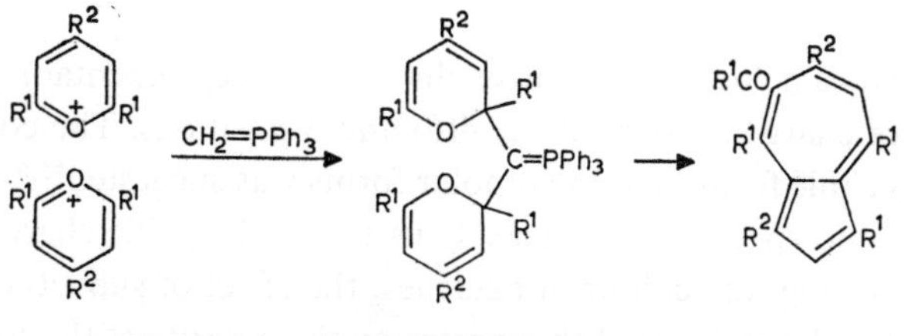

$R^1 = R^2 =$ phenyl or substituted phenyl group

The yield is about 30%.

Azulenes have also been prepared from other non-benzenoid aromatic compounds, for example from tropones[31], and from tropylium salts[32].

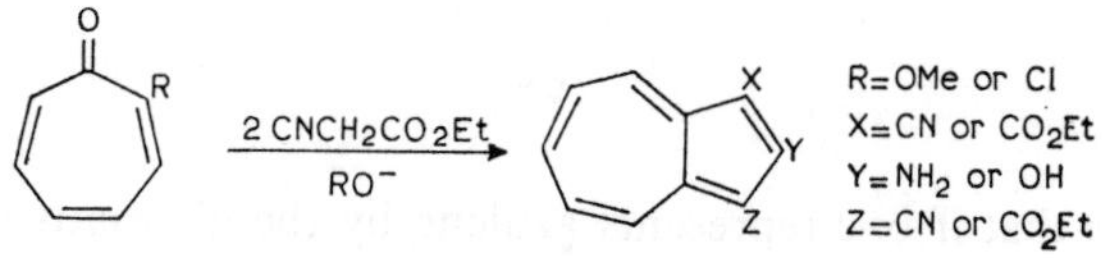

R = OMe or Cl
X = CN or CO_2Et
Y = NH_2 or OH
Z = CN or CO_2Et

(the ratio of these products formed depends largely on the reaction conditions)

References p. 212

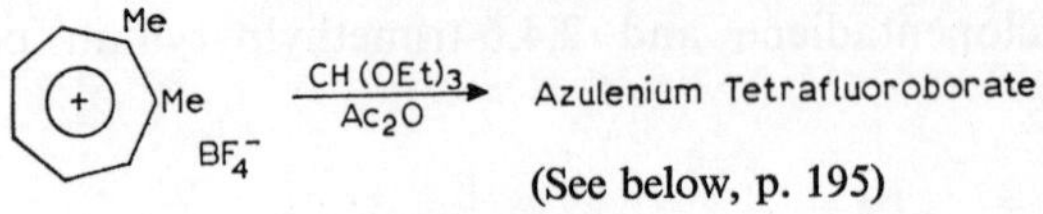

190   POLYCYCLIC COMPOUNDS

$$\text{[Me-substituted cycloheptatrienyl cation } BF_4^-] \xrightarrow{\frac{CH(OEt)_3}{Ac_2O}} \text{Azulenium Tetrafluoroborate}$$

(See below, p. 195)

They have also been prepared from fulvene aldehydes[33].

Structure of azulene

It was mentioned at the beginning of this chapter that azulene could be represented by Kekulé structures such as (IVa) and (IVb). It is also possible to draw dipolar structures such as (IVc)–(IVg). In structures (IVc), (IVd) the seven-membered ring resembles a tropylium ring and a negative charge appears at the 1(3)-position, while in structures (IVe)–(IVg) the five-mem-

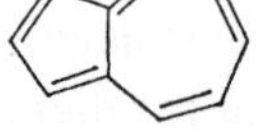

(IVa) 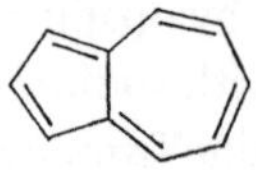   (IVb) 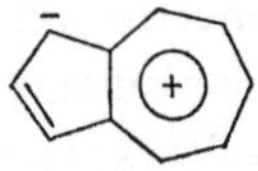   (IVc) 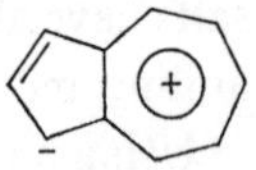   (IVd)

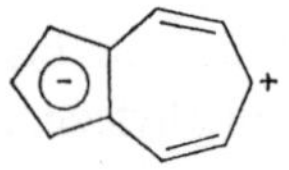

(IVe) 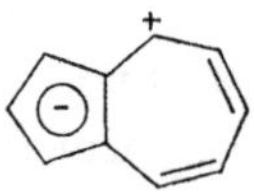   (IVf) 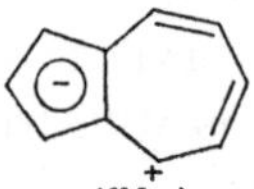   (IVg)

bered ring bears a delocalised negative charge as in a cyclopentadienide anion and a positive charge appears at the 4(6) and 8 positions. The contribution of both the Kekulé forms and the dipolar forms was suggested[34,35] to explain the reactions and spectra of azulene. Thus the nucleophilic character of the five-membered ring, the colour of azulenes, the effect of substituents on the spectra, and the dependence of the spectra on the polarity of the solvent were all reasonably explained by a contribution of a dipolar form to the overall structure of azulene.

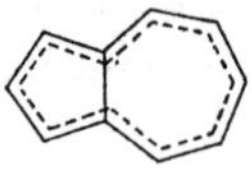

(V)

This means that if one represents azulene by the alternative formulation (V), which draws attention to the peripheral delocalised system of π-electrons it must be remembered that the π-electron system is not symmetrically disposed.

Physical evidence of the separation of charge is provided by dipole moment measurements[36-38]. The value of 1.08 D indicates that the contribution of the dipolar forms, though real, is not large. The direction of the dipole is confirmed by dipole moment measurements on 2-substituted azulenes such as 2-halo- and 2-cyano-azulenes[37]. The n.m.r. spectrum of azulene[39,144] also confirms the partial dipolar character, for it is necessary to invoke an a-symmetric distribution of electronic charge to explain the spectrum.

Azulene may thus be regarded as an aromatic hydrocarbon having a cyclic conjugated system of 10 π-electrons but this electron system is not evenly dispersed over the whole molecule. Its resonance energy (*ca.* 46 kcal/mole) indicates a considerable aromatic stabilisation[40,145]. Its ultra-violet spectrum is also in accord with this formulation[40].

In this picture of the azulene molecule the transannular bond serves to hold the molecule rigid and planar, but plays no part in the conjugated system. As will be discussed below it also plays a vital role in the stabilisation of transition states in substitution reactions[41]. In conformity with this interpretation an X-ray crystallographic examination of azulene[42] has shown that this transannular bond appears to be rather longer than the other carbon–carbon bonds in the molecule.

Quantum mechanical calculations on the distribution of electron density in the azulene molecule agreed that the five- and seven-membered rings should carry respectively negative and positive charges. They also suggested that electrophilic substitution should occur mainly at the 1 and 3 positions and nucleophilic substitution at the 4,6 and 8 positions[43]. Experimental evidence is entirely in accord with these suggestions.

In keeping with the nucleophilicity of the five-membered ring it was also suggested that the solubility of azulenes in strong acids was due to reversible protonation of this ring[44]. From calculations of the delocalisation energies of all the cations which might be formed by protonation of an azulene molecule it was further deduced that protonation should take place at the 1-position[45]. Confirmation came from studies of the basicity and the ultra-violet spectrum of the azulenium cation[46,152], and from its n.m.r. spectrum[47,48,144,152]. Conductivity measurements in 100% sulphuric acid confirm that only monoprotonation of azulene takes place[152]. In this azulenium cation the seven-membered ring is thus similar to a tropylium cation. Further evidence was provided by the fact (shown spectroscopically) that hydrogen-deuterium exchange takes place at the 1 and 3 positions in deuterophosphoric acid to give 1,3-dideuteroazulene[49]. In more recent years solid azulenium salts have actually been isolated[50-52].

An azulenium ion is formed as the transition state in electrophilic substitution reactions of azulene:

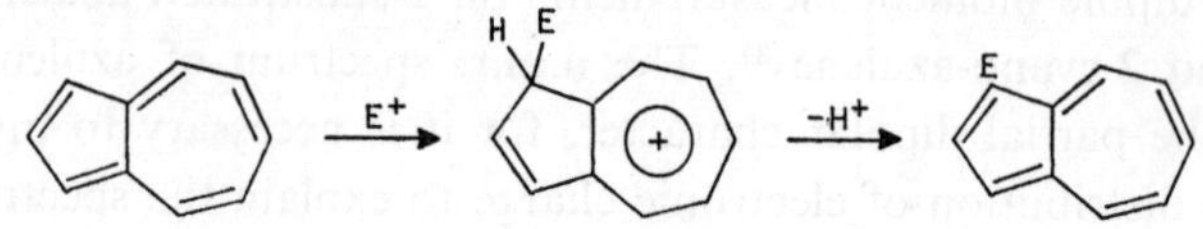

The transition state is stabilised by delocalisation of the positive charged over the seven-membered ring and its formation is thus energetically favoured. Similarly nucleophilic attack on the seven-membered ring involves a transition state in which the negative charge is delocalised over the five-membered ring:

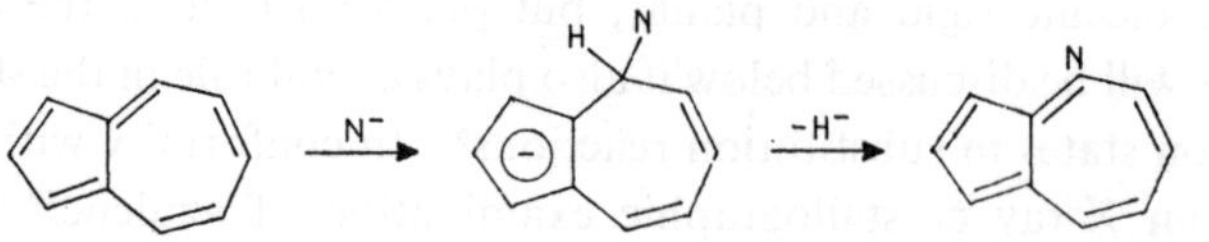

Thus, attack by a reagent on the appropriate ring is accompanied by the development of a delocalised sextet of π-electrons on the other ring in the transition state of the reaction. Rather therefore than an aromatic system being destroyed in the transition state, one such system (the peripheral) is exchanged for another (a tropylium cation or a cyclopentadienide anion). This means that the transition state in substitution reactions is energetically favoured and that the reactions are thereby facilitated.

The role of the transannular bond in these transition states should be noted[41]. If no such bond were present transition states involving aromatic sextets would not be possible, *e.g.*

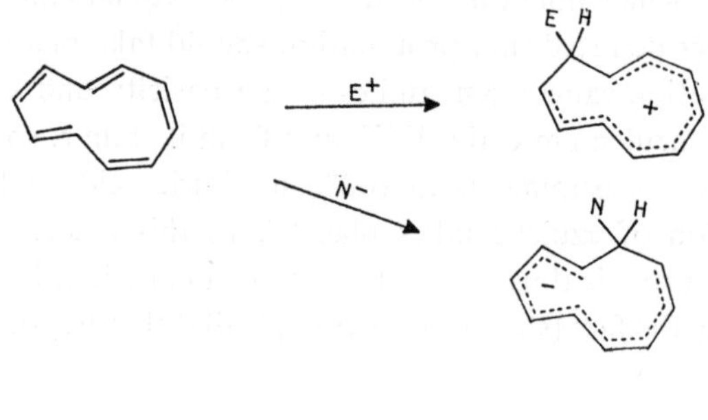

The fact that electrophilic or nucleophilic attack does result in substitution reactions indicates however that the change from the tropylium or cyclopentadienide type transition state to reform an azulene system is energetically favourable. In accord with this is the observation that 6-hydroxyazulene-1-aldehyde (VI) cannot be converted into its tautomeric fulveneotropone form[53]. Similarly the cycloheptatrienofulvene (VII) which is prepared by base cata-

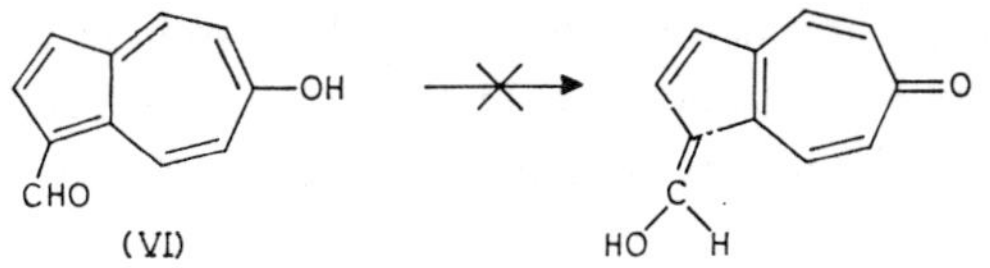

lysed condensation of a carbonyl compound with 1,6-dihydroazulene (see p. 194), although stable, can rearrange to an azulene derivative[54].

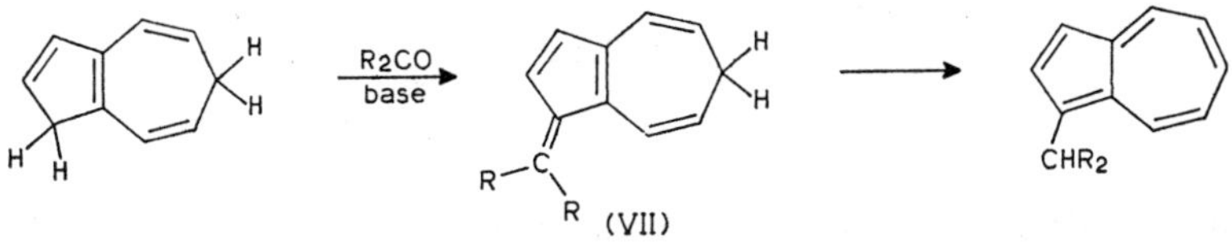

On the other hand the azulene π-electron system is readily disrupted if in the process a benzenoid system is formed in its place. For example on heating to $>350°$ in vacuo azulene undergoes a rearrangement reaction to give naphthalene in practically quantitative yield[55,56]. Similarly treatment of the azulene (VIII) with alkoxide results in the formation of the benzene derivative (IX)[57].

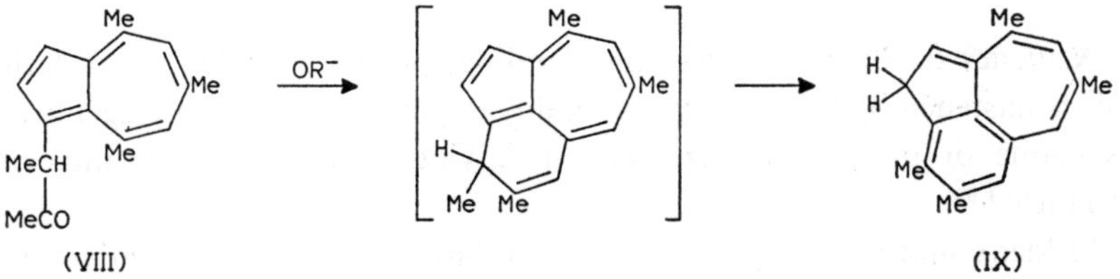

Compound (IX) reacts with strong bases to give a cycloheptatrienoindenide anion; it also forms a cyclopentadienobenzotropylium cation[57].

References p. 212

Reactions of azulenes

(a) Addition reactions; reduction and oxidation

Azulene will not react as a conjugated diene in a Diels–Alder reaction with maleic anhydride. Under forcing conditions they react together, to form, not a normal adduct, but instead 1-azulenylsuccinic anhydride[58].

Reduction of azulene by sodium in liquid ammonia gives 1,6-dihydroazulene[54]; azulenes are also readily reduced by sodium and ethanol or by catalytic hydrogenation to give polyhydroazulenes[5,6,59], Dehydrogenation of the latter products by means of sulphur results in reformation of the parent azulenes[6].

It is possible however to reduce an unsaturated side-chain catalytically without reducing the azulene ring-system[62], e.g.

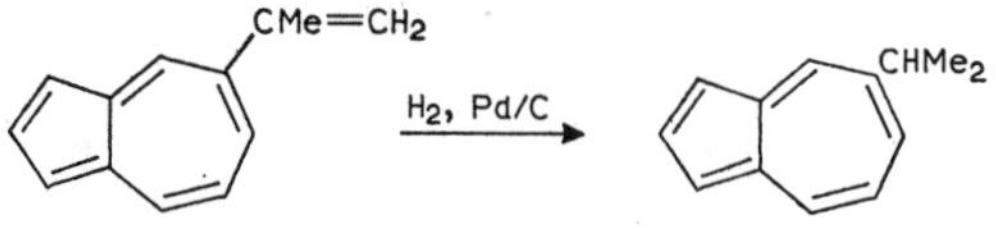

Similarly side-chain carbonyl groups can be reduced by means of the Wolff–Kishner reaction, while azulene carboxylic acids and esters have also been reduced to the corresponding alcohols by means of lithium aluminium hydride[63], e.g.

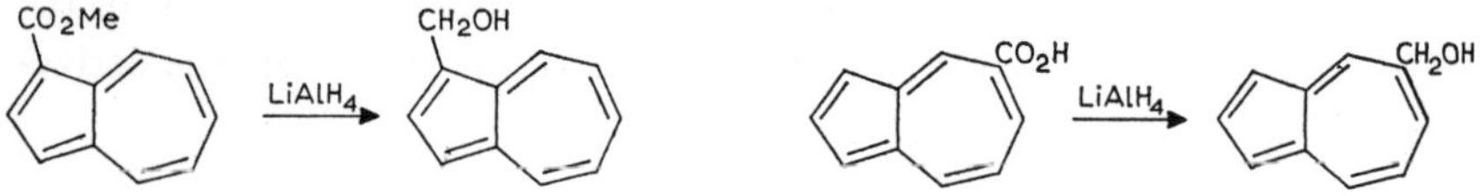

Azulenes decompose slowly on standing, presumably due to oxidation; this is hastened by light. They are readily degraded by oxidising agents such as ozone or potassium permanganate giving products of low molecular weight[6,59,60].

Milder conditions may result in the oxidation of azulene derivatives without destruction of the ring system. Thus guaiazulene (1,4-dimethyl-7-isopropylazulene) with selenium dioxide or with oxygen gave the ether (X) plus a small amount of an azulene aldehyde[61].

Azulenes readily form charge-transfer complexes with such agents as 1,3,5-trinitrobenzene, picric acid or iodine[65].

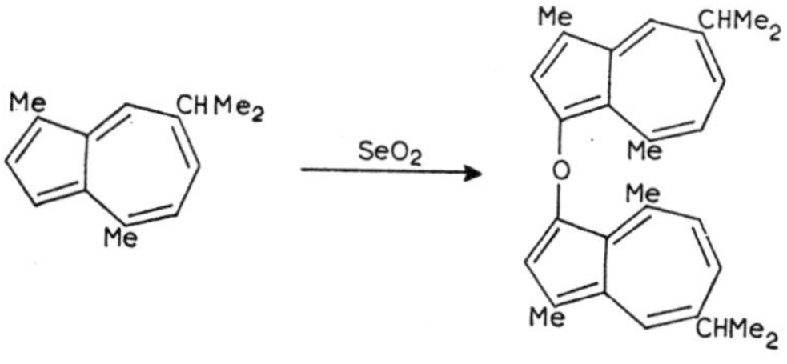

(b) Substitution reactions

Azulenes can undergo substitution reactions with electrophilic, nucleophilic and radical reagents. Some discussion of the mechanisms involved has already been given above (p. 192) and various examples of the different types of substitution are now considered.

(i) Electrophilic substitution. Electrophilic substitution of azulenes takes place at the 1- and 3-positions if these are vacant, reaction proceeding according to the path:

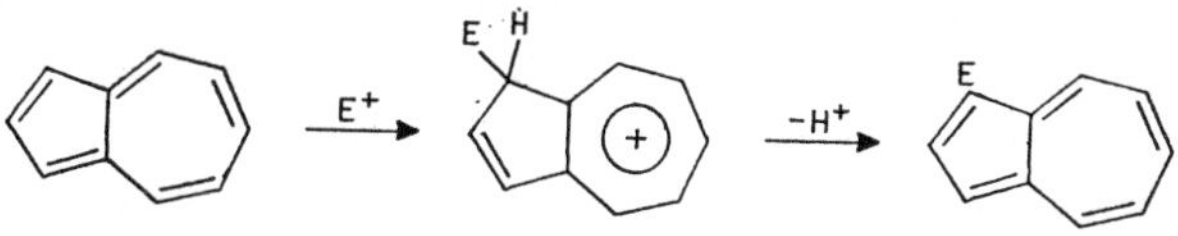

The simplest example of such a reaction is the protonation of azulenes to give azulenium salts; on dilution of the acid solution with water azulene is regenerated:

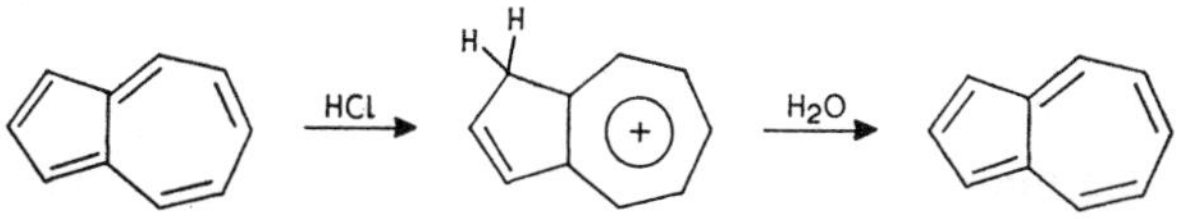

In azulenium ion formation the blue colour of the original azulene changes to yellow, the absorption maximum of the salt being at *ca.* 350 mμ compared with *ca.* 580 mμ in the azulene. Similarly azulene undergoes rapid deuterium exchange at the 1- and 3-positions; no such exchange occurs at other sites in the molecule[49]. The basic strength of azulene in fact approaches that of nitroaniline (pK in formic acid = 0.55)[64].

Azulenium salts have been isolated as crystalline compounds by the action of perchloric acid[50] and of tetrafluoroboric acid in ether[51] on azulene derivatives, *e.g.*

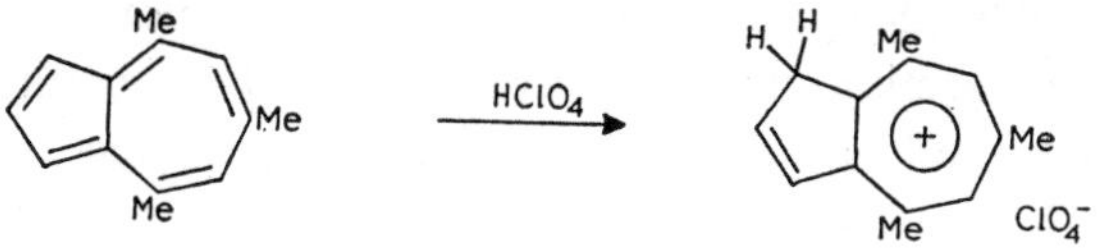

These salts are colourless or pale brown; addition of water instantaneously restores the blue colour of the parent azulene. Other 1-substituted azulenium salts, *e.g.* 1-triphenylmethylazulenium tetrafluoroborate have also been isolated as crystalline compounds[52]:

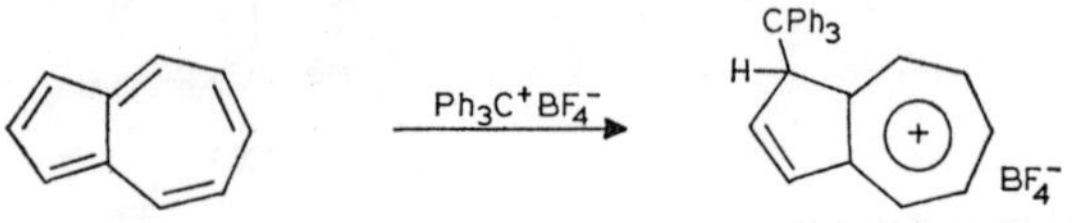

In addition to this reversible protonation on contact with strong acids, azulene also undergoes a slower irreversible transformation[8], which has been ascribed to electrophilic substitution of the azulene ring at the 1- and 3-positions by azulenium cations[147].

The first aromatic type substitution reaction of azulene to be described other than protonation was its coupling, in the 1-position, with diazonium compounds[34,52,73,74]. Only activated benzene derivatives couple with diazonium compounds, and the reaction with azulene is facilitated by the activation of the 1-position towards electrophilic attack. Reduction of 1-arylazoazulenes with sodium bisulphite gives 1-aminoazulenes which are unstable and can only be isolated as salts or as acetyl derivatives[75].

Azulenes are halogenated by *N*-bromosuccinimide or *N*-chlorosuccinimide to give mixtures of 1- and 1,3-substituted azulenes[34,38]. Pyridinium perbromide produces the dibromo-derivative[66]. Mono-iodo and di-iodo derivatives have been obtained by the action, respectively, of iodine monochloride[66] and iodosuccinimide[38]. 1-Bromoazulenes yield normal Grignard reagents with magnesium[76].

Nitration has been carried out under very mild conditions, using tetranitromethane[67,68], cupric nitrate in acetic anhydride[34,69], nitric acid in acetic anhydride[68] or urea nitrate[70]. Either the 1-nitro-[34,67–69] or 1,3-dinitro-compounds[68,70] are produced. The first method appears to be the most useful. 1-Nitroazulene has been converted into 1-acetylaminoazulene which was also obtained by Beckmann rearrangement of the oxime of 1-acetylazulene[34].

Mono-sulphonation of azulenes has been achieved using sulphur trioxide in dioxan[71,72].

With acetyl chloride in the presence of aluminium chloride diacetyl derivatives are formed; acetic anhydride gives a mixture of mono- and di-acetylated products. Under milder conditions, *e.g.* acetic anhydride and tetrachlorosilane in methylene chloride, the monoacetyl derivative is obtained[34,68].

Owing to the high reactivity of the 1- and 3-positions towards electrophilic reagents, azulenes can in fact be acylated without the presence of a catalyst. This has been achieved using acetyl bromide[77,67], benzoyl bromide[67,77], oxalyl chloride and bromide[67,77-79], malonyl chloride[80], phoshene[77,81], trifluoracetic anhydride[82,83] and even anhydrous acetic and formic acids[84]. The acylation reaction has been shown to be reversible in a number of cases, for example formyl, acetyl and benzoyl groups can be removed again by strong acids to regenerate the parent hydrocarbon[67,77]. Acetyl groups in the 1- and 3-positions can be oxidised to carboxyl groups by means of sodium hypoiodite[34]; the carboxyl group can also be got by treatment of 1-trifluoroacetylazulene with sodium hydroxide[83]. When azulene reacts with acid anhydrides in the presence of 74% perchloric acid, the 1-acylazulene which is first formed reacts with a further molecule of azulene to give a methine dye[158].

Formylation can be carried out either by means of the Gattermann reaction[67] or the Vilsmeier reaction[85-87]. In either case the 1-aldehyde is obtained. Ketones have also been prepared by means of the Vilsmeier reaction[85]. Azulene-1-aldehyde differs from benzenoid aldehydes in that it has reduced electrophilic character[77,89,108,109]. This is thought to be due to the contribution from a dipolar form:

Azulenes condense with benzenoid or heteroaromatic aldehydes in the presence of hydrochloric or perchloric acids to give azulenium salts which can react with nucleophiles to give a variety of 1-substituted derivatives[88,89], *e.g.*

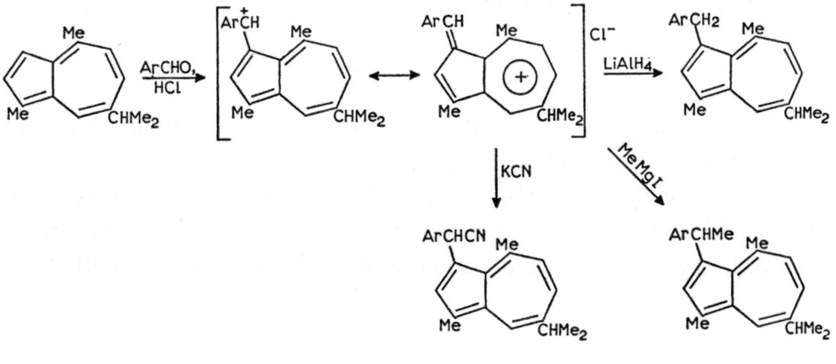

In the presence of tetrafluoroboric acid azulene has been shown to react in a similar way with aliphatic and benzenoid aldehydes and ketones and with orthoesters[90]. Azulenium salts similarly condense with aldehydes at the 1-position[51].

Cations such as the triphenylmethyl or tropylium ions react by substitution at the 1- and 3-positions[52,91]. 4,6,8-Trimethylazulene would not react in this way with the triphenylmethyl ion, very probably owing to steric hindrance[52,91]. Ethylation can similarly be achieved by means of triethyloxonium tetrafluoroborate[52].

Among other electrophilic substitutions undergone by azulenes are mercuration[72,92] and aminomethylation[93].

If the 1- and 3-positions are blocked substitution by electrophilic reagents can take place at the 5-position, since this is the position of next highest electron density[94], *e.g.*

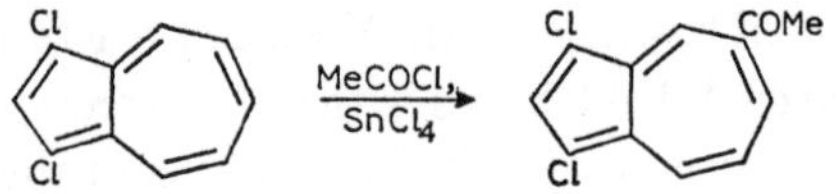

Acylation[94,95], formylation[95], halogenation[96] and nitration[95] have been achieved at the 5-position in similar circumstances. In the acetylation and formylation of 1,3-di-isopropyl- and 1,3-di-t-butyl-azulenes, alternative products were also obtained wherein replacement of the alkyl groups by formyl or acetyl groups had taken place[95], *e.g.*

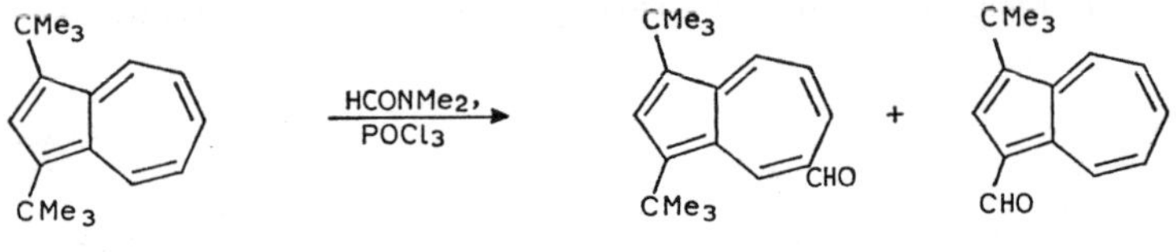

It has been shown that it is similarly possible to replace a *t*-butyl group by means of a *p*-nitrophenylazo group[52].

(ii) Nucleophilic substitution. Nucleophilic substitution of azulenes occurs preferentially at the 4- and 8-positions or if these positions are already occupied, at the 6-position. Reaction proceeds by the following paths:

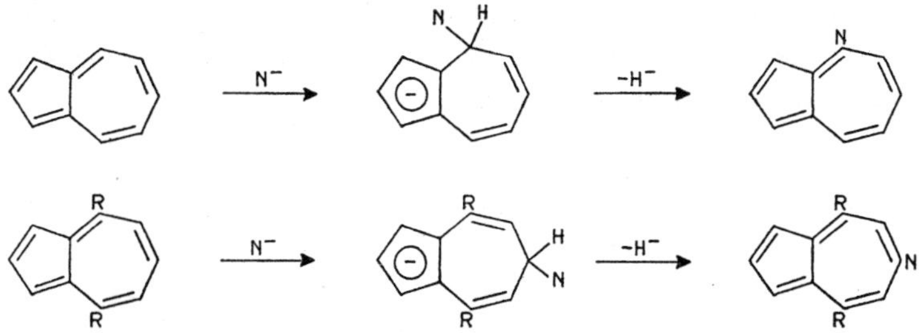

The study of nucleophilic substitution reactions has been hampered by the relative lack of stability of azulene towards bases. For example it is rapidly decomposed on heating with alcoholic alkali.

With alkali metal alkyls azulenes give alkylazulenide anions. The latter react with water to give dihydroazulenes which are readily dehydrogenated by means of chloranil to azulenes; alternatively the azulenide anions may be converted directly to azulenes by heating:

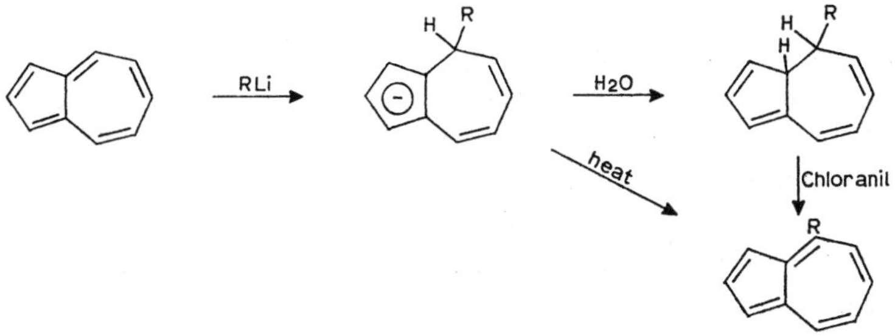

Similarly azulene reacts with sodamide in liquid ammonia to give a somewhat unstable red basic product which is probably the 4-aminoazulene[99,72]. It is interesting to note that this amine is colourless in acid solution which suggests that protonation takes place in the five-membered ring rather than on the amino-group[99].

Nucleophilic replacement of groups attached to the 4-position takes place readily. Thus it proved easy to replace a 4-methoxy group by an ethoxy or a hydroxyl group[99]. Similarly nucleophilic substitution of a halogen atom attached to the seven-membered ring proceeds readily; in contrast if the halogen atom is attached to the five-membered ring substitution takes place only to a small extent[66].

Since the 4-, 6- and 8-positions on azulenes are the positions of lowest electron density, methyl groups at these positions are activated and weakly

References p. 212

acidic and react with strong bases with loss of a proton. The anion so formed reacts with electrophilic reagents, the overall result being a substitution reaction at the methyl group[100], e.g.

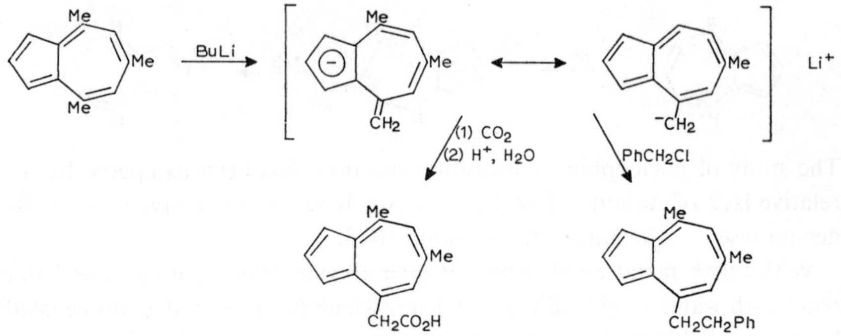

It is possible to isolate the intermediate alkali-metal salts[100]. The formation of sodium and potassium derivatives of azulenes from natural sources had in fact been observed many years ago[59,102,103], the blue azulenes becoming brown or grey on treatment with these metals in an inert solvent. Water, ethanol or exposure to air restored the blue colour. The reaction with carbon dioxide was also observed[102].

Another interesting reaction of such anions which has been reported[104] is that with ferric chloride, when a ferrocene-like sandwich compound is obtained.

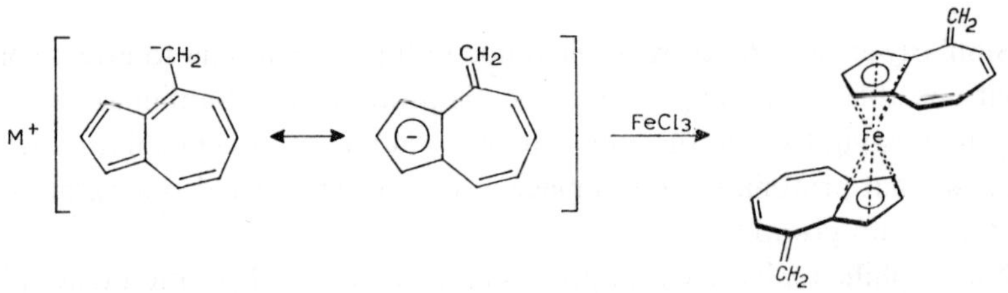

(iii) Radical substitution. Radical substitution of azulene appears to take place preferentially at the 1-position. Thus azulene reacts with *N*-nitrosoacetanilide in benzene to give 1-phenylazoazulene and 1-phenylazulene[105,106], while benzoyl peroxide gives a product which is assumed to be 1-azulenyl benzoate[106]. Benzyl radicals react with azulene to give a mixture of 1-benzylazulene, 2-benzylazulene and some pentabenzylazulene[107].

(c) Rearrangement reactions

Azulene rearranges on heating *in vacuo* at temperatures above 350° to give naphthalene in almost quantitative yield[55,56]. In the dehydrogenation of hydrogenated azulenes, some naphthalene derivatives are formed as by-products[60,110]. These naphthalene derivatives could of course be formed either by rearrangement of the hydroazulene before or during its dehydrogenation or by rearrangement of the resultant azulene at the high temperature of the reaction. Similarly alkyl groups may migrate during the dehydrogenation of hydroazulenes, *e.g.* in the treatment of 1,4-dimethyl-7-isopropylideneoctahydroazulene with selenium at about 300°, 2,4-dimethyl-7-isopropylazulene is formed[110]. Side-chain migration takes place much more commonly in the five-membered ring than in the seven-membered ring and usually involves migration of a group from the 1-position to the 2-position.

Spectra of azulenes

The infra-red spectrum of azulene is typical of an aromatic system[112]. There is a C—C in-plane vibration band at *ca.* 1570 cm^{-1} (6.37 μ) *cf.* naphthalene, 1597 cm^{-1}), and C—H stretching bands at 3030 and 3086 cm^{-1} (3.30 and 3.24 μ) (*cf.* naphthalene, 2994 and 3067 cm^{-1}). A discussion of the infra-red spectra of substituted azulenes is given in ref. 112.

The ultra-violet spectra of azulenes are highly complex (see Fig. 1)[18].

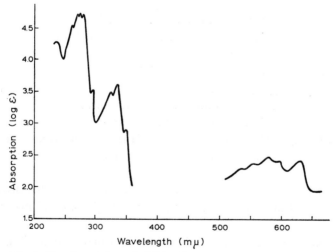

Fig. 1. The ultra-violet and visible absorption spectrum of azulene.

The spectrum of azulene, like that of many other non-alternant hydrocarbons which comply with the conditions of Hückel's rule, consists of weak, moderate and strong bands at progressively shorter wavelengths, corresponding to the α-, p- and β-band systems of alternant aromatic hydrocarbons[114]. Absorption takes place at longer wavelengths than in the case of the alternant isomer, naphthalene.

The low intensity bands in the visible portion of the spectrum, which are responsible for the blue colour, have been of particular value in the identification of azulenes. These bands suffer definite displacements by substitution of alkyl groups at different positions in the molecule. But whereas in the case of benzenoid hydrocarbons introduction of an alkyl group always causes a bathochromic shift, in the case of the long-wavelength absorption bands of azulene, the shift may be to longer or shorter wavelengths. The direction and extent of the shift depends on the position of the substituent alkyl group rather than on the nature of the group. An average displacement is associated with each position as shown in Fig. 2.

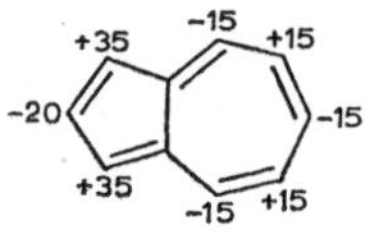

Fig. 2

Moreover in polyalkylazulenes these shifts are additive, *e.g.* azulene, 1,3-dimethylazulene and 1,8-dimethylazulene have principal maxima at 580 mμ, 638 mμ and 599 mμ respectively. In consequence, extensive use has been made of visible spectra in the determination of structures of azulenes.

Although alkyl groups affect these low-intensity bands differently depending on the position of substitution, they always displace the high intensity bands in the ultra-violet portion of the spectrum to longer wavelengths. As might be expected the shift is greatest if the substituents are attached to the five-membered ring since the electron density is greatest in this part of the molecule. Of other substituted azulenes, those with substituent groups in the 1-position have been studied in most detail[38]. For further discussions of the effects of different substituent groups refs. 38, 40, 115 may be consulted.

The n.m.r. spectra of azulene[39,144] and the azulenium cation[47,48] have been recorded and analysed (see also above, p. 191).

SESQUIFULVALENES

Another molecule made up from a five-membered ring and a seven-membered ring which might be expected to show some aromatic character is sesquifulvalene (XI)

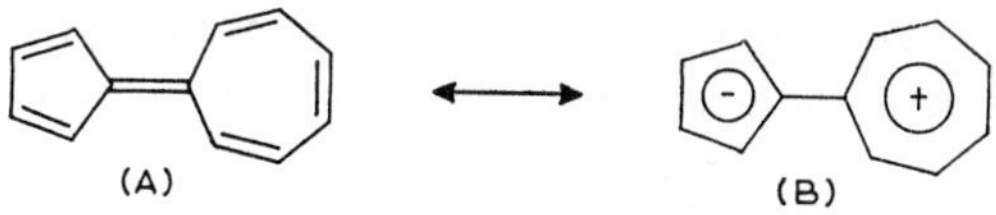

It is possible to draw a purely covalent form (A) of this molecule and also a dipolar form (B) and it might be expected to exist as a hybrid of these forms. Theory[36,116] predicts that sesquifulvalene should be stabilised by the contribution of the dipolar form and have a high delocalisation energy. In view of this it is surprising that it has not yet been isolated. There is evidence that it has been obtained in solution[119].

To date the only known sesquifulvalenes which show the expected thermal stability are benzo- and polyphenyl-derivatives[117,118,150].

A tetraphenylsesquifulvalene was prepared as follows[117]:

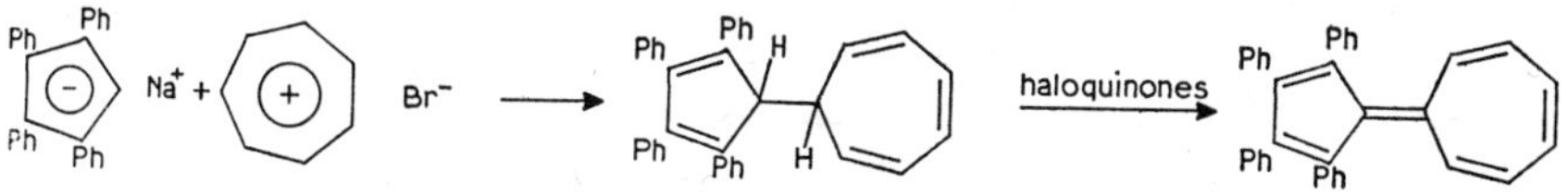

The product consisted of dark green crystals with a metallic lustre. Its u.v., i.r. and n.m.r. spectra were all in accord with its proposed structure. It reacts readily with lithium aluminium hydride and can be reduced smoothly catalytically.

A benzosesquifulvalene has also been prepared[118]:

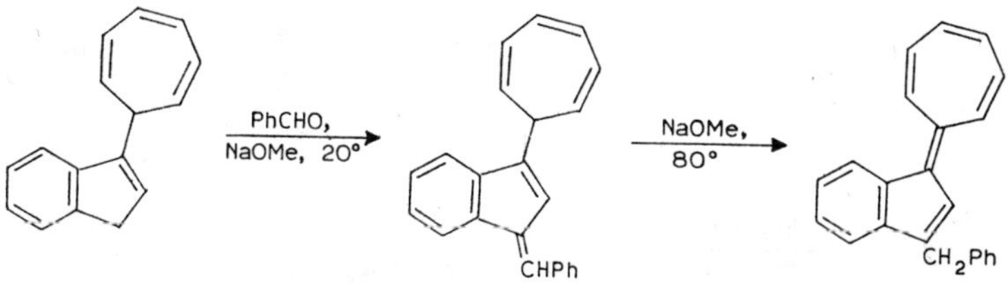

Both of these sesquifulvalenes take part in Diels–Alder reactions with tetracyanoethylene.

References p. 212

Another example of a benzosesquifulvalene has been prepared by condensing 4,5-benzotropone with 1,2,3,4-tetrachlorocyclopentadiene [150]. The structure of the product, which is violet, has been confirmed by X-ray crystallographic examination. It has a dipole moment of 5.20 D.

In contrast to these compounds the benzylsesquifulvalene (XII)[119] has poor stability and polyenoid character. This infers that the parent hydrocarbon may also lack the expected stabilisation; it has been suggested that this might be due to its being non-planar.

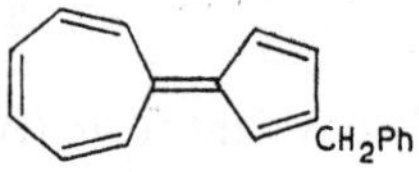

(XII)

Various heterocyclic cyclopentadienylidene derivatives, which are analogues of sesquifulvalenes have been prepared (see Chapter IV, p. 75).

CALICENES

Closely related to sesquifulvalene is cyclopropenylidenecyclopentadiene or *calicene* (from *calix*, a cup):

Like sesquifulvalene calicene should exist as a hybrid of dipolar and covalent forms. The parent hydrocarbon has not yet been prepared but a benzocalicene has been obtained as follows[155]:

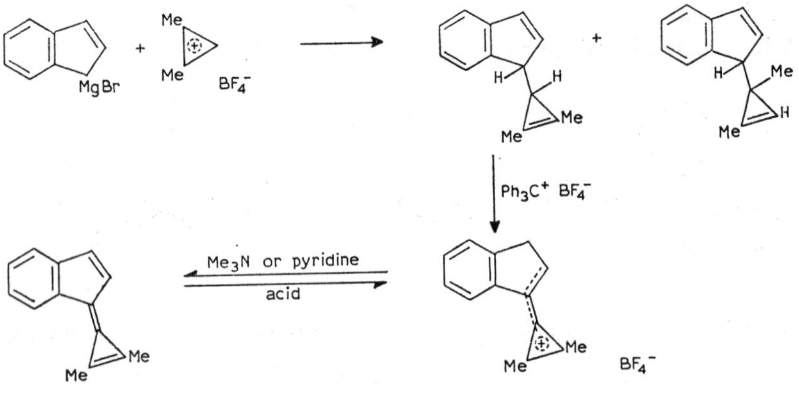

This benzocalicene has only been obtained in dilute solution. It is a strong base.

Other benzo- and dibenzo-calicenes have been prepared by the same method but starting from a diphenylcyclopropenium salt and the lithium derivative of indene or of fluorene[148]. In this case also the benzocalicene was only obtained in solution; evaporation of the solution to dryness or exposure of the solution to air resulted in decomposition of the benzocalicene. The dibenzocalicene on the other hand was isolated as a crystalline material. Indication of significant contribution from the dipolar forms is provided by measurements of the visible and ultra-violet spectra in solvents of different polarity. Decrease in the polarity of the solvent causes a shift of the long wavelength absorption to longer wavelengths; *e.g.* in the case of the dibenzocalicene there is a bathochromic shift of 41 mμ from a solution in acetonitrile to one in cyclohexane[148]. A similar dibenzocalicene has been obtained starting from a dimethylcyclopropenium salt[159].

An alternative approach to the preparation of calicene derivatives involved the condensation of a cyclopropanone with an indene derivative[149,160]:

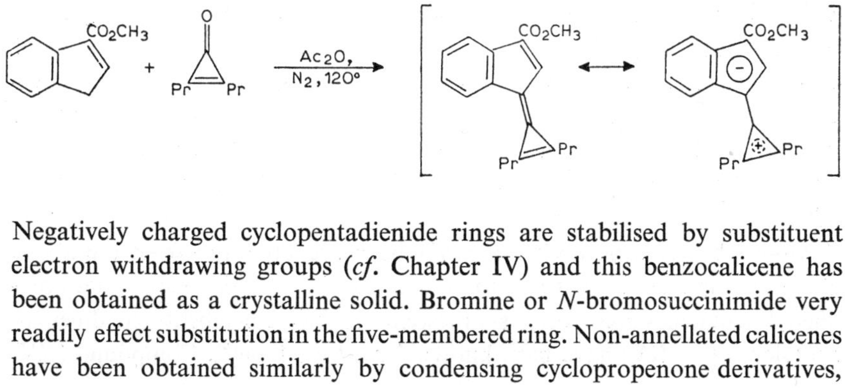

Negatively charged cyclopentadienide rings are stabilised by substituent electron withdrawing groups (*cf.* Chapter IV) and this benzocalicene has been obtained as a crystalline solid. Bromine or *N*-bromosuccinimide very readily effect substitution in the five-membered ring. Non-annellated calicenes have been obtained similarly by condensing cyclopropenone derivatives, with 2,3-dibenzoylcyclopentadiene[149] or with tetrachlorocyclopentadiene[161].

PENTALENES

The possibility that pentalene (XIII) might be an aromatic hydrocarbon was first considered more than forty years ago[120] but the same authors pointed out[121] that this was unlikely since aromatic sextets of electrons could not develop in each ring. The peripheral system of double bonds is completely

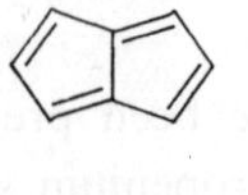

(XIII) 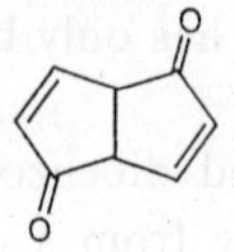   (XIV)

conjugated but is only made up of 8 π-electrons, *i.e.* it does not comply with Hückel's rule (but see below, p. 209).

All attempts to prepare pentalene have so far failed. For example attempted dehydrogenation of *cis*-octahydropentalene over platinum[122] or of tetrahydropentalenes over platinised asbestos[123] have given none of the desired product. Other preparative routes have also proved unsuccessful[124]. The diketone (XIV) has been prepared[125] and neither shows any detectable enol content nor forms any stable enol derivatives.

It has been pointed out[126] that the eight π-electrons in pentalene occupy four bonding orbitals and that the nearest unoccupied level proves to be one of zero energy. Because of this unoccupied level of zero energy the pentalene molecule should be converted comparatively readily from the ground singlet state to an excited triplet (diradical) state and consequently it should be extremely reactive. Furthermore, since the molecule can accept a pair of electrons into this unoccupied level without appreciable consumption of energy it can react readily with various nucleophilic reagents. These considerations demonstrate clearly the way in which the stability of aromatic systems is determined not only by a high conjugation energy of π-electrons but also by the way in which the molecular orbitals of a particular compound are filled.

Although pentalene itself has never been prepared and from the above discussion appears likely to be too reactive to be isolable, benzo- and polyphenyl-pentalenes have been synthesised; in fact 3,6-substituted dibenzopentalenes (see XV) have been known for a long time[127]. Compound (XV)

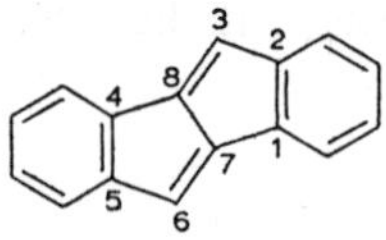

(XV) 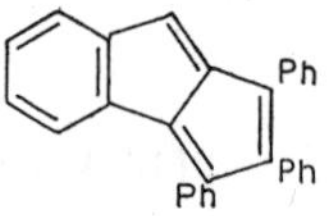   (XVI) 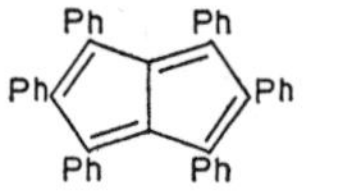   (XVII)

itself was prepared more recently[128]. It is a bronze coloured solid which is easily polymerised and behaves as a conjugated diene, rapidly absorbing four atoms of hydrogen in the presence of a catalyst and reacting vigorously with bromine by addition. The benzopentalene (XVI)[129] and the hexaphenylpentalene (XVII)[130] have also been prepared. Compound (XVI), which is a green crystalline material is stable to heat and air and to moderately strong acids. It is slowly destroyed by stronger acids and readily attacked by nucleophiles. Hexaphenylpentalene forms stable green-brown needles but in solution it is sensitive to air. It forms a blue solution in trifluoracetic acid. The tribenzopentalene fluoradene is strongly basic and forms a sodium salt[156] (see Chapter IV, p. 60).

A most interesting pentalene derivative which has been isolated in recent years is the dianion (XVIII), which is obtained by the action of butyl lithium on a dihydropentalene[131]. This salt forms white crystals which are stable in solution in tetrahydrofuran at room temperature.

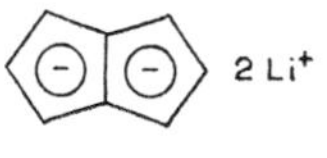

(XVIII)   2 Li$^+$ (XIX)

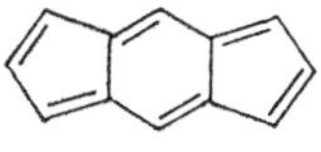

It has also been shown that 3,6-dihydro-1,2,4,5-dibenzopentalene can form a stable dianion on treatment with strong base[153].

Indacene (XIX), which is closely related to pentalene, has been prepared in recent years[136]. It is readily formed but its n.m.r. spectrum and chemical reactions indicate that it has limited π-electron delocalisation. The indacene derivative (XX) has also been obtained as a deep red-green crystalline compound[146]. With potassium fluoride it forms a yellow-orange dipotassium salt. The structure of this salt follows from analysis and spectral evidence[146].

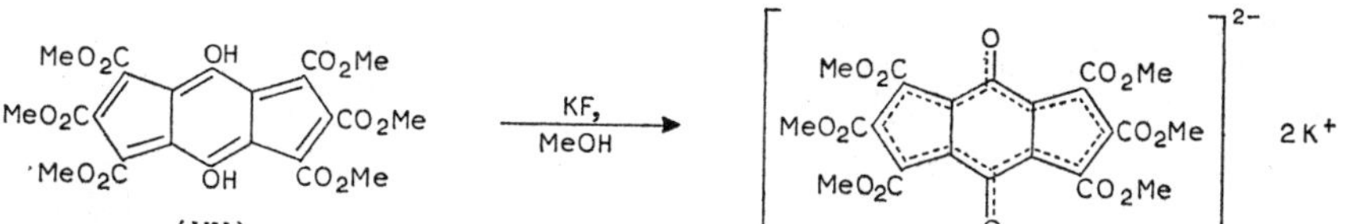

(XX) $\xrightarrow{\text{KF, MeOH}}$ [product] 2 K$^+$

A non-linear isomer of indacene is possible but it is likely to be even less stable than indacene itself. It is still unknown but its dianion has been prepared recently as follows[151]:

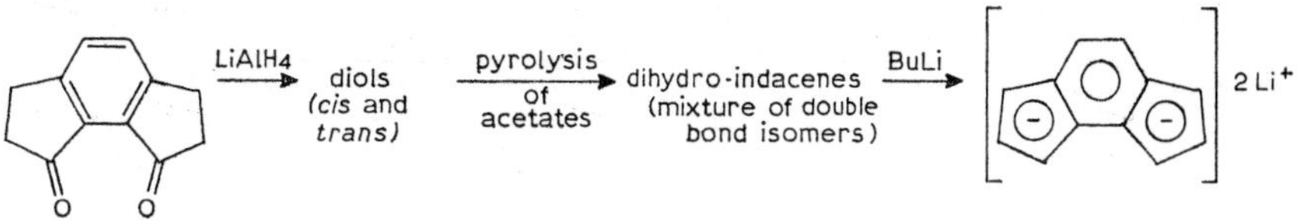

[diketone] $\xrightarrow{\text{LiAlH}_4}$ diols (cis and trans) $\xrightarrow[\text{acetates}]{\text{pyrolysis of}}$ dihydro-indacenes (mixture of double bond isomers) $\xrightarrow{\text{BuLi}}$ [dianion] 2 Li$^+$

This anion may be compared with the dianion derived from pentalene (see above). It reacts with ferric chloride to form a "sandwich" compound (*cf.* ferrocene, Chapter IV).

HEPTALENES

In heptalene (XIX) there is also a system of alternating double and single bonds, but 12 π-electrons are involved which again does not fit Hückel's requirements. Of these twelve electrons ten are in bonding orbitals and two in a non-bonding level of zero energy[126]. Consequently heptalene may be

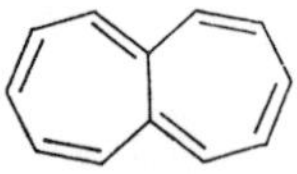

(XIX)

expected to give up a pair of electrons under the influence of even weak nucleophilic reagents. This is confirmed by the fact that it forms heptalenium salts (see below).

Attempts to dehydrogenate a decahydroheptalene[132] and also other hydroheptalenes[133] proved unsuccessful but heptalene was eventually prepared in 1961 by the following synthetic route[134]:

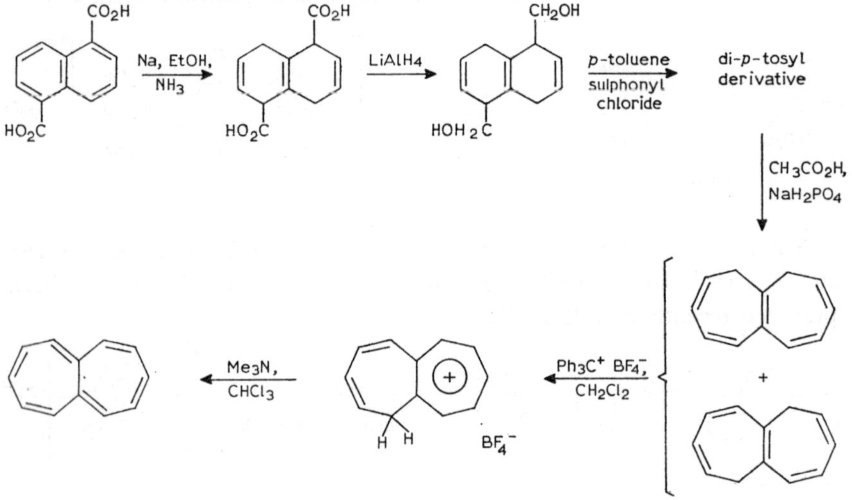

Heptalene proved to be a dark yellowish- or reddish-brown liquid which was stable at $-78°$ but polymerised on warming or on exposure to oxygen. It is readily reduced to bicyclo[5,5,0]dodecane. With 96% sulphuric acid it forms a positively charged heptalenium ion; this ion is apparently non-planar[135].

GENERAL COMMENTS ON POLYCYCLIC NON-BENZENOID AROMATIC HYDROCARBONS

For aromaticity in a cyclic polyolefin it is necessary to have a closed electron shell, *i.e.* a system of π-electrons such that either removal of electrons from it or addition of further electrons to it increase the total energy of the system and consequently diminish its stability[126]. Those hydrocarbons which fit Hückel's rule have such a closed shell.

Hückel's rule strictly applies only to monocyclic hydrocarbons but the results of much experimental work indicate that it is also applicable to bicyclic fused systems. This is readily understood if it is assumed that the central bridging bond introduces no appreciable change in the aromatic character of the system and that its function is only to maintain the planarity of the molecule. This argument is made more admissible by the fact that calculations show that the perturbation of the electron system caused by the bridging bond in both azulene and naphthalene is not excessive and that it cannot affect the general aromatic character of the system[126].

If the bridge bond is discounted in this way the peripheral π-electron systems in both naphthalene and azulene have 10 π-electrons and so these compounds should possess aromatic character. Pentalene and heptalene on the other hand have peripheral systems of 8 and 12 π-electrons respectively and should not be aromatic.

This approach can also be extended to a number of polycyclic hydrocarbons such as anthracene and phenanthrene each of which have peripheral systems of 14 π-electrons. In its general form the rule may be stated as: "Any planar (or nearly planar) fused system containing no electrons common to more than two rings will be aromatic if the number of π-electrons n it equals $(4n + 2)$ where n is an integer"[126].

In assessing the stability or otherwise of such a molecule however, lowering of stability due to angle strain or non-planarity must also be taken into account. For instance, such factors might greatly influence the stability of hypothetical molecules such as (XXI) (excessive angle strain) or (XXII)

References p. 212

(which might not be planar), although they meet the requirements of the general rule for aromaticity quoted above. Calculations have been made for a variety of fused ring systems[137] and it is possible that some as yet unprepared bicyclic or tricyclic systems may possess aromatic character.

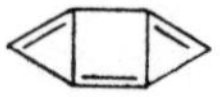

(XXI) 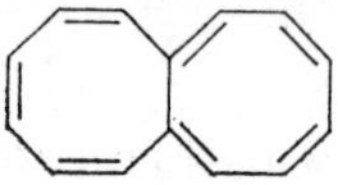   (XXII)

Matters are more complicated in the case of tri- and poly-cyclic systems having atoms common to three rings, and in these cases the simple $(4n + 2)$ rule no longer operates. It would seem that with polycyclic compounds each system has to be investigated individually by the molecular orbital method to

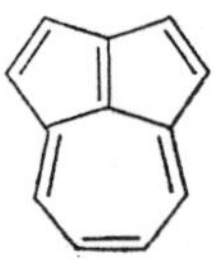

(XXIII) 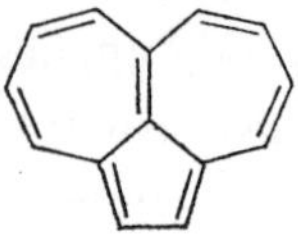   (XXIV)

determine the numbers of bonding, non-bonding, and anti-bonding orbitals.

Such calculations have been made for the systems (XXIII) and (XXIV)[138]. These systems have, respectively, 6 bonding orbitals and 12 π-electrons, and 7 bonding orbitals and 14 π-electrons. On these calculations both systems should possess aromatic stability.

Methyl derivatives of each of these systems have now been prepared, but only the derivatives of (XXIV) show the expected aromatic stability and reactions.

The dimethyl derivative of (XXIII)[139] behaves like a polyolefin. It adds dienes or dienophiles onto one of the five-membered rings to generate an azulene derivative, e.g.

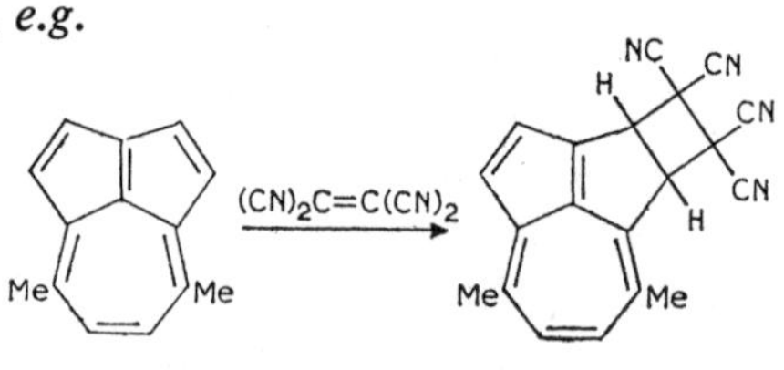

Other derivatives of (XXIII) behave similarly[140]. This lack of aromaticity has been attributed to angle strain[139].

On the other hand the dimethyl derivative of (XXIV)[139–142] shows the characteristics of an aromatic compound. It is a stable compound which forms dark red plates. It dissolves in sulphuric acid and can be reprecipitated unchanged from this solution. It readily undergoes electrophilic substitution in the five-membered ring, *e.g.* aminoethylation, acetylation by a Friedel–Crafts reaction, and formylation by means of a Vilsmeier reaction. These properties resemble those of an azulene as does its spectrum, and molecular orbital calculations[143] suggest that this ring system should be considered primarily as an azulene system with a diene bridge linking the 1,8-positions, with, however, electronic interaction between this diene system and the azulene part of the molecule. In keeping with this suggestion it is also found that the diene system of one of the seven-membered rings undergoes a Diels–Alder reaction with tetracyanoethylene or with maleic anhydride, *e.g.*

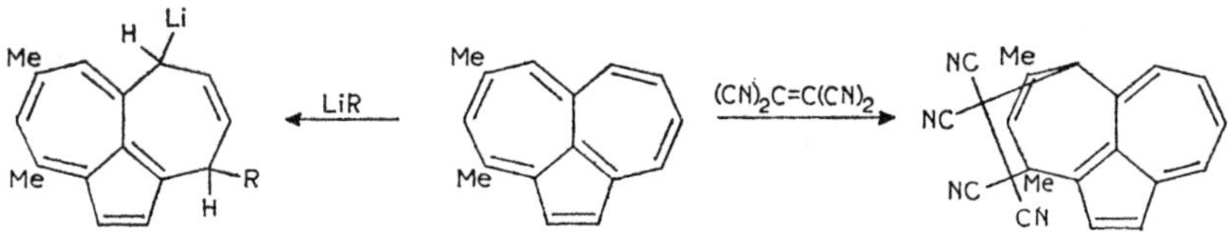

Similarly it reacts with lithium alkyls to form a blue azulene-type derivative.

The pentaleno[2,1,6-def]heptalene derivative (XXV) has also been prepared[154]. It forms black needles, is thermostable and dissolves reversibly in perchloric acid. It can be considered to consist structurally of two fused azulene units or of a pentalene unit fused to a heptalene. Also present in its structure are the elements of sesquifulvalene and of heptafulvene.

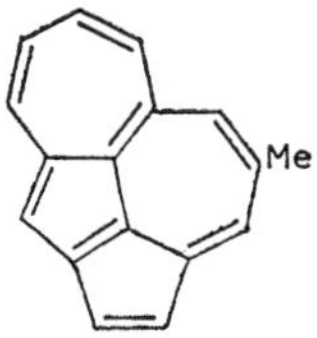

(XXV)

REFERENCES

1. P. BARBIER AND L. BOUVEAULT, *Compt. rend.*, 119 (1894) 281.
2. F. W. SEMMLER, *Aetherische Oele, Vol. III*, Veit and Co., Leipzig, 1906, p. 260.
3. D. PIESSE, *Compt. rend.*, 57 (1864) 1016.
4. A. ST. PFAU AND P. A. PLATTNER, *Helv. Chim. Acta.*, 19 (1936) 858.
5. A. E. SHERNDAL, *J. Am. Chem. Soc.*, 37 (1915) 167, 1537.
6. L. RUŽICKA AND E. A. RUDOLPH, *Helv. Chim. Acta.*, 9 (1926) 131.
7. M. GORDON, *Chem. Rev.*, 50 (1952) 127.
8. P. A. PLATTNER AND A. ST. PFAU, *Helv. Chim. Acta.*, 20 (1937) 224.
9. A. G. ANDERSON AND J. A. NELSON, *J. Am. Chem. Soc.*, 73 (1951) 232.
10. E. KOVÁTS, H. H. GÜNTHARD AND P. A. PLATTNER, *Helv. Chim. Acta*, 37 (1954) 2123.
11. P. A. PLATTNER AND J. WYSS, *Helv. Chim. Acta*, 22 (1939) 202; 23 (1940) 907.
12. R. R. COATS AND J. W. COOK, *J. Chem. Soc.*, (1942) 559.
13. W. VON E. DOERING, J. R. MAYER AND C. H. DEPUY, *J. Am. Chem. Soc.*, 75 (1953) 2386; S. DEV., *J. Indian. Chem. Soc.*, 30 (1953) 729; K. ALDER AND P. SCHMITZ, *Chem. Ber.*, 86 (1953) 1539.
14. H. ARNOLD, *Ber.*, 76 (1943) 777.
15. P. A. PLATTNER AND G. BÜCHI, *Helv. Chim. Acta.*, 29 (1946) 1608.
16. P. A. PLATTNER, A. FÜRST AND K. JIRASEK, *Helv. Chim. Acta*, 29 (1946) 730, 740.
17. E. A. BRAUDE AND W. F. FORBES, *Nature* 168 (1951) 874; *J. Chem. Soc.*, (1953) 2208. *Cf.* A. M. ISLAM AND R. A. RAPHAEL, *J. Chem. Soc.*, (1953) 2247.
18. D. LLOYD AND F. ROWE, *J. Chem. Soc.*, (1953) 3718.
19. D. LLOYD AND F. ROWE, *J. Chem. Soc.*, (1954) 4232.
20. P. A. PLATTNER AND A. STUDER, *Helv. Chim. Acta*, 29 (1946) 1432.
21. L. RUŽICKA AND V. PRELOG, U.S. Pat. 2838560, *Chem. Abs.*, 52 (1958) 17224.
22. W. HENTZSCHEL AND J. WISLICENUS, *Ann.*, 275 (1893) 312; *cf.* P. SABATIER AND A. MAILHE, *Compt. rend.*, 158 (1914) 985.
23. S. J. ASSONY AND N. KHARASCH, *Chem. and Ind.*, (1954) 1388; *J. Am. Chem. Soc.*, 80 (1958) 5978.
24. K. HAFNER, *Angew. Chem.*, 70 (1958) 419.
25. W. KÖNIG AND H. RÖSLER, *Naturwiss.*, 42 (1955) 211.
26. K. ZIEGLER AND K. HAFNER, *Angew. Chem.*, 67 (1955) 301.
27. K. HAFNER, *Ann.*, 606 (1957) 79.
28. K. HAFNER, *Angew. Chem.*, 67 (1955) 301.
29. K. HAFNER, *Angew. Chem.*, 69 (1957) 393; K. HAFNER AND H. KAISER, *Ann.*, 618 (1958) 140.
30. K. DIMROTH, K. H. WOLF AND H. WACHE, *Angew. Chem.*, 75 (1963) 860.
31. T. NOZOE, S. MATSUMURA, Y. MURASE AND S. SETO, *Chem. and Ind.*, (1955) 1257; T. NOZOE, S. SETO, S. MATSUMURA AND T. ASANO, *Proc. Japan. Acad.*, 32 (1956) 339.
32. K. HAFNER, H. W. RIEDEL AND M. DANIELISZ, *Angew. Chem.*, 75 (1963) 344.
33. K. HAFNER, K. H. VÖPEL, G. PLOSS AND C. KÖNIG, *Ann.*, 661 (1963) 52.
34. A. G. ANDERSON, J. A. NELSON AND J. TAZUMA, *J. Am. Chem. Soc.*, 75 (1953) 4980.
35. W. H. STAFFORD AND D. H. REID, *Chem. and Ind.*, (1954) 277.
36. G. W. WHELAND AND D. E. MANN, *J. Chem. Phys.*, 17 (1949) 264.
37. Y. KURITA AND M. KUBO, *J. Am. Chem. Soc.*, 79 (1957) 5460.
38. A. G. ANDERSON AND B. M. STECKLER, *J. Am. Chem. Soc.*, 81 (1959) 4941.
39. W. G. SCHNEIDER, H. J. BERNSTEIN AND J. A. POPLE, *J. Am. Chem. Soc.*, 80 (1958) 3497.
40. See E. HEILBRONNER in D. GINSBURG (Editor) *Non-benzenoid Aromatic Compounds*, Chapter V, Interscience, New York, 1959.

41 D. H. REID, *Chem. Soc. Special Publ.*, 12 (1958) 69.
42 J. M. ROBERTSON, H. M. M. SHEARER, G. A. SIM AND D. G. WATSON, *Acta Cryst.*, 15 (1962) 1.
43 C. A. COULSON AND H. C. LONGUET-HIGGINS, *Rev. Sci. Instr.*, 85 (1947) 929; R. D. BROWN, *Trans. Faraday Soc.*, 44 (1948) 984; A. PULLMAN AND G. BERTHIER, *Compt. rend.*, 227 (1948) 677; B. PULLMAN, M. MAYOT AND G. BERTHIER, *J. Chem. Phys.*, 18 (1950) 257; A. JULG, *Compt. rend.*, 239 (1954) 1498; F. H. SUMNER, *Trans. Faraday. Soc.*, 51 (1955) 315.
44 P. A. PLATTNER, E. HEILBRONNER AND S. WEBER, *Helv. Chim. Acta*, 32 (1949) 574.
45 E. HEILBRONNER AND M. SIMONETTA, *Helv. Chim. Acta*, 35 (1952) 1049.
46 W. SIMON, G. NAVILLE, H. SULSER AND E. HEILBRONNER, *Helv. Chim. Acta*, 39 (1956) 1107.
47 H. M. FREY, *J. Chem. Phys.*, 25 (1956) 600.
48 S. S. DANYLUK AND W. G. SCHNEIDER, *J. Am. Chem. Soc.*, 82 (1960) 997.
49 A. BAUDER AND H. H. GÜNTHARD, *Helv. Chim. Acta*, 41 (1958) 889.
50 E. C. KIRBY AND D. H. REID, *Chem. and Ind.*, (1960) 1217.
51 K. HAFNER AND H. PELSTER, *Angew. Chem.*, 72 (1960) 781.
52 K. HAFNER, A. STEPHAN AND C. BERNHARD, *Ann.*, 650 (1961) 42.
53 H. KAISER, unpublished work, quoted in ref. 24.
54 K. HAFNER AND W. KLINNER, unpublished work, quoted in ref. 24.
55 E. HEILBRONNER, P. A. PLATTNER AND K. WIELAND, *Experientia*, 3 (1947) 70.
56 E. HEILBRONNER AND K. WIELAND, *Helv. Chim. Acta*, 30 (1947) 947.
57 K. HAFNER AND H. SCHAUM, *Angew. Chem.*, 75 (1963) 90.
58 W. TREIBS, *Naturwiss.*, 47 (1960) 156.
59 R. E. KREMERS, *J. Am. Chem. Soc.*, 45 (1923) 717.
60 L. RUŽIČKA AND A. J. HAAGEN-SMIT, *Helv. Chim. Acta*, 14 (1931) 1104.
61 W. TREIBS, *Chem. Ber.*, 90 (1957) 761.
62 P. A. PLATTNER, A. FÜRST, A. MÜLLER AND W. KELLER-SCHIERLEIN, *Helv. Chim. Acta*, 37 (1954) 271.
63 W. TREIBS, H. ORTTMANN, R. SCHLIMPER AND C. LINDIG, *Chem. Ber.*, 92 (1959) 2152.
64 P. A. PLATTNER, E. HEILBRONNER AND S. WEBER, *Helv. Chim. Acta*, 35 (1952) 1036.
65 inter alia, R. S. MULLIKEN, *J. Am. Chem. Soc.*, 72 (1950) 600; 74 (1952) 811.
66 K. HAFNER, H. PATZELT AND H. KAISER, *Ann.*, 656 (1962) 24.
67 D. H. REID, W. H. STAFFORD AND W. L. STAFFORD, *J. Chem. Soc.*, (1958) 1118.
68 A. G. ANDERSON, R. SCOTONI, E. J. COWLES AND G. FRITZ, *J. Org. Chem.*, 22 (1957) 1193.
69 A. G. ANDERSON AND J. A. NELSON, *J. Am. Chem. Soc.*, 72 (1950) 3824.
70 W. TREIBS, *Angew. Chem.*, 67 (1955) 76.
71 W. TREIBS AND W. SCHROTH, *Ann.*, 586 (1954) 202.
72 A. G. ANDERSON, D. J. GALE, R. N. MCDONALD, R. G. ANDERSON AND R. C. RHODES, *J. Org. Chem.*, 29 (1964) 1373.
73 P. A. PLATTNER, *Angew. Chem.*, 62 (1950) 513.
74 W. TREIBS AND W. ZIEGENBEIN, *Ann.*, 586 (1954) 194.
75 K. G. SCHEIBLI, Dissertation, Zürich, 1952, quoted by W. KELLER-SCHIERLEIN AND E. HEILBRONNER in D. GINSBURG (Editor) *Non-benzenoid Aromatic Compounds*, Chapter VI, Interscience, New York, 1959.
76 K. HAFNER, *Angew. Chem.*, 70 (1958) 413.
77 W. L. GALLOWAY, D. H. REID AND W. H. STAFFORD, *Chem. and Ind.*, (1954) 724.
78 W. TREIBS, *Chem. Ber.*, 92 (1959) 2152.
79 W. TREIBS AND H. ORTTMANN, *Naturwiss.*, 45 (1958) 85.
80 W. TREIBS, *Naturwiss.*, 47 (1960) 179.

81 W. TREIBS, H. J. NEUPERT AND J. HIEBSCH, *Chem. Ber.*, 92 (1959) 1216.
82 A. G. ANDERSON, R. G. ANDERSON AND L. L. REPLOGLE, *Proc. Chem. Soc.*, (1960) 72.
83 A. G. ANDERSON AND R. G. ANDERSON, *J. Org. Chem.*, 27 (1962) 3578.
84 W. TREIBS, *Naturwiss.*, 45 (1958) 336.
85 K. HAFNER AND C. BERNHARD, *Angew. Chem.*, 69 (1957) 533.
86 W. TREIBS, J. HIEBSCH AND H. J. NEUPERT, *Naturwiss.*, 44 (1957) 352.
87 W. TREIBS, H. J. NEUPERT AND J. HIEBSCH, *Chem. Ber.*, 92 (1959) 141.
88 D. H. REID, W. H. STAFFORD, W. L. STAFFORD, G. MCLENNAN AND A. VOIGT, *J. Chem. Soc.*, (1958) 1110.
89 E. C. KIRBY AND D. H. REID, *J. Chem. Soc.*, (1960) 494.
90 K. HAFNER, H. PELSTER AND J. SCHNEIDER, *Ann.*, 650 (1961) 62.
91 E. C. KIRBY AND D. H. REID, *Tetrahedron Letters*, No. 27 (1960) 1.
92 A. G. ANDERSON, E. J. COWLES, J. TAZUMA AND J. A. NELSON, *J. Am. Chem. Soc.*, 77 (1955) 6321.
93 W. TREIBS, M. MÜHLSTÄDT AND K. D. KÖHLER, *Naturwiss.*, 45 (1958) 336.
94 A. G. ANDERSON AND L. L. REPLOGLE, *J. Org. Chem.*, 25 (1960) 1275.
95 K. HAFNER AND K. L. MORITZ, *Ann.*, 656 (1962) 40.
96 A. G. ANDERSON AND L. L. REPLOGLE, *J. Org. Chem.*, 28 (1963) 2578.
97 K. HAFNER AND R. WELDES, *Ann.*, 606 (1957) 90.
98 K. HAFNER, C. BERNHARD AND R. MÜLLER, *Ann.*, 650 (1961) 35.
99 D. H. REID, W. H. STAFFORD AND J. P. WARD, *J. Chem. Soc.*, (1958) 1100.
100 K. HAFNER, H. PELSTER AND H. PATZELT, *Ann.*, 650 (1961) 80.
101 V. PRELOG AND K. SCHENKER, *Helv. Chim. Acta*, 36 (1953) 1181.
102 J. MELVILLE, *J. Am. Chem. Soc.*, 55 (1933) 3288.
103 K. S. BIRRELL, *J. Am. Chem. Soc.*, 57 (1935) 893.
104 K. HAFNER, H. PELSTER AND W. AUS DER FÜNTEN, unpublished work, quoted by K. HAFNER, *Angew. Chem.*, 75 (1963) 1041.
105 H. ARNOLD AND K. PAHLS, *Chem. Ber.*, 89 (1956) 121.
106 A. G. ANDERSON AND G. M. C. CHANG, *J. Org. Chem.*, 23 (1958) 151.
107 J. F. TILNEY-BASSETT AND W. A. WATERS, *J. Chem. Soc.*, (1959) 3123.
108 K. HAFNER AND C. BERNHARD, *Ann.*, 625 (1959) 108.
109 E. C. KIRBY AND D. H. REID, *J. Chem. Soc.*, (1961) 163.
110 P. A. PLATTNER, *Helv. Chim. Acta*, 24 (1941) 283E.
111 C. GIANNOTTI, *Compt. rend.*, 258 (1964) 5225.
112 H. H. GÜNTHARD AND P. A. PLATTNER, *Helv. Chim. Acta*, 32 (1949) 284.
113 P. A. PLATTNER AND E. HEILBRONNER, *Helv. Chim. Acta*, 30 (1947) 910; 31 (1948) 804.
114 E. CLAR, *J. Chem. Soc.*, (1950) 1823; *Aromatische Kohlenwasserstoffe*, Springer Verlag, Berlin, 1952; *Polycyclic Hydrocarbons*, Springer Verlag, Berlin and Academic Press, New York, 1964.
115 S. F. MASON, *Quart. Rev.*, 15 (1961) 287.
116 G. BERTHIER AND B. PULLMANN, *Trans. Faraday Soc.*, 45 (1949) 484; E. D. BERGMANN, E. FISCHER, D. GINSBURG, Y. HIRSHBERG, D. LAVIE, M. MAYOT, A. PULLMAN AND B. PULLMAN, *Bull. Soc. chim. France*, 18 (1951) 684; J. F. TINKER, *J. Chem. Phys.*, 19 (1951) 981; A. JULG, *J. Chim. phys.*, 52 (1955) 50; A. JULG AND B. PULLMAN, *J. Chim. phys.*, 52 (1955) 481.
117 H. PRINZBACH, *Angew. Chem.*, 73 (1961) 169.
118 H. PRINZBACH AND D. SEIP, *Angew. Chem.*, 73 (1961) 169.
119 H. PRINZBACH AND W. ROSSWOG, *Angew. Chem.*, 73 (1961) 543.
120 J. W. ARMIT AND R. ROBINSON, *J. Chem. Soc.*, 121 (1922) 827.
121 J. W. ARMIT AND R. ROBINSON, *J. Chem. Soc.*, 127 (1925) 1604.
122 J. W. BARRATT AND R. P. LINSTEAD, *J. Chem. Soc.*, (1936) 611.

123 J. D. ROBERTS AND W. F. GORHAN, *J. Am. Chem. Soc.*, 74 (1952) 2278.
124 See, *e.g.* C. T. BLOOD AND R. P. LINSTEAD, *J. Chem. Soc.*, (1952) 2255; L. HORNER, H. G. SCHMELZER, H. U. ELTZ AND K. HABIG, *Ann.*, 661 (1963) 44.
125 H. J. DAUBEN, S. H.-K. JIANG AND V. R. BEN, *Hua Hsüeh Hsüeh Pao*, 23 (1957) 411; *Chem. Abs.*, 52 (1958) 16309h.
126 M. E. VOL'PIN, *Russ. Chem. Rev.*, 29 (1960) 129.
127 *inter alia*, W. ROSER, *Ann.*, 247 (1888) 153; K. BRAND, *Ber.*, 45 (1912) 307.
128 C. T. BLOOD AND R. P. LINSTEAD, *J. Chem. Soc.*, (1952) 2263.
129 E. LEGOFF, *J. Am. Chem. Soc.*, 84 (1962) 1505.
130 E. LEGOFF, *J. Am. Chem. Soc.*, 84 (1962) 3975.
131 T. J. KATZ AND M. ROSENBERGER, *J. Am. Chem. Soc.*, 84 (1962) 865.
132 G. O. ASPINALL AND W. BAKER, *J. Chem. Soc.*, (1950) 743.
133 D. H. S. HORN AND W. S. RAPSON, *J. Chem. Soc.*, (1949) 2421.
134 H. J. DAUBEN AND D. J. BERTELLI, *J. Am. Chem. Soc.*, 83 (1961) 4659.
135 E. HEILBRONNER, W. MEIER AND D. MEUCHE, *Helv. Chim. Acta*, 45 (1962) 2628.
136 K. HAFNER, K. H. HÄFNER, C. KÖNIG, M. KREUDER, G. PLOSS, G. SCHULZ, E. STURM AND K. H. VÖPEL, *Angew. Chem.*, 75 (1963) 35; K. HAFNER, *Angew. Chem.*, 75 (1963) 1041.
137 J. D. ROBERTS, A. STREITWIESER AND C. M. REGAN, *J. Am. Chem. Soc.*, 74 (1952) 4579.
138 M. E. DYATKINA AND E. M. SHUSTOROVICH, *Dokl. Akad. Nauk S.S.S.R.*, 117 (1957) 1021; M. E. DYATKINA, S. N. DOBRYAKOY AND E. M. SHUSTOROVICH, *Dokl. Akad. Nauk S.S.S.R.*, 123 (1958) 308.
139 K. HAFNER AND J. SCHNEIDER, *Ann.*, 624 (1959) 37.
140 K. HAFNER AND K. F. BANGERT, *Ann.*, 650 (1961) 98.
141 K. HAFNER AND J. SCHNEIDER, *Angew. Chem.*, 70 (1958) 702.
142 K. HAFNER AND G. SCHNEIDER, *Ann.*, 672 (1964) 194.
143 M. ASGAR ALI AND C. A. COULSON, *Molec. Phys.*, 4 (1961) 65; A. ROSOWSKY, H. FLEISCHER, S. T. YOUNG, R. PARTCH, W. H. SAUNDERS AND V. BOEKELHEIDE, *Tetrahedron*, 11 (1960) 121.
144 D. MEUCHE, B. B. MOLLOY, D. H. REID AND E. HEILBRONNER, *Helv. Chim. Acta*, 46 (1963) 2483
145 R. B. TURNER, W. R. MEADOR, W. VON E. DOERING, L. H. KNOX, J. R. MAYER AND D. W. WILEY, *J. Am. Chem. Soc.*, 79 (1957) 4127.
146 E. LEGOFF AND R. B. LA COUNT, *Tetrahedron Letters*, (1964) 1161
147 P. C. MYHRE AND R. D. ANDERSEN, *Tetrahedron Letters*, (1965) 1497.
148 W. M. JONES AND R. S. PYRON, *J. Am. Chem. Soc.*, 87 (1965) 1608.
149 A. S. KENDE AND P. T. IZZO, *J. Am. Chem. Soc.*, 87 (1965) 1609.
150 Y. KITAHARA, I. MURATA AND S. KATAGIRI, *Angew. Chem.*, 77 (1965) 345.
151 T. J. KATZ AND J. SCHULMAN, *J. Am. Chem. Soc.*, 86 (1964) 3169.
152 J. SCHULZE AND F. A. LONG, *J. Am. Chem. Soc.*, 86 (1964) 322.
153 A. J. SILVESTRI, *Tetrahedron*, 19 (1963) 855.
154 K. HAFNER, R. FLEISCHER AND K. FRITZ, *Angew. Chem.*, 77 (1965) 43.
155 H. PRINZBACH, D. SEIP AND U. FISCHER, *Angew. Chem.*, 77 (1965) 258.
156 H. RAPOPORT AND G. SMOLINSKY, *J. Am. Chem. Soc.*, 82 (1960) 934.
157 K. HAFNER AND K.-D. ASMUS, *Ann.*, 671 (1964) 31.
158 F. N. STEPANOV AND A. G. YURCHENKO, *Zhur. obshch. Khim.*, 34 (1964) 901.
159 H. PRINZBACH AND U. FISCHER, *Angew. Chem.*, 77 (1965) 621.
160 A. S. KENDE AND P. T. IZZO, *J. Am. Chem. Soc.*, 87 (1965) 4162.
161 M. UENO, I. MURATA AND Y. KITAHARA, *Tetrahedron Letters*, (1965) 2967.

Subject Index

Aminoiminocycloheptatrienes, 150
Aminothiotropones, 150
Annulenes, 10, 165 ff.
  aromaticity, 170
  crystallographic analysis, 172
  preparation, 167
  reactions, 173
  spectra, n.m.r., 170
  spectra, u.v., 172
Annulenes, bridged ring, 176
Annulenes, dehydro, 174
"Aromatic", definition, 1 ff., 11, 12
Aromatic character, 13
Aromatic compounds,
  formulae, 10
  n.m.r. spectra, 12
  orbitals, 6 ff.
Aromaticity, 11
  and molecular orbital theory, 6
Aromatic sextet, 4
Azulenes, 57, 182 ff.
  preparation, 24, 183 ff.
  reactions, 194
    addition, 194
    electrophilic substitution, 192, 195
    nucleophilic substitution, 192, 198
    oxidation, 194
    radical substitution, 200
    rearrangement, 201
    reduction, 194
  spectra, 201
  structure, 190 ff.
Azulenium ions, 191, 192

Benzene,
  formulae, 2
  orbitals, 6
Benzobiphenylenes, 50
Benzocyclobutadienes, 42 ff.
Benzocyclobutadienylides, 44
Benzotropolones, 151, 152
Benzotropones, 151
Benzotropylium salts, 113
Biphenylene, 45
  preparation, 45
  properties, 47 ff.
  reactions,
    complex formation, 48
    oxidation, 48
    reduction, 48
    substitution, di-, 49
    substitution, electrophilic, 48
  spectra, 47
  structure, 46
Biphenylene, benzo, 50
Biphenylenequinone, 50
Borazole, 9
Bridged-ring annulenes, 176

Calicenes, 33, 204
Colchicine, 118
Croconic acid, 42
Cyclobutadiene, 8, 35 ff.
  formation, 35, 37
  instability, 37
  orbitals, 8
  preparation, attempted, 35 ff.
Cyclobutadiene, benzo, 42 ff.
Cyclobutadiene, metal complexes, 37, 38 ff., 44
Cyclobutadiene, naphtho, 43
Cyclobutadienylides, benzo, 44
Cyclobutane, methylene derivatives, 38
Cyclobutene, methylene derivatives, 38
Cyclobutenediones, 41
Cyclobutenium salts, 40
Cycloheptatrienes, aminoimino, 150
Cycloheptatrienolones, see Tropolones
Cycloheptatrienones, see Tropones
Cycloheptatrienylium salts, see
    Tropylium salts
Cyclononatetraenide ion, 10, 163
Cyclooctatetraene, 5, 9
  bond lengths, 9
  orbitals, 9
  properties, 162
Cyclooctatetraenide dianion, 10, 162
Cyclopentadiene derivatives, 55 ff.

Cyclopentadiene, diazo, 61 *ff.*
  preparation, 61
  reactions, 63
    carbene formation, 65
    electrophilic substitution, 63
    reduction, 63
  spectra, 62
Cyclopentadiene, diazo, benzo
    derivatives, 69
Cyclopentadienide ions, 5, 7, 55, 58 *ff.*
  acidity, 57
  n.m.r. spectra, 58
  orbitals, 7
  preparation, 57, 60
  reactions, 58
  resonance energy, 58
  substituted, 59
Cyclopentadienylidene derivatives, 56, 75 *ff.*
Cyclopentadienylidenecycloheptatrienes
    see Sesquifulvalenes
Cyclopentadienylidenedihydropyridines,
  preparation, 76, 77
  properties, 78
  benzo derivatives, 79
Cyclopentadienylidenepyrans,
  preparation, 77
  properties, 78
  benzo derivatives, 80
Cyclopentadienylidenethiapyrans, 77
Cyclopentadienylides, 55, 66 *ff.*
  dipole moments, 67
  preparation, 66
  reactions, 68
  spectra, 67
  stability, 68
Cyclopentadienylides, benzo derivatives, 69 *ff.*
Cyclopolyolefins, *see* Annulenes
Cyclopropene, aromatic derivatives, 16 *ff.*
Cyclopropenes, methylene, 31 *ff.*
Cyclopropenium salts, 10, 16
  bond lengths, 17
  metal complexes, 24
  pK values, 21
  preparation, 16, 18 *ff.*
  reactions, 22 *ff.*
  solubility, 20
  spectra, i.r. and u.v., 20
  spectra, n.m.r., 16
  stability, 21

Cyclopropenones, 17
  dipole moments, 26
  preparation, 17, 24 *ff.*
  reactions, 26 *ff.*
  spectra, 26
  stability, 17

Dehydroannulenes, 174
Diazocyclopentadienes, 61 *ff.*
  preparation, 61
  reactions, 63 *ff.*
  spectra, 62
Diazocyclopentadiene, benzo
    derivatives, 69
Diazofluorene, 69 *ff.*
Dipole moments,
  azulene, 191
  benzotropones, 151
  cyclopentadienylidenedihydropyridines, 78
  cyclopentadienylides, 67
  cyclopropenones, 26
  fluorenylides, 74
  fulvenes, 81, 86
  tropolone, 129
  tropone, 129
Dithiafulvalenes, 77

Ferrocene, 56, 86 *ff.*
  bond lengths, 88
  preparation, 87
  reactions, 89 *ff.*
  spectra, 88
  stability, 89
  stereochemistry, 88
  structure, 88
Ferrocene analogues, 91
Fluoradene, 60
Fluorene, anion, 60
Fluorene, diazo, 69
Fluorenylidene derivatives, 79
Fluorenylides, 72 *ff.*
  dipole moments, 74
Formulae of aromatic compounds, 10
Fulvalenes, dithia, 77
Fulvenes, 56, 60, 81 *ff.*
  dipole moments, 81
  preparation, 81
  reactions, 83
  spectra, 82
  structure, 81

SUBJECT INDEX

Fulvenes, 6-substituted, 84
Furan, 5

Heptafulvenes, 153
Heptalene, 208
Heterocyclic compounds, aromaticity, 5, 9
Hinokitiol, 117
Hückel's rule, 6, 10

Indacene, 207
Indene, anion, 60
Indenylidene derivatives, 79
Indenylides, 69
Infra-red spectra,
  azulenes, 201
  cyclopropenium salts, 20
  cyclopropenones, 26
  diazocyclopentadiene, 63
  ferrocene, 88
  tropolones, 133
  tropones, 132
  tropylium ion, 104
Inscribed circle formulae, 4, 10

Methylenecyclobutanes, 38
Methylenecyclobutenes, 38
Methylenecycloheptatrienes, 153
Methylenecyclopentadiene, *see* Fulvene
Methylenecyclopropenes, 31 *ff*.
Molecular orbital theory and aromaticity, 6

Naphthalene, bond lengths, 5
Naphthocyclobutadiene, 43
Nuclear magnetic resonance spectra,
  annulenes, 170
  aromatic compounds, 12
  azulene, 191
  biphenylene, 47
  bridged-ring annulenes, 176, 177
  cyclobutadiene metal complexes, 39
  cyclopentadienide ion, 58
  cyclopropenium salts, 16
  cyclopropenones, 26
  diazocyclopentadiene, 63
  ferrocene, 88
  fulvenes, 82
  naphthocyclobutadiene, 43
  tropolone, 133
  tropone, 132
  tropylium ion, 104

Orbitals
  aromatic compounds, 6
  benzene, 6
  cyclobutadiene, 8
  cyclooctatetraene, 9
  cyclopentadienide ion, 7
  tropylium ion, 8

Pentalenes, 205
Pentalene dianion, 60, 207
Polycyclic compounds, 181, 209
Pseudoaromatic compounds, 13
Purpurogallin, 118
Pyrrole, 5

Quasiaromatic compounds, 14
Quinocyclopropenes, 29 *ff*.

Rhodizonic acid, 42

Sesquifulvalenes, 80, 203 *ff*.
Spectra, *see* i.r., n.m.r. and u.v. spectra
Squaric acid, 41
Stipitatic acid, 117

Thiotropones, 150
Thujaplicins, 117, 127, 128
Tropenium salts, *see* Tropylium salts
Tropolones, 117 *ff*.
  aromatic character, 120
  dipole moment, 129
  nomenclature, 118
  preparation, 126
  properties, 139
  reactions,
    addition, 142
    electrophilic substitution, 143
    hydroxyl group, 141
    oxidation, 139
    rearrangement, 146
    reduction, 140
    replacement of substituent groups, 145
  reactivity, 120, 121
  spectra, 132
  structure, 119, 129
Tropolones, benzo, 151, 152
Tropolones, metal complexes, 141
Tropolones, thio, 150
Tropolonium ion, 130

Tropones, 119 ff.
  aromatic character, 120
  dipole moment, 129
  preparation, 29, 122 ff.
  properties, 133
  reactions,
    addition, 134
    carbonyl group, 133
    conversion into tropolones, 128
    rearrangement, 138
    reduction, 133
    replacement of substituent groups, 136
    substitution, 135
  reactivity, 120
  spectra, 131
  structure, 129
Tropones, alkoxy, reactions, 136, 137
Tropones, aminothio, 150
Tropones, benzo, 151
Tropones, halo, reactions, 136, 137
Tropones, 3- and 4-hydroxy,
  preparation, 147
  properties, 149
Tropylium ion, orbitals, 8
Tropylium salts, 98 ff.
  pK values, 108, 114
  preparation, 98
  properties, 104
  reactions,
    conversion into tropones, 123
    electrophilic attack, inertia towards, 112
    nucleophiles, reaction with, 107
    oxidation, 106
    reduction, 105
  spectra, 104
  structure, 103
Tropylium salts, alkyl substituted, reactions, 112
Tropylium salts, benzo, 113
Tropylium salts, metal complexes, 113

Ultra-violet spectra,
  annulenes, 172
  azulenes, 201
  benzotropylium ions, 114
  biphenylene, 47
  cyclopentadienylidenedihydropyridines, 78
  cyclopentadienylides, 67
  cyclopropenium salts, 20
  cyclopropenones, 26
  diazocyclopentadiene, 62
  fulvenes, 82
  tropolones, 132
  tropones, 131
  tropylium ion, 104

DATE DUE

Randall Library – UNCW
QD331 .L58   NXWW
Lloyd / Carbocyclic non–benzenoid aromatic compoun

3049001776734